Future Health Scenarios

Digital technologies is a major emerging area to invest and research in new models of health management. Future health scenarios are constituted by technologies in health and clinical decision-making systems. This book provides a unique multidisciplinary approach for exploring the potential contribution of AI and digital technologies in enabling global healthcare systems to respond to urgent twenty-first-century challenges. Deep analysis has been made regarding telemedicine using big data, deep learning, robotics, mobile and remote applications.

Features:

- Focuses on prospective scenarios in health to predict possible futures.
- Addresses the urgent needs of the key population, socio-technical and health themes.
- Covers health innovative practices as 3D models for surgeries, big data to treat rare diseases, and AI robot for heart treatments.
- Explores telemedicine using big data, deep learning, robotics, mobile and remote applications.
- Reviews public health based on predictive analytics and disease trends.

This book is aimed at researchers, professionals and graduate students in computer science, artificial intelligence, decision support, healthcare technology management, biomedical engineering and robotics.

Advances in Smart Healthcare Technologies
Editors: Chinmay Chakraborty and *Joel J. P. C. Rodrigues*

This book series focus on recent advances and different research areas in smart healthcare technologies including Internet of Medical Things (IoMedT), e-Health, personalised medicine, sensing, Big Data, telemedicine, etc. under the healthcare informatics umbrella. Overall focus is on bringing together the latest industrial and academic progress, research, and development efforts within the rapidly maturing health informatics ecosystem. It aims to offer valuable perceptions to researchers and engineers on how to design and develop novel healthcare systems and how to improve patient's information delivery care remotely. The potential for making faster advances in many scientific disciplines and improving the profitability and success of different enterprises is to be investigated.

Blockchain Technology in Healthcare Applications
Social, Economic and Technological Implications
Bharat Bhushan, Nitin Rakesh, Yousef Farhaoui, Parma Nand Astya and Bhuvan Unhelkar

Digital Health Transformation with Blockchain and Artificial Intelligence
Chinmay Chakraborty

Smart and Secure Internet of Healthcare Things
Nitin Gupta, Jagdeep Singh, Chinmay Chakraborty, Mamoun Alazab and Dinh-Thuan Do

Practical Artificial Intelligence for Internet of Medical Things
Emerging Trends, Issues, and Challenges
Edited by Ben Othman Soufiene, Chinmay Chakraborty, and Faris A. Almalki

Intelligent Internet of Things for Smart Healthcare Systems
Edited by Durgesh Srivastava, Neha Sharma, Deepak Sinwar, Jabar H. Yousif, and Hari Prabhat Gupta

Future Health Scenarios
AI and Digital Technologies in Global Healthcare Systems
Edited by Maria José Sousa, Francisco Guilherme Nunes, Generosa do Nascimento and Chinmay Chakraborty

For more information about this series, please visit: www.routledge.com/Advances-in-Smart-Healthcare-Technologies/book-series/CRCASHT

Future Health Scenarios
AI and Digital Technologies in Global Healthcare Systems

Edited by
Maria José Sousa, Francisco Guilherme Nunes,
Generosa do Nascimento and Chinmay Chakraborty

CRC Press is an imprint of the
Taylor & Francis Group, an **informa** business

Designed cover image: © Shutterstock

First edition published 2023
by CRC Press
6000 Broken Sound Parkway NW, Suite 300, Boca Raton, FL 33487-2742

and by CRC Press
4 Park Square, Milton Park, Abingdon, Oxon, OX14 4RN

CRC Press is an imprint of Taylor & Francis Group, LLC

© 2023 selection and editorial matter, Maria José Sousa, Francisco Guilherme Nunes, Generosa do Nascimento and Chinmay Chakraborty; individual chapters, the contributors

Reasonable efforts have been made to publish reliable data and information, but the author and publisher cannot assume responsibility for the validity of all materials or the consequences of their use. The authors and publishers have attempted to trace the copyright holders of all material reproduced in this publication and apologize to copyright holders if permission to publish in this form has not been obtained. If any copyright material has not been acknowledged please write and let us know so we may rectify in any future reprint.

Except as permitted under U.S. Copyright Law, no part of this book may be reprinted, reproduced, transmitted, or utilized in any form by any electronic, mechanical, or other means, now known or hereafter invented, including photocopying, microfilming, and recording, or in any information storage or retrieval system, without written permission from the publishers.

For permission to photocopy or use material electronically from this work, access www.copyright.com or contact the Copyright Clearance Center, Inc. (CCC), 222 Rosewood Drive, Danvers, MA 01923, 978-750-8400. For works that are not available on CCC please contact mpkbookspermissions@tandf.co.uk

Trademark notice: Product or corporate names may be trademarks or registered trademarks and are used only for identification and explanation without intent to infringe.

ISBN: 9781032131498 (hbk)
ISBN: 9781032131504 (pbk)
ISBN: 9781003227892 (ebk)

DOI: 10.1201/9781003227892

Typeset in Times
by Newgen Publishing UK

Contents

Foreword ix
Preface..xi
List of Contributors xiii
About the Editors xvii

Chapter 1 Scenarios for the Future of Healthcare: Moving towards Smart Hospitals 1

  Maria José Sousa, Francesca Dal Mas and Paul Barach

Chapter 2 People Management in Healthcare: The Challenges in the Era of Digital Disruption 21

  Alzira Duarte, Generosa do Nascimento and Francisco Nunes

Chapter 3 Managing Acute Patient Flows in Hospitals 37

  Svante Lifvergren and Axel Lifvergren

Chapter 4 Influence of Commercial Excellence and Digital in Healthcare Professionals Relationships Management: A Pharmaceutical Focus Group Study 61

  António Pesqueira

Chapter 5 Psychology on the Edge of Health Tech Challenges 85

  Lourdes Caraça and Vera Proença

Chapter 6 Perceptions of Clients on Quality of Health Services 97

  Maria Carolina Martins Rodrigues, Luciana Aparecida Barbieri da Rosa, Maria José Sousa and Waleska Yone Yamakawa Zavatti Campos

Chapter 7 Innovation Management Applied to Primary Care: An Integrative Review 113

  Patricia Gesser da Costa, Guilherme Agnolin, João Paulo da Silveira, Andreia de Bem Machado, Gertrudes Aparecida Dandolini, João Artur de Souza and Maria José Sousa

Chapter 8 The Spinner Innovation and Knowledge Flow for Future
Health Scenarios Applications.. 129

*Ronnie Figueiredo, Marcela Castro, Pedro Mota Veiga and
Raquel Soares*

Chapter 9 Reshaping Flows in Healthcare Systems? Digital Technologies
Are the Perfect Ally 149

José Crespo de Carvalho and Teresa Cardoso-Grilo

Chapter 10 Work with Me, Don't Just Talk at Me: When "Explainable"
Is Not Enough... 169

Brian Pickering

Chapter 11 Application of Mobile Technologies in Healthcare During
Coronavirus Pandemic Lockdown.. 193

*Edeh Michael Onyema, Nwafor Chika Eucheria,
Ugboaja Samuel Gregory, Nneka Ernestina Richard-Nnabu,
Akindutire Opeyemi Roselyn, Emmanuel Chukwuemeka Edeh and
Ifeoma Ugwueke*

Chapter 12 Application of Pattern Recognition in Taste Perception for
Healthcare: An Exploratory Study 207

Dipannita Basu, Anusruti Mitra and Ahona Ghosh

Chapter 13 New Perspectives for Knowledge Management in
Telemedicine: Preliminary Findings from a Case Study
in the COVID-19 Era.. 231

*Francesca Dal Mas, Helena Biancuzzi, Maurizio Massaro,
Lorenzo Cobianchi, Rym Bednarova and Luca Miceli*

Chapter 14 Prioritization of Quality of Public Health Services in the
Sector of Graphic Methods: University Hospital 241

*Roger da Silva Wegner, Leoni Pentiado Godoy,
Taís Pentiado Godoy, Maria José Sousa and
Luciana Aparecida Barbieri da Rosa*

Chapter 15 Telework and Conflict (Work–Family and Work–Family):
What Is the Effect on Occupational Stress? 257

*Ana Moreira, Mónica Salvador, Alexandra de Jesus,
Catarina Furtado and Madalena Lopez-Caño*

Index 269

Foreword

The Future of the Health Sector

The health sector represents a relevant factor in economic growth and human development. Modern societies recognise the importance of scientific research and technological innovation in health. The crisis associated with the COVID-19 pandemic proved the relevance of health systems in the protection of populations. As stated by the World Health Organization, it is essential that countries implement universal and general health coverage.

The future of the health sector will be strongly linked to the development of science. One of the most important aspects is related to the digital transition. In addition to digital tools, technological innovation can help to bring care closer to people's health needs. The ongoing demographic transition in the more developed countries will bring in a profound reform of health systems. It will be necessary to invest more in proximity health care in view of increased longevity. Health systems will need to be increasingly integrated to ensure responses to the growing burden of chronic diseases. High-level healthcare based on advanced technological support will have to coexist with long-term care. Hospitals will increasingly have technological characteristics of acute care while the role of primary health care and home care will be strengthened.

The future of the health sector will be closely linked to economic growth and will contribute to a more sustained global development. The health professions will be increasingly differentiated and prepared to work in multidisciplinary teams. Technical skills will be ever more sophisticated and supported by a growing role of robotics, artificial intelligence, and genomics. Citizens will be increasingly informed by acquiring new skills in health literacy. Individual choices will be made with more information, allowing a shared decision process in the individual health path throughout the life cycle.

Health in the world, in the post-pandemic period, will be more global. Governments will tend to give the health sector a strategic dimension not only for the impact on people's lives, but also for contributing to wealth creation and well-being. The energy transition to a more sustainable and greener world such as the digital transition will be part of a broader process of transformation of the sector and health systems. The pursuit of the Sustainable Development Goals will require greater integration of sectoral policies in the context of which health will play a decisive role. As stated more than two decades ago by the World Health Organization, a sustainable and secure future for humanity requires that the objective of HiP (Health in all Policies) be kept in mind.

Adalberto Campos Fernandes
PhD, MPH, MD Professor NOVA ENSP

Preface

The development of knowledge is a major force in the progress of all facets of human life and the provision of care to those who need it is not an exception. Encapsulating multiple and often interrelated streams of technology, the knowledge potentially useful to improve health care has grown exponentially. This book represents an effort to bring together distinct accounts of these formidable changes we observe in the delivery of care, especially those engendered by knowledge accumulation and expansion, often codified in technologies.

Emerging technologies are challenging how organised health providers govern and manage themselves, and most entities, namely hospitals, are now experimenting with new models of care provision, offering extensive learning opportunities, and selecting the most promising among them (Sousa, Dal Mas & Barach, 2022; Basu, Mitra & Ghosh, 2022). In this ever-changing context, special attention must be given to professionals, those who, in practice, deliver care, and whose abilities, motivation, and opportunities must be addressed in a distinctive but integrated way (Duarte, Nascimento & Nunes, 2022). Studying the relationship between managing people and the quality of services as it happens in specific contexts can provide critical inputs for changing how employees can perform activities with excellence (Wegner et al., 2022).

If virtually, all facets of human resource management can be optimised as a consequence of using a digital approach (Figueiredo et al., 2022), professionals must be able to look at advanced technologies, like machine learning, artificial intelligence, and predictive modelling, as supportive partners instead of black boxes, which requires progress not only in terms of technology acceptance but also regarding the corresponding changes in ethical standards that guide professionals relationships with patients (Pickering, 2022).

As a consequence, among other factors, of more effective health systems, the population is ageing, and a growing number of people seek more and more sophisticated care, increasing the pressure over European health systems, which requires a better understanding and management of hospitals' patient flows (Lifvergren & Lifvergren, 2022). To consider the distinction between patients and goods flows is relevant if we want to take advantage of different digital technologies (e.g. Internet of Things, artificial intelligence, big data, blockchain, cloud services, and 3D printing) in reshaping healthcare logistics (Carvalho & Cardoso-Grilo, 2022).

The pandemic context also prompted the adoption of technologies to support clinical practices. More precisely, mobile technologies became an important tool for facilitating the delivery of care, because its use enhances patient access to care, case reporting, contact tracing, examinations, information sharing about patients, and facilitates research and education (Onyema et. al., 2022). In the context of the COVID-19 outbreak, technology-enabled the online delivery of mental health services. Although potentially effective in supporting patients' needs, to be sustained, the practice of online care for mental health requires additional reflections, including legal matters and the training of professionals (Caraça & Proença, 2022). More generally, in this context, due to social distance requirements, telemedicine becomes an increasingly common practice, and specific programmes are offering opportunities to learn more about the dynamics of knowledge translation, transfer, and sharing (Dal Mas et al., 2022).

Increased telework was one of the solutions for reducing social contacts as a measure for reducing SARS-COV-2 contagions. However, this change can increase both work–family and family–work conflicts, which usually leads to higher levels of stress and compromises work performance. Understanding the dynamics of the relationships between these variables is of paramount importance for both individual well-being and economic recovery (Moreira et al., 2022). Complementarily, and on the grounds of economic sustainability, in the context of the COVID-19 pandemic, pharmaceutical companies are facing different challenges (and opportunities as well), which calls for a greater reliance on commercial excellence and digital frameworks as supports for dealing with future scenarios pertaining sales, medical affairs, and marketing operations (Pesqueira, 2022).

The centrality of patients in determining the quality of healthcare is widely acknowledged, but in the context of increased knowledge being produced, it becomes relevant to service providers to understand the dynamics of knowledge transitions between patients and staff (Rodrigues et al., 2022). Stressing the criticality of the knowledge in addressing future health scenarios, namely in designing public health policies for effective primary healthcare, can benefit from the extensive knowledge accumulated about innovative solutions for service delivery to the population (Sousa et al., 2022).

As a whole, this book celebrates the diversity of a topic that is burgeoning and is complex but, at the same time, exerts a powerful attraction upon those who, on the grounds of distinct knowledge bases, strive for a healthier world.

Maria José Sousa
Francisco Guilherme Nunes
Generosa do Nascimento
Chinmay Chakraborty

Contributors

Ahona Ghosh Department of Computer Science and Engineering, Maulana Abul Kalam Azad University of Technology, West Bengal, India

Akindutire Opeyemi Roselyn Department of Statistics, Ekiti State University, Ekiti, Nigeria

Alexandra de Jesus ISMAT-Instituto Superior Manuel Teixeira Gomes

Alzira Duarte Instituto Universitário de Lisboa (ISCTE-IUL)

Ana Moreira ISMAT-Instituto Superior Manuel Teixeira Gomes e ISPA- Instituto Universitário

Andreia de Bem Machado Universidade Federal de Santa Catarina, Brazil

António Pesqueira Independent Researcher, Zug, Switzerland

Anusruti Mitra Department of Information Technology, Maulana Abul Kalam Azad University of Technology, West Bengal, India

Axel Lifvergren Department of Orthopaedic Surgery, Capio St Göran Hospital, Sweden, Department of Medical Sciences, Uppsala UniversityDepartment of Management and Engineering, Linköping University

Brian Pickering Electronics and Computer Science, University of Southampton

Waleska Yone Yamakawa Zavatti Campos PUC-Rio, Rio Janeiro, Brazil

Catarina Furtado ISMAT-Instituto Superior Manuel Teixeira Gomes

Da Rosa, L.A.B. UAB/UFSM, Santa Maria, Brazil

Dipannita Basu Department of Information Technology, Maulana Abul Kalam Azad University of Technology, West Bengal, India

Edeh Michael Onyema Department of Vocational and Technical Education, Faculty of Education, Alex Ekwueme Federal University, Ndufu-Alike, Abakaliki, Nigeria / Adjunct Faculty, Saveetha School of Engineering, Saveetha Institute of Medical and Technical Sciences, Chennai, India

Emmanuel Chukwuemeka Edeh Faculty of Education, University of Benin, Benin, Nigeria

Francesca Dal Mas Department of Management, Ca' Foscari University, Venice, Italy

Francisco Nunes Instituto Universitário de Lisboa (ISCTE-IUL)

Generosa do Nascimento Instituto Universitário de Lisboa (ISCTE-IUL)

Gertrudes Aparecida Dandolini Universidade Federal de Santa Catarina, Brazil

Guilherme Agnolin Universidade Federal de Santa Catarina, Brazil

Helena Biancuzzi, Department of Pain Medicine, National Cancer Institute CRO, Aviano, Italy

Ifeoma Ugwueke Department of Biological Sciences, Coal City University, Enugu, Nigeria

João Artur de Souza Universidade Federal de Santa Catarina, Brazil

João Paulo da Silveira Primary Care Directorate, State Department of Health, Government of Santa Catarina, Brazil

José Crespo de Carvalho Instituto Universitário de Lisboa (ISCTE-IUL)

Leoni Pentiado Godoy Universidade Fderal de Santa Maria – UFSM

Lorenzo Cobianchi Department of Clinical, Diagnostic and Paediatric Science, University of Pavia, Italy. IRCCS Policlinico San Matteo Foundation, General Surgery, Pavia, Italy

Lourdes Caraça Health Department, Grouping of Health Centers of the Estuary Tagus, Polytechnic Institute of Santarem, Health College

Luca Miceli Department of Pain Medicine, IRCCS C.R.O. National Cancer Institute of Aviano, Aviano, Italy

Luciana Aparecida Barbieri da Rosa Instituto Federal de Rondônia – IFRO – Zona Norte

Madalena Lopez-Caño ISMAT-Instituto Superior Manuel Teixeira Gomes

Marcela Castro Universidade Europeia, Portugal

Maria José Sousa ISCTE-Instituto Universitário de Lisboa– Universidade de Aveiro: Lisboa, Lisboa, Portugal

Maurizio Massaro Department of Management, Ca' Foscari University Venice, Italy

Mónica Salvador ISMAT-Instituto Superior Manuel Teixeira Gomes

Nneka Ernestina Richard-Nnabu Department of Computer Science/Informatics Alex Ekwueme Federal University Ndufu Alike-Ikwo, Abakaliki, Nigeria

Nwafor Chika Eucheria Department of Science Education, EBonyi State University, Abakaliki, Nigeria

Patricia Gesser da Costa Universidade Federal de Santa Catarina, Brazil

Paul Barach Jefferson University, Philadelphia, PA, USA

Pedro Mota Veiga NECE (Research Center in Business Sciences), University of Beira Interior (UBI) & Polytechnic Institute of Viseu, Portugal

Contributors

Raquel Soares UNIDCOM-IADE (Design and Communication Research Unit). Cinturs (Research Centre for Tourism, Sustainability and Well-being). Universidade Europeia, Portugal

Maria Carolina Martins Rodrigues Universidade do Algarve (CinTurs) Faro, Portugal

Roger da Silva Wegner Instituto Federal de Educação Ciência e Tecnologia Farroupilha – (IFFAR) – Júlio de Castilhos

Ronnie Figueiredo NECE (Research Center in Business Sciences), University of Beira Interior (UBI), Spinner Innovation Centre, Portugal

Rym Bednarova Department of Pain Medicine, Hospital of Latisana (ASUFC) Latisana, Italy

Svante Lifvergren Centre for Healthcare Improvement, Chalmers University of Technology, Sweden

Taís Pentiado Godoy Universidade Federal de Santa Maria – Campus Cachoeira o Sul

Teresa Cardoso-Grilo Instituto Universitário de Lisboa (ISCTE-IUL), Business Research Unit (BRU-IUL) Centre for Management Studies of Instituto Superior Técnico, Universidade de Lisboa

Ugboaja Samuel Gregory Department of Computer Science, Michael Okpara University of Agriculture, Nigeria

Vera Proença ISPA- Instituto Universitário

About the Editors

Maria José Sousa (PhD in Management) is a University Professor with Habilitation and a research fellow at ISCTE/Instituto Universitário de Lisboa. Her research interests currently are public policies, information science, innovation and management issues. She is a best-selling author in Research Methods, ICT and People Management and has co-authored over 100 articles and book chapters and published in several scientific journals (i.e. *Journal of Business Research, Information Systems Frontiers, European Planning Studies, Systems Research e Behavioral Science, Computational and Mathematical Organization Theory, Future Generation Computer Systems*, and others). She has also organised and peer-reviewed international conferences and is the guest-editor of more than five Special Issues from Elsevier and Springer. She has coordinated several European projects of innovation and is also External Expert of COST Association – European Cooperation in Science and Technology and is former President of the ISO/TC 260 – Human Resources Management, representing Portugal in the International Organisation for Standardisation.

Francisco Guilherme Nunes, before joining ISCTE-IUL in a full-time position (2005), worked as HRM consultant for five years and Marketing Researcher for ten years. After 2005, he taught several courses in HRM and Applied Research Strategies. It was the manager of several programmes in the HRM field (undergraduate, master and Ph.D. levels). He was the Director of the Human Resource and Organisational Behaviour Department for four years. More recently, he shifted his career to a more research orientation. His research interests are the determinants of organisational performance, organisational identity, and leadership development, especially in highly institutionalised contexts like healthcare services or public and non-profits.

Generosa do Nascimento is a PhD in Management, specialising in Human Resources and Organisational Behaviour, by ISCTE-IUL. She is Assistant Professor at the Department of Human Resources and Organisational Behaviour at ISCTE Business School. She is Director of the Executive Master in Strategic People Management and Leadership, of the Executive Master in Healthcare Services Management and of the Post-Graduation at ISCTE Executive Education. The main areas of research are healthcare management and strategic people management. She has been a member of boards of directors (or adviser) and has extensive consulting experience in strategic people management, healthcare services management and organisational change in multiple organisations, private, public and social economy, across different business sectors.

Chinmay Chakraborty, SMIEEE, is an Assistant Professor in Electronics and Communication Engineering, Birla Institute of Technology, Mesra, India, and postdoctoral fellow of Federal University of Piauí, Brazil. His main research interests include the Internet of Medical Things, wireless body sensor networks, wireless networks, telemedicine, m-health/e-health, and medical imaging.

1 Scenarios for the Future of Healthcare
Moving towards Smart Hospitals

Maria José Sousa, Francesca Dal Mas and Paul Barach

CONTENTS

1.1 Introduction 1
1.2 Conceptualization of Smart Hospitals 3
1.3 Methodological Considerations 6
1.4 Interdisciplinary Approach of Smart Hospitals 6
1.5 International Health Innovation Activities 9
1.6 Future Scenarios 12
1.7 Contributions of the Research 14
1.8 Final Considerations 16
References 17

1.1 INTRODUCTION

Healthcare needs to make structural changes; even if there have been noticeable improvements in the efficiency and effectiveness of care in some settings, patients continue to suffer unacceptable harm and frequently struggle to have their voices heard; processes aren't as efficient as they could be; and costs continue to rise at alarming rates while quality issues persist (Phelps and Barach, 2014). New digital technologies have the potential to be revolutionary in the long run by involving citizens and transforming European society and the places we live in, but progress has been gradual. Individual patients could help shape the healthcare and social system by participating in health and social policymaking, organization, design, and delivery, ideally with the backing of their representative organizations. Technology's role in value-based healthcare must be founded on the improvement of patient safety and hospital performance.

How can we leverage connected systems, novel technologies, and clinical solutions to optimize the patient experience while improving patient safety, quality, and financial performance? If healthcare is to be founded on value creation, the service provided should anticipate the needs of the user and should add value to the perception of the user. Patient experience (Cortese and Smoldt, 2007) is increasingly recognized as an important pillar of quality in European healthcare alongside clinical effectiveness and patient safety (Foglino, 2016).

The purpose of this chapter is to define future scenarios for the European (EU) healthcare sector in this setting. How prepared are European healthcare systems for such dangers in the twenty-first century? How can the EU improve its readiness and commitment to achieve the triple goal? The core purpose is to discover strategies to improve healthcare system preparation, to investigate levers to encourage the healthcare sector to innovate, and to propose new techniques to engage the private sector in developing a more diverse and resilient health-tech industry in Europe. According to our early study, the scope of these problems is just too great for immediate, comprehensive national solutions. Progress must be done in small steps, gradually increasing systemic knowledge and awareness. To do so, we will examine the hospital of the future using several vectors: management, skills development, technologies, and health professionals' relationships and practices with patients and their careers. In this context, the impact of new digital technologies (robots, AI, platforms) on the health sector will be wide-ranging.

Besides the research to develop management models, and technological scenarios, we need to instigate a more robust debate in Europe across the different scientific fields. Job demands and resources model (Bakker, 2010), psychological safety (Edmondson, 2005), and job crafting approaches can help hospitals, governance leaders, and teams to (re)design their workflows, processes, and protocols to improve learning, performance, and added value and reduce burnout and others psychosocial risks in health teams. To achieve this, new research is needed to capture the new digital transformations and their potential in the health sector. It meets these scientific needs and goes beyond what is currently available. It will necessitate cutting-edge research to provide new scientific insights, as well as the development and implementation of strategies to prepare for specific threats to patient safety, quality, and reliability of healthcare services in the EU. Public–private partnerships will be necessary as foundations for more extensive resilience and preparedness on a manageable scale while also ensuring a much longer-term and multiparty set of obligations.

The innovation potential of this research is enormous as follows:

- A new way of conceptualizing technological transformation in the health sector and the impacts on the management model, skills, health professionals, and patient relationships, based on empirical real-world healthcare delivery platforms.
- A theoretical framework of Smart Hospitals giving insights to policymakers to create a strategic European health sector, based on the expertise of health professionals, physicians, health networks, and patients.
- Competencies development model for health professionals to use emergent technologies such as augmented reality, simulation, and gamification in health activities.
- An integrative approach to assess the true impact of technological innovations, with new task measured skills, and also measuring management models, and technology integration in work processes.

1.2 CONCEPTUALIZATION OF SMART HOSPITALS

The healthcare landscape in Europe is diverse, with service providers ranging from big healthcare delivery organizations, such as general or specialized hospitals, to local primary care units or health centers, each with its own organizational culture, size, and structure (Glasby, 2003). Empirical research on the best strategies to address hospital governance and change management is critical for supporting informed policy decisions that can improve health care in European countries, particularly in Member States experiencing health system restructuring and reform. Hospitals of the future will be focused on patient safety and patient-centeredness – a true patient centricity model – with a pan-European and international health services priority (Currie, Waring, and Finn, 2008). The last decade has seen significant advances in research and policy in these areas. In an EU survey, over a quarter of respondents said they believed that they (or a member of their family) have experienced an adverse healthcare event in hospitals. (Access to Healthcare Survey Report, 2016).

A significant re-conceptualization of hospital clinical risk has occurred, highlighting how upstream 'latent factors', such as hospital architecture, permit, condition, or increase the potential for 'active errors and patient harm' (Cassin and Barach, 2012). As a result, a study to optimize the interactions between physical design, people, tasks, and dynamic environments is required to understand the features of a safe, resilient, and high-performing system (Lomas, 2007). According to the socio-technical approach, adverse incidents can be investigated from both an organizational and a technical perspective, which includes the concept of latent conditions such as physical and organizational design, as well as the cascading nature of the human error, which starts with management decisions and actions (Waring, 2010).

A 'systems approach' to healthcare emphasizes the larger organization, administration, and culture of the field. Inter-departmental interactions, attitudinal differences, and cultures that normalize risk are shaped by inter-departmental relationships, attitudinal differences, and cultures that normalize risk, according to the applicants' research (Galvan, et al. 2005; Hesselink, 2014). To date, however, most studies have concentrated on a specific clinical site or organization (Swan, 2001), such as primary or secondary care, operating rooms, or the emergency department. The risks to patient safety that arise when patients move between hospitals and healthcare systems, as well as from one EU member to another, receive little attention (Hesselink, 2012). To better comprehend the barriers and drivers to patient safety, it's necessary to think of them as complex and intertwined 'constellations' of factors that exist both within and between care procedures. Regulatory pressures, organizational boundaries, the impact of perverse financial incentives, and professional duty shifting are all examples (Chumer, 2000).

A very important challenge that can be mitigated by our research is related to the opportunities of integrative care approaches, namely the practices and common protocols for patient referral/transfer namely from hospitals to primary care and continuous units and vice-versa. A scientific understanding of the impact of new digital technologies on safe management models, practices, and processes in the health sector is one of the main goals of the Smart Hospitals research. The technological transformations (e.g., nano, biotech, robotics, artificial intelligence, human–computer interaction) have been emerging faster than organizations can adapt nor appreciate.

However, these technologies can be applied to all the clinical and managerial processes creating a European smart health sector, as follows:

a. Patient-facing technologies can be developed strongly dependent on user involvement. Fundamental to success is meaningful and productive end-user involvement and engagement in the process of co-creation with patients (Jacquinet, 2019; Jurgens-Kowal, 2012). This can be applied for better patient and family management in a collaborative system comprised of smartphones/tablets (Matias and Sousa, 2017; Matias and Sousa, 2016), automated scheduling mobile apps connected with appointment systems at hospitals, and with all concerned persons updated for a particular schedule (Farias et al., 2020).
b. Information on patient management can be sent to patients about their pre-tests for any surgery or pathology results directly. In the diagnostic process technologies such as biosensing, surface machines can be used to monitor and diagnose the patient at home such as is being done for COVID-19 patients in Korea. Touch robots can enable remote patient monitoring using telemedicine, and help hospitals follow up on patients after they have been discharged, helping patients follow instructions and avoid the need to be readmitted to the hospital (Broens, 2007).
c. Health professionals can use tablets to improve patient care processes, and they can be installed in each hospital room to make patient charting easier (Gonzalez et al., 2019). Nurses can more easily deliver treatment and update charts. Newer digital life-saving equipment, such as electrocardiograms (ECGs), ventilators, and digital sensors, such as wireless temperature counters, could be linked to smart applications on tablets that can be linked to nursing staff's smartphones. Smartphones equipped with life-saving emergency notifications can send and track disparities in postoperative physical activity between patients who have had a postoperative incident and those who have not. These data can be used to objectively assess patient-centered surgical recovery, which could improve and promote collaborative decision-making, recovery monitoring, and patient participation.
d. Laboratories can be automated by enabling doctors to send online investigation requests and delivering the results can be automated.

In this context, Smart Hospitals need to achieve an effective governance status to enable sustainability in the future, as rapidly evolving technologies are expected to disrupt healthcare and health-related services worldwide (Chen et al., 2019; Gordon et al., 2017; IBM, 2013). There is a growing demand for out-patient ambulatory services, leading fewer people to hospital facilities. However, complex and severe patients will still need in-patient services like the recent coronavirus pandemic is demonstrating (Gardner, 2020, Gordon et al., 2017), especially considering the challenges posed by an ageing population (Bowser et al., 2019) and the impact of chronic diseases (Dal Mas, Massaro, et al., 2019; Massaro et al., 2015; Siemens, 2018), which lead to an increasing healthcare expenditure (Siemens, 2018). The system is moving from a provider-centric model to a patient-centric model (Siemens, 2018), with a focus on the outcome- and value-based care (Dal Mas, Massaro, et al., 2019; Siemens, 2018), with higher expectations by informed stakeholders (IBM, 2013) also in terms of accountability (Chen et al., 2019). The healthcare system is evolving, turning into an open ecosystem in which several stakeholders must cooperate (Bowser et al., 2019; Secundo et al., 2019). In this system, researchers, providers, and payers can engage with patients, caregivers, and others. New technologies can facilitate such an engagement, and if based around trust, allow it to be affordable and scalable (Bowser et al., 2019). New value models are emerging, leading to an everyone-to-everyone (E2E) economy, which is digitally based (Bowser et al., 2019) and will lead to a more sustainable and entrusting healthcare system (IBM, 2013).

Hospitals are an integral part of the care delivery process (Siemens, 2018). Hospital facilities and infrastructures are getting old, especially in low- and middle-income countries (LMIC), but there is an increased demand for hospital beds. In coping with this issue, there is an open call for governments and policymakers to rethink the entire healthcare system, trying to optimize in-patient and out-patient settings, integrate digital technologies, and engage users more effectively leading to co-produced services (Gordon et al., 2017). Emerging technologies can help reduce inefficiencies and costs while improving at the same time care outcomes (Gordon et al., 2017).

Investments in people, technologies, processes, and premises will be required. Such deep and sustained investments may not provide immediate returns in the short run, but they will, in the long run, improve care delivery, increase operational efficiencies, and enhance the experiences of citizens, healthcare professionals, and patients. The outcomes sought would include higher quality of care are, improved efficiency, and better patient satisfaction (Gordon et al., 2017) through productivity (Butt et al., 2018), transparency, building efficiency, and flexibility, and safety and security (IBM, 2013; Siemens, 2018).

A smart hospital, defined as "a hospital that relies on optimized and automated processes built on an ICT environment of interconnected assets, particularly based on Internet of Things (IoT), to improve existing patient care procedures and introduce new capabilities," will be the hospital of the future (ENISA, 2016, p. 9). The smart hospital must be linked to the rest of the environment via the internet (Chen et al., 2019). Remote care systems, mobile client devices, identification systems (using wearables, biometric scanners, and other technologies), secure buildings, networking equipment, interconnected clinical information system (which includes labs, radiology, picture archiving, pharmacy, pathology, blood bank, and other services), networked medical devices (including stationery for life support machines), and data will be among the hospital's assets (ENISA, 2016, p. 15). When these technologies are incorporated into the hospital, they will reach a complete digital reinvention, which incorporates digital technology "to create revenues and results through innovative strategies, products, and experiences" (Bowser et al., 2019, p. 9).

In a micro-analysis, the digitization concepts are first explained in the context of the technical change in the organizations of the health sector and its impacts on the medical practices supported by new technological devices. The digitization can transform health organizations, impacting jobs, skills, and organizational systems (Sousa and Rocha, 2019a). There are two main forms of digitization: the smart industry (Industry 4.0) applied to manufacturing and services, using automation, and advanced robotics (based on artificial intelligence) changing production and services delivery. Secondly, the possibilities for generating new jobs using the native competencies of people as creativity, imagination, and the capacity to reconfigure the world are immense and can open up new avenues for future innovation. However, we should not overlook the potential negative perspectives of policymaking and society including the potential loss of millions of jobs worldwide due to smart automation (Frey and Osborne, 2013; and confirmed by EU-research), the possibilities for generating new jobs using the native competencies of people as creativity, imagination, and the capacity to reconfigure the world are immense and can open up new avenues for future innovation (Gonzalez, Sousa, and Pinto, 2017).

1.3 METHODOLOGICAL CONSIDERATIONS

Using a qualitative approach, we will develop an analysis of the dimensions of Smart Hospitals using an in-depth meta-analysis of the existing studies on organizational transformations as a consequence of the technological transformations in the health sector and will identify health policy directions, governance models and practices, skills, relationships between health professionals, patient practices and external stakeholders, and technologies.

1.4 INTERDISCIPLINARY APPROACH OF SMART HOSPITALS

The topic of Smart Hospitals is highly complex in its interrelations of technology, health sector, and society and its richness at multi-level, multi-jurisdictional issues. Instead of simple causality, there is a multitude of variables that are multi-causal and are mediating or have modifying relationships between those variables and are being mediated and modified by these and other variables. We will use a multi-disciplinary, multi-methods, systems approach (Lilford et al., 2010) to unpack and uncouple these relationships and study the phenomenon from different angles and perspectives. This includes the following:

a. New organizational capabilities (Gordon et al., 2017), namely, new practices forced by the emergence of new technologies, as artificial intelligence and robotics.
b. Create a culture for digital transformation (Sousa and Rocha, 2018a), as managers and executives must understand the need for a digital future, leading its implementation (Gordon et al., 2017) with a top-down approach (Renaudin et al., 2018), pursuing a new focus (Bowser et al., 2019). A cultural shift is needed (IBM, 2013).
c. Use of technologies that communicate with each other, being all devices and technologies used to talk to each other, in a highly interdependent way (Gordon et al., 2017). Process automation will be essential (IBM, 2013).
d. Consider technologies as they change over time, the chosen framework should be flexible and adaptable to add or modify existing solutions (Gordon et al., 2017).

e. Making good use of data, using data analytics (Sousa, 2019), can lead to great value (Dal Mas, Piccolo, et al., 2019; Dal Mas, Piccolo and Ruzza, 2020; Dal Mas, Piccolo, Edvinsson, et al., 2020; Presch et al., 2020). The digital hospital of the future should create a strong system-wide data infrastructure (Gordon et al., 2017).
f. Enhance the digital skills of healthcare professionals, as healthcare organizations will turn to digital leaders (Bowser et al., 2019; Sousa and Rocha, 2018b). Healthcare professionals will need to deal with a variety of technologies. Their future training should take this into high consideration (Dal Mas, Piccolo, Edvinsson, et al., 2020). Multidisciplinary approaches are needed (Siemens, 2018), as new ways of working will be established (Bowser et al., 2019).
g. Ensure cyber security and resilience, because cyber breaches may represent a massive threat for the hospitals of the future (ENISA, 2016). Both cyber security and cyber resilience should be enhanced as primary processes (ENISA, 2016; Gordon et al., 2017).

Major stakeholder groups are given in Table 1.1: (a) health technology networks; (b) policies and experts on Smart Hospitals networks; (c) European University

TABLE 1.1
Stakeholders' Networks

Stakeholders	Potential Contributions
European University Hospital Alliance	The European University Hospitals Alliance (EUHA) is a group of nine university hospitals. Their goal is to create a network of long-term healthcare ecosystems that produce the greatest potential results with the resources available. EUHA may collaborate with Smart Hospitals to share best practices and replicate some of Smart Hospitals' healthcare and education solutions.
European Hospital and Healthcare Federation	HOPE is a non-profit organization with members and representatives from 30 countries that monitors EU policies and legislation that may have an impact on the organization and operation of hospitals and healthcare services, as well as providing input and the viewpoint of healthcare providers in forums where these issues are discussed. Smart Hospitals and HOPE may form a partnership to share best practices and prototype some of Smart Hospitals' ideas.
European Reference Networks for Rare and Low Prevalence Complex Diseases	ERN's mission is to combat complicated or rare diseases and ailments that necessitate highly specialized treatment as well as a concentration of knowledge and resources. There are 24 ERNs that cover all main illness groups and span 25 European nations, with approximately 300 hospitals and 900 healthcare facilities. ERN and Smart Hospitals may collaborate to provide information regarding rare diseases and how they might be treated in a modern, digital healthcare setting, including the need for in-patient and out-patient services.

(continued)

TABLE 1.1 (Continued)
Stakeholders' Networks

Stakeholders	Potential Contributions
i~HD European Network of Excellence for Hospitals	i~HD is made up of several organizations, including the HOPE, several hospitals, and healthcare institutions, as well as pharmaceutical businesses. Its mission is to become the European reference organization for directing and catalyzing the greatest, most efficient, and trustworthy applications of health data and interoperability, to improve health and knowledge discovery. Its mission is to facilitate, organize, and accelerate the development and deployment of interoperable and seamless eHealth solutions and research techniques, to attain best practices and long-term integrated person-centered health care to improve health and wellness. Because iHD's goals are so closely matched with Smart Hospitals', iHD could be the right partner to talk about Smart Hospitals solutions and best practices.
European Health Management Association	The European Health Management Association is a non-profit membership organization with more than 100 members from 30 countries, dedicated to improving health management's capacity and capability to deliver high-quality healthcare through information sharing, training, and research. Given Smart Hospitals' goal of providing high-quality healthcare, EHMA is an appropriate partner for knowledge transfer and training.
European Hospital and Healthcare Employers' Association	HOSPEEM is the European hospital and healthcare employers' association. Given the relevance paid by Smart Hospitals in enhancing the healthcare professionals' wellbeing, HOSPEEM stands as a relevant potential partner to discuss and share best practices related to workflows and workloads.
European Association for Communication in Healthcare	European Association for Communication in Healthcare (EACH) is a non-profit organization dedicated to researching and improving how healthcare professionals, patients, and families communicate with one another. One of the goals of Smart Hospitals is to improve knowledge translation across the various players in the healthcare ecosystem. As a result, EACH's knowledge and the ability to communicate it may be relevant to Smart Hospitals.
International Alliance of Patients' Organizations (IAPO) European Patients Forum	IAPO: promote patients at the centre of healthcare and develop a patient-centered healthcare system around the world, promote patient-centered healthcare around the world. Smart Hospitals will benefit from their experience while also helping to expand the network and focus on the requirements of patients, therefore boosting the quality of connections between health professionals and patients.

Hospital Alliance, and Occupational, Safety and Health networks such as PEROSH and, but also to network such as R&D&I – innovation policy networks.

1.5 INTERNATIONAL HEALTH INNOVATION ACTIVITIES

The five major (inter)national innovation trends and developments with high relevance to the EU are (1) Smart Industry (Industry 4.0: Robotics, AI, IoT, augmented reality), (2) health digital platforms; (3) health digital skills; (4) big data and artificial intelligence (AI); and (5) health knowledge transfer systems. In this context, next, we present the key innovation areas and past and future core European projects related to Smart Hospitals (Table 1.2).

TABLE 1.2
Projects Analysis

Project	Main Results and Link with Our Project
TO-REACH – Transfer of Organizational innovations for Resilient, Effective, equitable, Accessible, sustainable and Comprehensive Health Services and Systems	These project goals are complementary to the Smart Hospitals project as they will contribute to the resilience, effectiveness, equity, accessibility, and comprehensiveness of health services and systems. Being one of the goals of Smart Hospitals is to develop a model for the Hospital of the Future integrating several dimensions: management, technology, skills development, and relationships between health professionals and patients. Their main tasks are mapping the health system challenges and priorities by synthesizing different materials and stakeholder inputs, and this information is important to the definition of our scenarios for a strategic health sector.
Constructing Healthcare Environments through Responsible Research Innovation and Entrepreneurship Strategies	The project's goal is about supporting healthcare research and innovation policy and pilot actions by interlinking RRI – Responsible Research and Innovation, demand-side policy, and territorial innovation models including smart specialization. RRI support interconnections between healthcare research, innovation policies, and testing practices, therefore, such topic can create synergies within Smart Hospitals.
SPRING – Socially Pertinent Robots in Gerontological Healthcare	The project aims to create a new paradigm and concept for socially aware robots, as well as develop novel methods and algorithms for computer vision, audio processing, sensor-based control, and spoken dialog systems that are based on modern statistical and deep-learning techniques to ground the required social robot skills. The project's goal is to develop a new generation of hospital-based robots that are flexible enough to adapt to the needs of their users. The project can develop major synergies with Smart Hospitals, given the importance of robotics in Smart Hospitals, notably in jobs such as drug distribution.

(*continued*)

TABLE 1.2 (Continued)
Projects Analysis

Project	Main Results and Link with Our Project
EURIPHI – European wide Innovation Procurement in Health and Care	The project aims at creating a CoP (Community of Practice) using innovative procurement methods. Unmet needs and shortcomings in the current health systems and care delivery are addressed by using innovation procurement using an EU co-funded instrument of pre-commercial procurement (PCP) or public procurement of innovative solutions (PPI). Smart Hospitals sees knowledge as the primary asset for the healthcare ecosystem. Therefore, knowledge management tools as CoPs using innovative methodologies are highly appreciated, and create, therefore, synergies.
ProACT – Integrated Technology Ecosystems for ProACTtive patient-centered Care	The goal of the project is to create and test an ecosystem that integrates a variety of new and existing technologies to improve and advance home-based integrated care for older adults with multimorbidity and associated co-morbidities, as well as improving patient engagement, workflow management, and healthcare models, both in-patient and out-patient. The project's findings can feed into Smart Hospitals' concept, as Smart Hospitals strive to reduce the number of hospitalized patients by providing out-patient care.
NIGHTINGALE – Connecting Patients and Carers using wearable sensor technology	By designing wearable technology that can continuously monitor patients' vital signs, including blood results and other clinical data, as well as ensuring early warning of actual deterioration in and out of the hospital, the project aims to develop innovative wireless and wearable technology that can be integrated into a clinical decision support system. Given the importance of data information and analytics in the philosophy and workflows of Smart Hospitals, wearable technology can help to build synergies within the Smart Hospitals' idea.
BigMedilytics – Big Data for Medical Analytics	By applying big data technologies to complicated datasets and maintaining the security and privacy of personal data, the initiative intends to improve patient outcomes and increase efficiency in the health sector. Cost reduction, improved patient outcomes, and improved access to healthcare facilities are all goals of the initiative, which considers both in-patient and out-patient treatments. Big data analytics, one of the most widely employed Industry 4.0 technologies in Smart Hospital dynamics, is at the heart of the project. As a result, the project's outcomes may be compatible with Smart Hospitals' concept and strategy.
In-demand – Digital Health solutions proposed and co-created with healthcare organizations	The project aims to develop a new model where healthcare organizations and companies co-create Digital Health solutions. Co-production is a central topic of the modern healthcare ecosystem and a key topic of Smart Hospitals. Therefore, this project's results can create synergies with Smart Hospitals.

TABLE 1.2 (Continued)
Projects Analysis

Project	Main Results and Link with Our Project
Impact HTA – Improved Methods and Actionable Tools for Enhancing Health Technology Assessment	The project examines new and improved methods in ten thematic areas with the goal of better understanding cost and health outcomes within and across countries while integrating clinical and economic data from various sources to improve economic evaluation methods in the context of health system performance measurement. Keeping expenses under control is critical in a patient-centered healthcare system where health outcomes are the primary goal. Smart Hospitals attempt to balance the benefits of outcomes with cost savings through technology and improved workflows, therefore there are meaningful synergies with this initiative.
DeepHealth – Deep-Learning and HPC to Boost Biomedical Applications for Health	The goal of the project is to combine HPC infrastructures with deep learning (DL) and artificial intelligence (AI) techniques to support biomedical applications that require the analysis of large and complex biomedical datasets, resulting in new and more efficient methods of disease diagnosis, monitoring, and treatment. Smart Hospitals are built on new industry 4.0 technologies, which can improve healthcare outcomes, streamline procedures, and save costs. As a result, the initiative has a lot of overlap with Smart Hospitals' goals and philosophy.
EU mHealth Hub Project	Early diagnosis of diseases and prevention can help the management of healthcare services to an ageing population, with increasing chronic diseases. The project aims to establish a 'Knowledge and Innovations Hub for mHealth' to enable and monitor mHealth innovation and adoption in Europe and to support the development of national mHealth interventions in selected members. Early diagnosis, prevention, the increase in healthcare outcomes, and the reduction of costs are key topics within Smart Hospitals philosophy and project plan, therefore, there can be synergies with such projects.
ImpleMentAll – Getting eHealth implementation right	Through the development, application, and evaluation of tailored implementation strategies in a natural laboratory of ongoing eHealth implementation initiatives, the project aims to provide an evidence-based answer to the issue of knowledge translation from scientific research into clinical practice. Knowledge translation from scientific research into clinical practice helps to the modern ecosystem's and Smart Hospitals' patient-centric goals. As a result, there are synergies between this project and the goals of Smart Hospitals.

(continued)

TABLE 1.2 (Continued)
Projects Analysis

Project	Main Results and Link with Our Project
THALEA II – Telemonitoring and Telemedicine for Hospitals Assisted by ICT for Life-saving co-morbid patients in Europe as part of a patient personalised care program of the EU	Through the development, application, and evaluation of tailored implementation strategies in a natural laboratory of ongoing eHealth implementation initiatives, the project aims to provide an evidence-based answer to the issue of knowledge translation from scientific research into clinical practice. Knowledge translation from scientific research into clinical practice helps to the modern ecosystem's and Smart Hospitals' patient-centric goals. As a result, there are synergies between this project and the goals of Smart Hospitals.
CROWDHEALTH – Collective wisdom driving public health policies	The project's goal is to combine large amounts of health-related heterogeneous data from many sources to help policymakers make better decisions. The translation of clinical practice and procedures into policy is an important aspect of modern healthcare, and it is also important to the philosophy of Smart Hospitals. Triangulation of data and analytics appear to be effective methods for promoting such dynamics. As a result, there are parallels with the notion of Smart Hospitals.
PatientDataChain – Blockchain approach to disrupt patient–provider medical records data exchange	The project uses Smart Contracts to map patient–provider relationships, thanks to the transparency and trust of blockchain technology. According to the literature, blockchain is the most disruptive new technology. There is a great potential in the use of the blockchain for the management of patients' data, ensuring transparency and reducing errors, and allowing to use of anonymised big data for research purposes. Given such aims, the project can be well integrated within Smart Hospitals' goals and philosophy.

1.6 FUTURE SCENARIOS

The main lines and dimensions to think about a definition of future scenarios for the health sector are based on several premises: (a) The introduction of cultural and organizational governance approaches focused on the leadership pipeline alignment with a new organizational "blue-print" – as managers and leaders are the first practitioners of the new practices; (b) In an assessment of resource utilization for patient services including the identification of the best and most efficient and effective practices in hospital planning and design; (c) On the identification of the main barriers and facilitators of end-user engagement and involvement in hospital planning, construction, and management; (d) on developing and producing content dealing with transforming a hospital to a hospital of the future: technologies, management models, new work models, new practices, employee engagement, professional ethics, citizens/patient relationships models; (e) in developing and implementing diversified training programs and pedagogical approaches based on real problems; (f) design, with the stakeholders from across the EU, including national coalitions (health professional bodies, universities, policymakers, and other stakeholders), an evidence-based framework for measuring intermediate and long-term outcomes of hospital planning, design, and management that federal and national governments can customize according to the context of their healthcare and political systems; (g) in developing, disseminating, and sustaining plans regarding methods, tools, best practices, and policies on hospital design, planning, and management through partnering with the key stakeholders such as the European Society of Quality Healthcare (ESQH), the European Hospital and Healthcare Federation (HOPE), the World Health Organization (WHO) Patients for Patient Safety Programme, the International Alliance of Patient's Organizations (IAPO), and the European Patient's Forum (EPF).

According to this premises, six forces for creating the scenarios were identified:

Force 1: Address the organizational impact of technological transformations and contribute to promoting the development of healthcare professionals.

Research generally neglects or makes unfounded predictions about the impacts of technological transformation at the national and EU health sector levels. European data are underexplored and extant research and concepts are scattered. A balanced European research consortium is needed to address this complex issue and is missing to date.

Force 2. Transforming the impact of technology on the health sector is a major value for health professionals to strengthen their skills.

Technologies are impacting the medical practice with robotization and artificial intelligence, also the management processes are being influenced, and also the relationships between professionals of the health sector and patients, which requires new skills development.

Force 3: Create a robust governance for the health sector, based on scientific and evidence-based knowledge.

This robust governance requires a balanced, integrative approach to health policies at a European and a national dimension.

Force 4: Policy for developing and empowering health professionals and technological penetration in health organizations.

Policy recommendations to deal with technological transformation impacts on skills and to promote the changes in health organizations.

Force 5: Strengthening the innovation capacity of health professionals regarding their relationships with the patients.

Health professionals require new skills development with the advent of new technologies to make the diagnosis, treatment, and follow-up of their patients.

Force 6: Creating research synergy with external stakeholders.

Smart Hospitals contribute to state-of-the-art research and knowledge and is also strongly committed to strengthening the European health networks, through case studies and good practices. This guarantees access to and uptake of knowledge by and for all. By doing so, we utilize, mobilize, and synergize all knowledge produced by the stakeholders and develop the needed skills for gaining maximum impact.

The interactions among the driving forces resulted in four scenarios (Figure 1.1).

1.7 CONTRIBUTIONS OF THE RESEARCH

This research contributes to the advancement of the theoretical and empirical knowledge on smart hospital concepts and in identifying scenarios for the future of the health sector. It delivers a state-of-the-art and analysis of the role of the stakeholders, an analysis of the innovation introduced by research projects during the past decade, and a proposal of scenarios for the future of hospitals.

It gives directions for creating new management models for hospitals, based on the European best practices and the stakeholder's insights, which will impact the way a hospital can be managed. It develops recommendations for policy regarding the promotion of the empowerment and skills development of the health professionals, and regarding the technological penetration of the technology in health organizations, giving directions for a European strategy for the health sector.

This research contributes also to altering the relationships among healthcare professionals and patients, as the new practices and technologies can change the way healthcare services will be delivered, preferentially with greater efficiency.

Finally, it strengthens the European research area on health management, technology, health management skills, and relationships between health professionals and patients, providing means for health scientists, health policymakers, and health organizations to optimally deal with the technological and digital transformations.

In the short term, this research will:

- Contribute and add to the work of existing state-of-the-art in Smart Hospitals.
- Give knowledge about the international networks concerned with Smart Hospitals development and challenges and set up relations for collaboration and exchange of knowledge.
- Support capacity building and continuing development on health skills that contribute to the efficiency and efficacy of the health sector.
- Provide the opportunity for international colleagues to address the smart hospital challenges and to build new relationships and future working partnerships on health technological challenges.
- Promote and expand knowledge in the European health sector spheres.

In the long-term, the impacts will continue in terms of:

- Creating knowledge about Smart Hospitals model development according to the technological evolution;

Moving towards Smart Hospitals

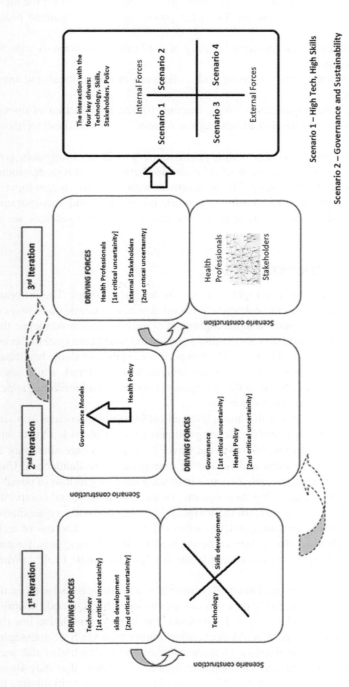

FIGURE 1.1 Scenarios model.

- Influencing policy at a national, European, and international level for the introduction of recommendations on Smart Hospitals, and a more strategic health sector;
- Create awareness of the capacity building of the health professionals with the skills development;
- Sharing knowledge, and expertise with colleagues at the national and international levels;
- Allow government agencies to create avenues for the introduction of disruptive technologies while also addressing the societal issues that these technologies bring;
- Develop innovative ways of providing public services while ensuring public governance (via the reconfiguration of relationships between health professionals and patients) and enhancing public involvement with disruptive technology;
- Developing new methods, optimizing work procedures, and incorporating evidence-based decision-making processes into health public services are all priorities.

1.8 FINAL CONSIDERATIONS

Smart Hospitals are strategic and highly relevant in the current times. The introduction of the latest technologies in healthcare, embedded in new processes and ways of working, can help the management of the nightmare of the coronavirus pandemic that is unfolding around the world and is ravaging patients and healthcare systems across Europe at the time of writing. The COVID-19 experience is pushing all the healthcare systems under great pressure. New healthcare professionals are hired, while others had to change their roles to cope with the emergency. Flexibility and resilience appear to be the top priority in managing the crisis.

According to the most recent literature, telemedical innovations such as allowing clinicians to see patients at home or allowing patients to schedule video visits with established or on-demand providers to avoid travel to in-person care sites may be an answer in managing disasters and healthcare emergencies (Hollander and Carr, 2020). Artificial Intelligence can assist doctors in quickly detecting lesions of possible coronavirus pneumonia, measuring their volume, shape, and density, and comparing changes of multiple lung lesions from the image, all of which provide a quantitative report to assist doctors in making quick decisions (McCall, 2020). The use of technology was positively used also in Taiwan during the COVID-19 emergency, for case identification, containment, and resource allocation to protect public health. (Wang et al., 2020).

New technology was employed to classify travellers' infectious risks based on the flight origin and travel history in the previous two weeks, including QR code scanning and online reporting of travel history and symptoms. Passengers classified as low risk (based on alert areas) received a health declaration border pass via SMS messaging on their phones for faster immigration clearance; those classified as higher risk were quarantined at home and tracked via their smartphones to ensure that they stayed at home during the incubation period (Wang et al., 2020). This study highlights the necessity of professional training in dealing with crises and activating emergency management mechanisms to address an emergent pandemic, but this can only be done correctly if people are trained on how to respond and apply their talent.

REFERENCES

Barach, P. and Phelps, G. (2013). Clinical sensemaking: A systematic approach to reduce the impact of normalised deviance in the medical profession. *Journal of the Royal Society of Medicine*, *106*(10): 387–390. http://dx.doi.org/10.1177/0141076813505045.

Bowser, J., Saxena, S., Fraser, H. and Marshall, A. (2019). *A Healthy Outlook: Digital Reinvention in Healthcare*. Armonk, NY: IBM Institute for Business Value, , available at papers3://publication/uuid/00C50A8C-9B9B-49D3-96B8-C66B960B2068

Broens, T.H.F, et al. (2007). Determinants of successful telemedicine implementations: A literature study. *Journal of Telemedicine and Telecare*, *13*(6): 303–309.

Butt, M.A., Nawaz, F., Hussain, S., Sousa, M.J. et al. (2018). Individual knowledge management engagement, knowledge-worker productivity, and innovation performance in knowledge-based organizations: The implications for knowledge processes and knowledge-based systems. *Computational and Mathematical Organizational Theory*, *25*(3): 336–356. https://doi.org/10.1007/s10588-018-9270-z

Cassin, B. and Barach, P. (2012). Making sense of root cause analysis investigations of surgery-related adverse events. *Surgical Clinics of North America*, *92*(1): pp. 101–115.

Chen, B., Baur, A., Stepniak, M. and Wang, J. (2019). *Finding the Future of Care Provision: The Role of Smart Hospitals*. McKinsey & Company (retrieved from: www.mckinsey.com/~/media/McKinsey/Industries/Healthcare%20Systems%20and%20S services/Our%20Insights/Finding%20the%20future%20of%20care%20provision%20the%20 role %20of%20smart%20hospitals/Finding-the-future-of-care-provision-the-role-of-smarthospitals.ashx).

Chumer, M., Hull, R., and Prichard, C. (2000). Introduction: Situating discussions about "knowledge". In: Pritchard, C., Hull, R., Chumer, M., and Willmott, H. (Eds), *Managing Knowledge: Critical Investigations of Work and Learning*. Basingstoke: Macmillan.

Cortese, D. and R. Smoldt (2007). A health system by design: The future of healthcare must be about competition for patients based on value. *Modern Healthcare*, *37*(38): 38.

Currie, G., Waring, J., and Finn, R. (2008). The limits of knowledge management for public sector modernisation: The case of patient safety and quality. *Public Administration*, *86*(2): 363–385.

Dal Mas, F., Massaro, M., Lombardi, R., and Garlatti, A. (2019). From output to outcome measures in the public sector: A structured literature review, *International Journal of Organizational Analysis*, *27*(5): 1631–1656.

Dal Mas, F., Piccolo, D., Cobianchi, L., Edvinsson, L., Presch, G., Massaro, M., Skrap, M., et al. (2019). The effects of Artificial Intelligence, Robotics, and Industry 4.0 technologies. Insights from the Healthcare Sector. Proceedings of the first European Conference on the impact of Artificial Intelligence and Robotics, Academic Conferences and Publishing International Limited, pp. 88–95.

Dal Mas, F., Piccolo, D., Edvinsson, L., Skrap, M., and D'Auria, S. (2020). Strategy innovation, intellectual capital management and the future of healthcare: The case of Kiron by Nucleode. In: Matos, F., Vairinhos, V., Salavisa, I., Edvinsson, L., and Massaro, M. (Eds.), *Knowledge, People, and Digital Transformation: Approaches for a Sustainable Future*. Cham: Springer, pp. 119–131.

Dal Mas, F., Piccolo, D., and Ruzza, D. (2020) Overcoming cognitive bias through intellectual capital management: The case of pediatric medicine. In: Ordonez de Pablos, P. and Edvinsson, L. (Eds.), *Intellectual Capital in the Digital Economy*. London: Routledge.

Edmondson A.C. and Mogelof J.P. (2005). Explaining psychological safety in innovation teams. In: Thompson L. and Choi H. (Eds.), Creativity and Innovation in Organizations, Mahwah, NJ: Erlbaum, pp. 109–136.

ENISA. (2016), Smart hospitals: Security and resilience for smart health service and infrastructures, (Security, E.U.A. for N. and I.,Ed.), ENISA, Heraklion, doi:10.2824/28801.

Farias, F.A.C. de, Dagostini, C.M., Bicca, Y. de A., Falavigna, V.F., and Falavigna, A. (2020). Remote patient monitoring: A systematic review. *Telemed J E-Health Off J Am Telemed Assoc.*, 26(5): 576–583. doi:10.1089/tmj.2019.0066

Foglino, S., Bravi, F., Carretta, E., Fantini, M.P., Dobrow, M.J. and Brown, A.D. (2016). The relationship between integrated care and cancer patient experience: A scoping review of the evidence. *Health Policy*, 120 : 55–63.

Frey, C. and Osborne, M. (2013). *The Future of Employment: How Susceptible Are Jobs to Computerisation? Working Paper.* Oxford: University of Oxford.

Galvan, C., Bacha, B., Mohr, J., and Barach, P. (2005). A human factors approach to understanding patient safety during pediatric cardiac surgery. *Progress in Pediatric Cardiology*, 20:13–20.

Gardner, L. (Update January 31, 2020). *Modeling the Spreading Risk of 2019-nCoV*. Johns Hopkins University.

Glasby, J. (2003). *Hospital Discharge: Integrating Health and Social Care*. Oxford: Radcliffe Publishing.

Gonzalez, Garcia, M., Fatehi, F., Bashi, N., et al. (2019). A review of randomized intervention trials utilizing telemedicine for improving heart failure readmission: Can a realist approach bridge the translational divide? *Clinical Medicine Insights in Cardiology*, 13:1179546819861396. doi:10.1177/1179546819861396; PMID: 31316270; PMCID: PMC6620724.

Gordon, R., Perlman, M., and Shukla, M. (2017). *The hospital of the future: How digital technologies can change hospitals globally, Deloitte*, available at:www2.deloitte.com/content/dam/Deloitte/global/Documents/Life-Sciences-Health-Care/us-lshchospital-of-the-future.pdf.

Hamel, L., Kirzinger, A., and Brodie, M. (2016). 2016 Survey of Americans on the U.S. Role in Global Health. Retrieved from www.kff.org/global-health-policy/poll-finding/2016-survey-of-americans-on-the-u-s-role-in-global-health/

Hesselink, G., Schoonhoven, L., Barach, P., Spijker, A., Gademan, P., Kalkman, C., Liefers, J., Vernooij-Dassen, M., and Wollersheim, W. (2012). Improving patient handovers from hospital to primary care: A systematic review. *Annals of Internal Medicine*, 157(6): 417–428.

Hesselink, G., Zegers, M., Vernooij-Dassen, M., Barach, P., Kalkman, C., Flink, M., Öhlen, G., Olsson, M., Bergenbrant, S., Orrego, C., Suñol, R., Toccafondi, G., Venneri, F., Dudzik-Urbaniak, E., Kutryba, B., Schoonhoven, L., and Wollersheim H. (2014). European HANDOVER Research Collaborative: Improving patient discharge and reducing hospital readmissions by using intervention mapping. *BMC Health Services Research*, 13(14): 389. doi: 10.1186/1472-6963-14-389.

Hollander, J.E. and Carr, B.G. (2020). Virtually perfect? Telemedicine for Covid-19. *New England Journal of Medicine*, 382(18): 1679–1681, DOI: 10.1056/NEJMp2003539

IBM. (2013). The digital hospital evolution: Creating a framework for the healthcare system of the future., Somers, available at: www.himss.eu/sites/himsseu/files/education/whitepapers/IBM Digital Hospital Evolution GBW03203-USEN-00.pdf.

Jacquinet, M., Nobre, A., Curado, H., Sousa, M.J., Pimenta, R., Arraya, M., and Martins, E. (2019). *Management of Tacit Knowledge and the Issue of Empowerment of Patients and Stakeholders in the Health Care Sector in Healthcare Policy and Reform: Concepts, Methodologies, Tools, and Applications*, Hershey, PA: IGI Global. DOI: 10.4018/ 978-1-5225-6915-2

Jurgens-Kowal, T. (2012). The power of co-creation. *Journal of Product Innovation Management*, 29(4): 683–683.

Lilford, R.J., Chilton, P.J., Hemming, K., Girling, A.J., Taylor, C.A., and Barach, P. (2010). Evaluating policy and service interventions: Framework to guide selection and interpretation of study end points. *British Medical Journal*, 341: c4413.

Lomas J. (2007). The in-between world of knowledge brokering. *BMJ*, 334(7585): 129–132.

Massaro, M., Dumay, J., and Garlatti, A. (2015). Public sector knowledge management: A structured literature review. *Journal of Knowledge Management*, 19(3): 530–558.

Matias, N. and Sousa, M.J. (2016). Mobile health as a tool for behaviour change in chronic disease prevention: A systematic literature review. 11th Iberian Conference on Information Systems and Technologies (CISTI16), 1–6.

Matias, N. and Sousa, M.J. (2017). Mobile health, a key factor enhancing disease prevention campaigns: Looking for evidences in kidney disease prevention. *Journal of Information Systems Engineering & Management*, 2(3): 1–15, ISSN, 2468-2071

McCall, B. (2020). COVID-19 and artificial intelligence: Protecting health-care workers and curbing the spread, *The Lancet*, February 20, DOI: https://doi.org/10.1016/ S2589-7500(20)30054-6

Phelps, G. and Barach, P. (2014). Why the safety and quality movement has been slow to improve care? *International Journal of Clinical Practice*, 68(8): 932–935.

Presch, G., Dal Mas, F., Piccolo, D., Sinik, M., and Cobianchi, L. (2020). The World Health Innovation Summit (WHIS) platform for sustainable development: From the digital economy to knowledge in the healthcare sector. In: Ordonez de Pablos, P. and Edvinsson, L. (Eds.), *Intellectual Capital in the Digital Economy*. London: Routledge, pp. 217–224.

Renaudin, M., Dal Mas, F., Garlatti, A. and Massaro, M. (2018). Knowledge management and cultural change in a knowledge-intensive public organization. In: Remeniy, D. (Ed.), 4th Knowledge Management and Intellectual Capital Excellence Awards, Academic Conferences and Publishing International Limited, Reading, pp. 85–96.

Ruzza, D., Dal Mas, F., Massaro, M. and Bagnoli, C. (2020). The role of blockchain for intellectual capital enhancement and business model innovation. In: Ordonez de Pablos, P. and Edvinsson, L. (Eds.), *Intellectual Capital in the Digital Economy*. London: Routledge.

Secundo, G., Toma, A., Schiuma, G., and Passiante, G. (2019). Knowledge transfer in open innovation: A classification framework for healthcare ecosystems. *Business Process Management Journal*, 25(1): 144–163.

Siemens. (2018). Smart hospitals – smart healthcare Creating perfect places to heal, Zug.

Sousa, M.J. (2019). Skills to boost innovation: In the context of public policies. *SWS Journal of Social Sciences and Art*, 1: 90–103 ISSN: 2664-0104

Sousa, M.J. and González-Loureiro, M. (2016). Employee knowledge profiles: A mixed-research methods approach. *Information Systems Frontiers*, 18: 1103–1117, ISSN: 1572-9419

Sousa, M.J., Pesqueira, A., Lemos, C., Sousa, M., and Rocha, Á. (2019). Decision-making based on big data analytics for people management in healthcare organizations. *Journal of Medical Systems*, 43(9): 1–10, ISSN: 01485598.

Sousa, M.J. and Rocha, A. (2018). Learning analytics measuring impacts on organisational performance. *Journal of Grid Computing*, 18: 563–571, ISSN: 1572-9184

Sousa, M.J. and Rocha, A. (2018a). Digital learning: Developing skills for digital transformation of organizations. *Future Generation Computer Systems* (JCR (Q1) IF 4.639/ Scopus) *91*: 327–334, ISSN: 0167739

Sousa, M.J. and Rocha, A. (2018b). *Corporate Digital Learning: Proposal of Learning Analytics Model*. In Rocha, A. (Ed.), AISC Series. Berlin: Springer.

Sousa, M.J. and Rocha, A. (2019a). Skills for disruptive digital business. *Journal of Business Research*, Elsevier, 94(C): 257–263 doi: https://doi.org/10.1016/j.jbusres.2017.12.051

Sousa, M.J. and Rocha, A. (2019b). Leadership styles and skills developed through game-based learning. *Journal of Business Research*, *94*(C): 360–366 https://doi.org/10.1016/j.jbusres.2018.01.057

Sousa, M.J. and Rocha, A. (2019c). Editorial strategic knowledge management in the digital age. *Journal of Business Research*, *94*(C): 223–226, ISSN: 0148-2963

Sousa, M. J., Rocha, A., and Sousa, M. (2019). Digital and innovation policies in the health sector. In Tomé, E., Cesario F. and Soares, R. (Eds), Proceedings of the European Conference on Knowledge Management, ECKM (Vol. 2, pp. 967–977).

Sousa, M.J. and Wilks, D. (2018). Sustainable skills for the world of work in the digital age. *Systems Research and Behavioral Science*, *35*(4): 399–405, ISSN 1099-1743

Swan, J. and Scarbrough, H. (2001). Knowledge management: Concepts and controversies. *Journal of Management Studies*, *38*(7): 913–921.

Wang, C.J., Chun, Y. N., and Brook, R. H. (2020). Response to COVID-19 in Taiwan: Big data analytics, new technology, and proactive testing. *JAMA*, 323(14):1341–1342, March 3, DOI:10.1001/jama.2020.3151

Waring, J., Rowley, E., Dingwall, R., Palmer, C., and Murcott, T. (2010). Narrative review of the UK Patient Safety Research Portfolio. *Journal of Health Services Research & Policy*, 15 Suppl 1: 26–32.

2 People Management in Healthcare

The Challenges in the Era of Digital Disruption

Alzira Duarte, Generosa do Nascimento and Francisco Nunes

CONTENTS

2.1 Introduction 21
2.2 Technological Evolution and Digital Disruption in Healthcare
 Organizations..22
 2.2.1 Perspectives of Human Resources Management............................24
 2.2.2 Challenges to the Integration of the Digital HRM in Healthcare
 Organizations..27
2.3 Proposal for an Integrated People Management Model 31
2.4 Conclusion...34
References..34

2.1 INTRODUCTION

In the current context, notably brittle, anxious, non-linear and incomprehensible (BANI), human and social capital, supported by cutting-edge technology, have become an essential competitive advantage for any organization in any sector. The COVID-19 pandemic and its effects have come in an unpredictable but undelayable way to demonstrate the importance of technologies and the capacity for adaptability of organizations and people as a condition for survival and success. However, this period came also to demonstrate the importance of the existence of solid management models for organizations that maximize results in their multiple aspects: for people, organizations, stakeholders, and society.

Schwab in 2016 warned of the emergence of the Fourth Industrial Revolution, marked by cyber-physical systems, and by the rapid advances that new technologies brought to society. Effectively, the information and communication technologies, in the form of artificial intelligence, robotics and advanced technologies, have evolved exponentially and with a strong impact on how to organize the services and the functions themselves, forcing a dynamic and permanent reinvention of the human resources management itself (Ancarani et al., 2019).

These new technologies have changed the operation models of organizations, imposing not only the emergence of new products and services, but above all forcing radical changes in the role of people and how they are managed in the organizational context (Vrontis et al., 2021). The automation of administrative processes, the virtualization of collaboration/communication and the speed of creating new services and products have imposed new forms of work and their organization (Bondarouk et al., 2017) bringing with them new challenges and threats.

We live nowadays what already in 2013 Scharmer and Kaufer designated as the age of disruption. A period strongly marked by:

- Talent scarcity – new market requirements, new skills and capabilities to the detriment of the traditional perspectives;
- The flourishing of automation – the largest expansion of automation replacing less sophisticated functions;
- Impermanent work market – the massive extinction of jobs with the emergence of new functions still unexpected;
- Increased patient capacity and requirements – by the patients who virtually accompany the innovations and demand them in their interaction with the services;
- Value-based healthcare volatility – The speed of adaptation to markets requirements is determinant for the survival of healthcare units;
- And we also add the dematerialization of processes and jobs themselves and the virtualization of services, as we traditionally knew them – with the ubiquitous that technology brings, it becomes imperative to enable users to use it (professionals, users, citizens) as a condition of preparation for a Society 5.0.

The relationship between technologies and people has become increasingly a symbiotic and complementary one, in which success and sustainability are based on the ability to adapt and integrate, with an impact not only on organizations but also upon society itself. In this way, it is essential to reflect on an integrated model for the management of people, which contemplates the individual/society complementarity, mediated by advanced technological contexts that ensure the necessary conditions for the success, well-being and development of professionals, and for organizational and sustainability results.

In this manner, we propose to make a synthetic reflection on technological evolution and digital disruption in healthcare services that guide the current management of people; the challenges associated with the integration of technologies in the management of people in the context of health organizations; and finally the presentation of a proposal for an Integrated People Management Model (IPM 4.0).

2.2 TECHNOLOGICAL EVOLUTION AND DIGITAL DISRUPTION IN HEALTHCARE ORGANIZATIONS

Throughout time, technology has evolved into different stages that marked significantly the models of production and the availability of goods and services, with consequent effects on the management of people, as illustrated in Table 2.1.

TABLE 2.1
Technological Development Stages

Stage 1. "Use of machines driven by steam power to replace labor in the transformation of raw materials into products" (Martin & Siebert, 2016, p. 313)

Stage 2. "Use of electrically powered machines to move materials between machines and to power moving assembly lines and flow production" (Martin & Siebert, 2016, p. 313)

Stage 3. "Use of electronics-based ICT to coordinate and control transfer of information and tasks" (Martin & Siebert, 2016, p. 313)

Stage 4. "Use of several related technologies, all of which reinforce each other and are based on the microchip, cheap and massive computing power and the Internet" (Martin & Siebert, 2016, p. 313)

Stage 5. Development of artificial intelligence (AI), nanotechnology, Internet of Things, and other intelligent technologies, with the potential for customization and the creation of unique and anticipatory responses to future needs.

Source: Adapted from Martin & Siebert (2016).

The first two stages mark the traditional work, first and second industrial revolutions (stage 1 – steam; stage 2 – electric), and have determined new forms of production with the change of manual production models to mechanization and the need for the implementation of new models of work organization.

The new technologies (stage 3 – computing and information systems; stage 4 – web technologies) identified with the digital revolution (Barley, 2020) or 4.0 revolution (Schwab, 2016) came to create conditions of change at the infrastructure level, the optimization of production conditions and the provision of services, giving rise to new forms of doing work.

It is, however, with the emergence of smart technologies (artificial intelligence, nanotechnology and IoT (stage 5)) and of ubiquitous computing, with its immense transformative potential, that the conditions for the design of new services and new ways to make them available have emerged, transforming economies and organizations, and imposing new functions and ways of working in the direction of "super smart society – Society 5.0".

Although they are of an incremental character, the verified changes associated with the evolution of technology have often brought new challenges that have the tendency to intensify as the systems that mold them become more complex.

Health systems are high complexity adaptative systems (Begun et al., 2003). They are intensive knowledge systems, with high technological dependence and are determined by the ability to adapt, and by inter-complementary, self-organization and by co-evolution, all framed by an ethical climate.

As we witness technological evolution, we are confronted with questions that refer to the funding, development and maintenance of knowledge in various resources and decision contexts. At present, and in the field of health, the development of healthcare data analytics models and the integration with artificial intelligence, potentiate the emergence of new equipment, processes and communication models. New practices, many of them supported by mobile and remote devices, by wearable and implantable medical devices, telemedicine, have come to respond to the needs of efficiency and accessibility, underlying the success of what health organizations will be in the future.

However, and being associated with the potential for more directed and accurate responses, technological development has also brought new challenges that can be analyzed in three distinct dimensions:

1. *For the patients*, technological advancement has shown the differences in the conditions of access and in the familiarity with the resources/processes. For example, telemedicine, which favors real-time communication between professionals and patients, with relevant information sharing, presupposes knowledge and familiarity with the internet and also connectivity conditions that are still limited in many regions and for many patients (e.g. older populations or the more vulnerable).
2. *For health professionals*, the speed and type of skills required by new technologies, demand that, allied to specialized training of high technicality, it should associate all in a relational pattern and it also requires permanent and constant learning throughout life, which goes beyond the domain of expertise. For example, the use of healthcare data analytics and artificial intelligence models presupposes the development and validation of data synthesis algorithms, where many of them are based on statistical principles and on high requirements in terms of collection and systematization of information for clinical decisions.
3. *For the management of healthcare services*, in addition to the creation and to the development of the conditions for timely responses to the needs of the patients (in terms of human and physical resources), concerns have come up, and they are associated with the development of integrating policies, and with cost control (efficiency and optimization of resources), with security and confidentiality, and with the satisfaction of the patients and the professionals. Technology, by providing conditions for streamlining processes, and for the compensation for human insufficiencies, has equally brought in some challenges, such as the legitimacy of data collection, the validity of decisions supported by artificial intelligence or even questions related to the knowledge and intentions of different potential users (cybersecurity issues, such as exposing data in cyberattacks contexts).

It is in this context that human resources management asserts itself as an essential condition in providing integrated management models for support staff, healthcare professionals and patients, who have been evolving and adapting over time.

2.2.1 Perspectives of Human Resources Management

When considering the recent history of the evolution of human resources management, three major perspectives are clearly identified:

1. The first perspective, characterized as traditional human resources management (HRM), that focuses essentially on the fulfillment of activities and tasks of an operational/ administrative nature, that is, the dominant idea is control and execution.
2. The second perspective, the strategic human resources management (SHRM), which emerged in the 1980s, focuses on the influence of HRM practices on relevant individual and organizational results, such as well-being or performance (Beer et al., 1984), emphasizing the link between HRM practices and strategic alignment. The main conceptual approaches, as mentioned by Wight and Ulrich (2017), are based on the theory of human capital, and on the theory of social exchange, as well as on the resource-based approach. As mentioned by Armstrong and Brown (2019), this implies the need for "adjustment" between organizational strategy and HR practices, in a multi-stakeholder approach, while also developing a culture of trust. By considering the SHRM as a social system, this puts stakeholders' relationships at the center of action. SHRM takes a broader approach than HRM, focusing on the elaboration of a long-term strategy (Alfawaire and Atan, 2021).
3. Finally, the third perspective is the digital HRM or HRM 4.0, whose dominant idea is the efficiency and rationality, arising from technological evolution and focusing on the performance potential of the employees, leveraged by the development and adoption of new technologies, namely process automation and artificial intelligence. Verlinden (2018) states that the use of advanced mobile, social, analytic and cloud technologies favors management of people as well as it improves the efficiency and optimization of functions.

In short, in Table 2.2, there are five comparison vectors between the three approaches.

Considering the analysis of Table 2.2, we would be led to consider that digital HRM does not constitute a real alternative to the two approaches usually considered (the traditional perspective of HRM and the strategic HRM). This is not true and needs further reflection.

Effectively, analyzing the evolution of technology and its integration in people management practices and models, two great moments can be seen that, although consonant, mark deep differences in terms of philosophy, practices, purposes and results in an organizational context.

In line with the first perspective of people management (the traditional HRM, previously mentioned), we have the adoption of technology marked by an operational application of digital technologies (Strohmeier, 2020). This conception of HRM digital foresees the use of information and communication technology essentially as a support for HRM operational practices (e.g. in procedural operations related to attendance control systems and administrative management, recruitment and selection, and rewards among others). As stated by Strohmeier (2020, p.355) "this support is realized by automating HR practices with the aim of increasing their speed, decreasing their costs, and improving their quality (e.g. Snell et al., 1995; Strohmeier, 2007). Strategic purposes are not supported". This perspective of operational application of digital technologies tends to become more sophisticated, creating conditions to move from a phase of strict

TABLE 2.2
A Comparison of HRM, SHRM and Digital HRM

Dimensions	HRM (Alfawaire and Atan, 2021, p.34)	SHRM (Alfawaire and Atan, 2021, p.34)	Digital HR
Focus	People emotions or motivational requirements	The connection between organizational strategies and HRM practices affecting individual and organizational performance	The use of technology in HRM to improve its effectiveness and collection, exchange, and control of information, inside and outside the organization
Scope	Fragmented and micro vision (aiming at the local optimization of HRM individual practices)	Comprehensive and macro vision (aiming at overall optimization through vertical and horizontal HRM practices	Micro vision (aiming at the local optimization of HRM individual practices with emphasis on efficiency and rationality) and transition to macro vision (integrative)
Duration	Short term	Long and medium term	Long, medium, and short term
Roles and functions	Not associated with organizational goals and systems	Motivators for reinforcement and self-governance	Greater administrative quality and flexibility in HRM, and new types of work
Relating HR practices with non-HR practices	Peripheral, support	Nuclear, collaborative	Integrative, support

Source: Adapted from Alfawaire and Atan (2021).

execution to a mixed phase of execution and support for the implementation of human resources strategies, without determining or influencing the strategy itself.

In a second phase, when we consider the perspective of digital HRM, we identified a significant change in the way technology and people management are articulated. Literature and reflection from practice allow us to identify a second phase of true digital disruption with a high impact on people management and which directs us toward an HRM 4.0. Essentially characterized by the power of transformation in the design of new services and products, in the way organizations are structured and managed, namely through the adoption of self-service tools aimed at workers and the community. As defended by Strohmeier (2020, p.349) "digital transformation means a strategic opportunity based on the potential for digital technologies to create innovative business opportunities as expressed by 'digital business strategies'" (Bharadwaj et al., 2013). It involves the fundamental strategic change of the entire organization due to the business potential of digital technologies (e.g. Hanelt et al., 2018; Hausberg et al., 2018; Ismail et al., 2017 cited by Strohmeier, 2020).

In this perspective, people management no longer integrates technological evolution as a mere support for the optimization of HRM processes (traditional perspective), or as a mere alignment between the facilitating potential of digital technologies and previously defined people management strategies (SHRM). on the contrary. It constitutes a real change in the way technologies are understood, what their role and impact is on the creation of new services. The appropriation of the transforming potential of technologies, in their more sophisticated forms of Artificial Intelligence, Cloud Computing and above all, "ubiquitous computing", through HRM, has repositioned the management of people in organizations not only for its strategic value but above all for its potential to create value for the organizations themselves. In the words of Strohmeier (2020, p. 351), "Fully transferring the concept to HRM implies not merely aligning digital technologies to pre-formulated HR strategies but formulating and executing HR strategies that are directly based on the potential for digitization to create value for an organization".

2.2.2 CHALLENGES TO THE INTEGRATION OF THE DIGITAL HRM IN HEALTHCARE ORGANIZATIONS

The above allows us to consider that we are facing a real revolution based on digital disruption and with an impact on society and on the way in which healthcare organizations are understood. As mentioned by Buchelt et al. (2020, p.2) in their analysis of the role of HRM in the management of health professionals "the effects of that revolution are progressing in two different ways – in both medicine itself and the management of healthcare organizations" and "it has dramatically revolutionized the design and delivery of care, and also creates implications for the performance of healthcare entities" with permanent implications on:

- "Strategy creation and reconstruction of organizational structures;
- The judicious management of healthcare resources aimed at greater efficiency in the delivery of care and improved quality of healthcare services;
- Better allocation of resources;
- A new patient-centered philosophy of care resulting in co-production and the co-creation of value in healthcare services;
- Increasing the ability to monitor and manage the flow of activities performed by healthcare professionals;
- Personalized medicine;
- The permanent development of knowledge within the healthcare sector" (Buchelt et al., 2020, p.2).

The management of people in organizations that provides healthcare services have also sensed this evolution. However, its integration has demanded and will demand, in the future, specific attention.

In these organizations, and due to the inherent complexity of their specialization and also due to the type of services provided, the outcome results they produce are ensured by a highly qualified workforce capable of mobilizing professional knowledge.

Mintzberg (1982) calls it those of professional bureaucracy. This high qualification enhances autonomous work, where roles are performed according to requirements arising from professional values, norms and codes with high bargaining power, tending to low hierarchical levels.

Along these lines, Empson (2021) suggests four fundamental dimensions for these professional organizations services: primary activity, knowledge, governance and identity.

Citing Empson (2021), each of the main characteristics of a professional service firm is defined as:

i. Primary activity – applying specialist knowledge to creating customized solutions to clients' problems;
ii. Knowledge – Core assets are professionals' specialist technical knowledge and in-depth knowledge of clients;
iii. Governance – Extensive autonomy and contingent authority; core producers own or control core assets;
iv. Identity – Core producers recognized as professionals by clients, competitors and each other.

The author refers to the fact that these organizations, by favoring the mobilization on in-depth and technical knowledge to solve patients' problems, with high autonomy, create an identity in which providers recognize each other and are also recognized by other as professionals. They have such staking features of governance, organizational model, orientation of identity and desired organizational practices that have the possibility of directly influencing organizational outcomes.

In the field of healthcare provision, the disruption associated with artificial intelligence and advanced technologies create disruptive conditions with the traditional way in which organizations, professionals, patients and stakeholders interact and communicate. Greater scrutiny, demand and speed forces the flexibility hitherto unknown or ignored.

These types of requirements force the integration of new technologies and their full potential for change, in order to enhance people management, their satisfaction, motivation, well-being and results.

In addition to the potential gains in efficiency and productivity, as defended by Martin and Siebert (2020), it is essential to reflect on the value of technologies and how they can be integrated with the objective of creating the response to the needs of professionals and patients, of intellectual capital and/or innovation for the organizations, social and environmental value for society.

With the intention of responding to the challenges associated with the age of disruption (Scharmer and Kaufer, 2013), this chapter introduces a reflection on the potential of new technologies and how they can be integrated into people management.

Challenge 1 – The threat of talent scarcity: The health sector, by its dynamism and its need for evolution and permanent adaptation, imposes new requirements, new skills and competencies, which are not always aligned with the traditional academic perspective. These permanently seek and incorporate tacit and explicit knowledge with both internal and external origin. The way it is used and its integration into daily actions allows the creation of new knowledge and thus generates intellectual capital in the form of human, structural and relational capital.

Technologies, either in terms of facilitation of access to information, or in terms of its diffusion, have a high potential of empowerment to internal resources organizations, but at the same time, the potential for attraction and diffusion of new competencies. The strategies for analysis, control and development of skills associated with operations for enabling and attracting new assets can bridge the risk of shortage or inadequacy of talent. Nowadays, there are technological solutions for knowledge management issues in organizations with high potential, taking into account the objectives and organizational characteristics (culture, values, technologies, etc.) and the cost–benefit relationship.

The proposal then passes through the development of skills and the sharing of knowledge, using technology and electronic empowerment of human resources.

Challenge 2 – Increment of automation: As in numerous sectors, in the provision of healthcare, the replacement of less sophisticated functions by automation or robotization processes has become increasingly more evident. In a cost containment and optimization of resources logic, routine, repetitive or administrative operations tend to be developed by equipment and/or applications that quickly and successfully respond to the current requirements for people management.

Guetal and Stone, in 2005, demonstrated the change potential for HRM resulting from the use of standard technology in an operational context. Fields as distinct as candidate management, rewards, onboarding, performance management, time and attendance, or even employee self-service tools, provide technology-based services that are fast, efficient and safe. Thus setting free human resources professionals for more valuable functions, while simultaneously fostering a culture of self-sufficiency and self-development.

As mentioned by Towers Perrin (2002) cited by Martin e Siebert (2016, p.343) "It represents a significant internal change for most organizations, and has to be supported with education, communication, the right tools and processes, and frequent and consistent reinforcement."

The proposal to integrate the challenge of cost containment and job extinction, by the domain of computerization, involves the conversion of administrative processes through e-HR methodologies and through the conversion and requalification of replaced professionals to more productive and enriching functions.

Challenge 3 – Impermanent Labor Market: The massive extinction of jobs with the emergence of unanticipated new functions, associated with technological development and competitive movements between organizations, have made the health sector market less predictable. If HRM policies, oriented toward attracting and retaining talents, are a condition for organizations to have professionals, they need in order to ensure the necessary answers, on the other hand, the technological evolution itself has threatened the desirable stability. Lots of functions, due to technological evolution have become obsolete, leading to these posts being extinct and thus threatening the established psychological contract of organizational commitment in exchange of the exercising of a given function. Paradoxically, it turns out that it is the development of technology that imposes the creation of new jobs in the labor market and very often, the creation of new functions that we have not even anticipated.

In this way, the development of a culture of openness and learning by the organization has become essential in order to enhance existing resources and proactively integrate new challenges.

However and at the same time we need to foster a culture of individual learning, so that all the professionals are able to follow and lead along the challenges of the market and meet the conditions to be active members in the organizations. This task is facilitated by the integration of e-learning as a personal development methodology. E-learning means the practices of sharing information and knowledge by the internet or intranet, for the purpose of learning development. Other authors, more broadly, define it as any system that "generates and disseminates information and is designed to improve performance" (Rosenberg, 2001, p.11). The integration of e-learning as a human resources practice allows organizational and individual development and thus to prepare in advance the response to future needs.

> The proposal in a context of extinction of jobs due to low labor demand and the increasingly accentuated emergence of new functions, some not even identifiable, past to update knowledge and overcome the risk of outdatedness should be based on the use of technology to support organizational and individual learning.

Challenge 4 – Greater patient's capacity and exigencies: Greater access and scrutiny by the patients that virtually accompany the innovations and demand them in their interaction with the services has been verified. New ways of providing health services, such as telemedicine, e-health, m-health or participated forms of distance interaction have been gaining increasing importance. In a logic of accessibility and greater proximity to patients, the services tend to incorporate the potential of the new information and communication technologies to ensure, not only the punctual responses, but also the monitoring, in situations where travelling is difficult or impossible. This greater proximity is the guarantee of access to the services, but also it is strongly conditioned by the availability of resources and familiarity with the technologies by stakeholders. In the same way, the ease of communication arising from the new technologies sometimes requires an emergency character in the contact, which is not always justifiable or easily attended. For the relation with patients, to be successful, the adaptation of new technologies must attend the expectations of these users. It is also fundamental to create the internal conditions for these expectations to be met. To this end, it is essential to prepare the patient or a significant person, with the knowledge to use of the technology. It is also fundamental for this purpose to equip the professionals with the needed skills to correspond and thus strength the relationship and confidence in the service.

The proposal to respond to the greater scrutiny and exigencies from patients and society in general, is that healthcare organizations must direct their efforts towards new products and services that are accessible to different users, but above all to ensure management models that favor innovation and social cohesion and development. The proposal involves the integration of technology to support innovative business models and participative organizational cultures.

Challenge 5 – Value-based healthcare volatility: The speed of adaptation to market requirements are determinants for the survival and adequacy of healthcare organizations. Recent challenges, resulting from COVID-19, are illustrative of the way, where in a short period of time, organizations have seen themselves in the contingency of adapting their plans and services to excessive demand, new disease manifestations and new treatment alternatives. Being a field in which the space for experimentation is reduced and has unattainable costs, the ability to develop cohesive teams in full interaction and with effective communication is central. In this context, the development of communities of practice, virtual teams and the use of networks facilitate the communication and the work flexibility (e.g. by remote telecommuting work), but also the creation of social capital. The communities of practice and virtual teams, based on digital technologies, has different nature but the same goal. While the communities of practice are targeted to the joint team results, the virtual teams are sets of professionals who fulfill organizational tasks, although in dispersed spaces – both inside or outside the organizations. Both allow the continuity of the exercising of some functions in contexts of geographic dispersion or the impossibility of permanence in the usual work post.

The proposal to the challenge of creating cohesive teams with the potential to create value despite profound dynamic and unpredictable contexts, often with demands higher than desirable, refers to the creation of teams with solid values and high resilience and flexibility to respond to new challenges. To overcome this challenge, the use of technology to create new forms of community at work is proposed.

This is the context that justifies the presentation of a human resources management model that integrates the different challenges and solutions discussed above.

2.3 PROPOSAL FOR AN INTEGRATED PEOPLE MANAGEMENT MODEL

To respond to the identified challenges in healthcare services, we propose an Integrated People Management Model (IPM 4.0) that integrates the main challenges and concerns presented, and also constitutes a guiding framework for action in the future.

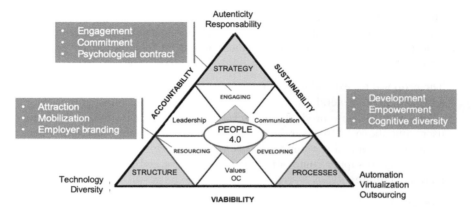

FIGURE 2.1 Integrated People Management Model – IPM 4.0 (Duarte et al., 2019).

The model (Figure 2.1) is synergistically articulated around *three structural vectors*, with the human component (people and their experiences) as its epicenter, as stated in Nascimento and Duarte (2021). The following are the structuring vectors:

i. *The structure* is defined as the organizational design that emerges through the identification, analysis, ordering and grouping of the activities and resources of the organizations. It contains all of the necessary resources (material, relational, capital, informational, and so on) as well as their interconnections. The structure should be agile and flexible enough to respond to sudden and unpredictable changes. It should also increase new collaborative models and partnerships with the stakeholders (internal and external) formed in a logic of responsibility and confidence, hence improving organizational outcomes and action predictability.

ii. *The processes* are a set of actions, procedures and routines that are special to health organizations and strive to maximize the value of health and well-being.

iii. *The strategy* is defined as the mediating force between organizations and their environment, characterized by a pattern or a plan that integrates the organization's key goals, policies and tactics in order to achieve its objectives and goals. Anchored in a vision of the future, it allows the allocation of resources (structure) and the assimilation of processes that guarantee in advance a differentiation in the face of contextual premises. It is the strategy that ensures long-term viability and sustainability conditions of the organization.

There are *three essential domains* of people management that should be highlighted in conjunction with the three organizational structure vectors. These are the three domains:

i. *Resourcing* – aligned with the Structure vector, its goal is to plan and prospect, as well as attract, identify and select talent that is critical to the delivery of health services. It supports employer branding dimensions, recruitment and selection, and talent mobility management.

ii. *Developing* – connected with the Processes vector, it focuses on human development and empowerment, as well as knowledge generation and dissemination. The features of this new generation of digital-knowledge workers necessitate a paradigm of lifelong learning and reinforcement. "The acquisition and diffusion of knowledge, the management and evaluation of performance all have requirements for personalization and sharing, and come with expectations of a quick and permanent feedback" (Nascimento & Duarte, 2021, p.130). It should serve as a vehicle for progress, recognition and motivation, as well as providing a pleasurable experience. Incorporating technologies to aid managerial processes and standard administrative procedures is no longer sufficient for human management. A new type of self-regulation of learning and assessment, which is personalized and self-managed, emphasizes taking advantage of opportunities presented by social, mobile and cloud technologies, as well as the virtualization of several processes and procedures.

iii. *Engaging* – this domain, which is associated with the Strategy vector, is responsible for building and managing employees' commitment to the organization. Committed employees not only contribute more to the organization's success, but they also manage to benefit from and utilize their talents, competencies and abilities more effectively. It is critical that People Management 4.0 considers and incorporates the drivers of organizational engagement into its policies. The importance of meaningful work, the ability to engage and form satisfying connections, and the establishment of inclusive policies connected with a culture of trust and empowerment will have to be allied with remuneration policies and appealing careers.

To guarantee the alignment between the structuring vectors (structure, processes and strategy) and the key domains of intervention of people management (Resourcing, Developing and Engaging), it is important to consider the *three determinants of success* (Nascimento & Duarte, 2021, p.131). The following are the determinants of success:

i. *Organizational culture and values* – mediating the structure and the processes, it is defined by two unique and inseparable perspectives: the individual and the organizational. Regarding individual terms, it's critical to emphasize ethical issues and societal commitment (social and environmental), all of which are anchored in patterns and standards of autonomy, responsibility and development. The sharing of these values generates new ways of being in organizations. The organizations, in their turn, combine these values and cultivate cultures of flexibility, inclusivity and creativity, all based on trust and commitment principles. This combination produces sustainability, which is the organization's ultimate goal and service to society.

ii. *Communication* – influenced by strategy and processes, communication plays a key role in people management as well as the success and consolidation of the organization's identity. As a result, heightened concern and care for both formal and informal internal communication, as well as external communication, becomes crucial and extremely important.

iii. *Leadership* – it serves as a link between strategy and structure. Leadership must elicit and assimilate change in order to establish strategies that promote innovation and active learning throughout the organization, with a focus on employee's engagement. "This challenge calls for new leadership competencies and skills that are structured in relational terms (e.g. development of a culture of trust, authentic communication, promotion of individual and team development, and ensuring the inclusion of individual interests and values)" (Nascimento & Duarte, 2021, p.131). Regarding technical terms, leadership must be able to challenge the status quo and foster creativity and innovation, as well as build benchmarking and networking activities that assure continuous updating and responsiveness in a constantly changing environment. Leaders in the twenty-first century will distinguish themselves by adaptable and disruptive behaviors, value integration and growth strategy implementation, and value creation for society.

2.4 CONCLUSION

It was our intention in this chapter to reflect on the evolution of technology, its impact on healthcare organizations and on people management.

Since health organizations are high-complex intensive knowledge systems, it is deciphered that, while essential, technologies by themselves are not the assurance of the desired quality.

The value of technology is boosted by correct alignment with people management and business strategy, thus emerged the proposal of an integrated people management model (IPM 4.0), which stems from the guidance according to three principles: sustainability, viability and accountability.

With the challenge of managing people in 2030, we believe the core will be effective in a perspective that is consistent with Society 5.0's goals (Super Smart Society). A society founded on the principles of equity, responsibility, accountability and long-term sustainability. A world in which (re)thinking and (re)positioning technologies to improve humanity's quality of life will be critical.

Based on the learning resulting from the management of people in the turbulent context of the COVID-19 pandemic, we believe that, more than ever, human and social capital will be the determinant of organizational success, and a competitive advantage for organizations, in general, and for health, in particular.

(Nascimento & Duarte, 2021, p.131)

REFERENCES

Alfawaire, F., & Atan, T. (2021). The effect of strategic human resource and knowledge management on sustainable competitive advantages at Jordanian universities: The mediating role of organizational innovation. *Sustainability*, 13(15), 8445. https://doi.org/10.3390/su13158445

Ancarani, A., di Mauro, C., & Mascali, F. (2019). Backshoring strategy and the adoption of Industry 4.0: Evidence from Europe. *Journal of World Business*, *54*(4), 360–371. https://doi.org/10.1016/j.jwb.2019.04.003

Armstrong, M., & Brown, D. (2019). Strategic human resource management: Back to the future. Institute for Employment Studies reports, 1–36.

Barley, S. R. (2020). *Work and Technological Change (Clarendon Lectures in Management Studies)*. Oxford University Press.

Beer, M., Spector, B., Lawrence, P., Quinn Mills, D., & Walton, R. (1984). *Human Resource Management: A General Manager's Perspective*. Free Press.

Begun, J. W., Zimmerman, B., & Dooley, K. (2003). Health care organizations as complex adaptive systems. In: Mick, S. S., & Wyttenbach, M. E. (Eds). *Advances in Health Care Organization Theory* (1st ed.). Jossey-Bass, pp. 253–288.

Bharadwaj, A., El Sawy, O., Pavlou, P., et al. (2013). Digital business strategy: Toward a next generation of insights. *MIS Quarterly*, *37*(2), 471–482.

Bondarouk, T., & Brewster, C. (2016). Conceptualising the future of HRM and technology research. *The International Journal of Human Resource Management*, *27*(21), 2652–2671. https://doi.org/10.1080/09585192.2016.1232296

Buchelt, B., Frączkiewicz-Wronka, A., & Dobrowolska, M. (2020). The organizational aspect of human resource management as a determinant of the potential of Polish hospitals to manage medical professionals in Healthcare 4.0. *Sustainability*, *12*(12), 5118. https://doi.org/10.3390/su12125118

Duarte, A., Nascimento, G., & Almeida, F. (2019). Gestão de Pessoas 4.0 – Entre a continuidade e a reinvenção. In: Carolina Machado and J. Paulo Davim (Ed.), *Organização e Políticas Empresariais*. Conjuntura Actual Editora, pp. 15–53.

Empson, L. (2021). Researching the post-pandemic professional service firm: Challenging our assumptions. *Journal of Management Studies*, *58*(5), 1383–1388. https://doi.org/10.1111/joms.12697

Gueutal, H., Stone, D. L., & Salas, E. (2005). *The Brave New World of e-HR: Human Resources in the Digital Age* (1st ed.). Pfeiffer.

Hanelt, A., Bohnsack, R., Marz, D., et al. (2018). Same, same, but different!? A systematic review of the literature on digital transformation. In: 78th annual meeting of the academy of management, Chicago, IL, 10–14 August 2018.

Hausberg, J., Liere-Netheler, K., Packmohr S., et al. (2018). Digital transformation in business research: A systematic literature review and analysis. In: DRUID18, Copenhagen Business School, Copenhagen, Denmark, 11–13 June 2018.

Ismail, M. H., Khater, M., Zaki, M. (2017). Digital business transformation and strategy: What do we know so far? Working Paper, University of Cambridge, Cambridge, November.

Martin, G., & Siebert, S. (2016). *Managing People and Organizations in Changing Contexts* (2nd ed.). Routledge.

Mintzberg, H. (1978). *The Structuring of Organizations*. Pearson.

Nascimento, G. & Duarte, A. (2021). Healthcare people management – Preparing today's professionals for tomorrow. *HealthManagement.org – The Journal*, 21 (3), 126–131.

Rosenberg, M. J. (2000). *E-Learning: Strategies for Delivering Knowledge in the Digital Age* (1st ed.). McGraw-Hill Education.

Scharmer, O., & Kaeufer, K. (2013). *Leading from the Emerging Future: From Ego-System to Eco-System Economies* (1st ed.). Berrett-Koehler Publishers.

Schwab, K. (2016, January 14). The fourth industrial revolution: What it means, how to respond. WEForum.org. Retrieved from www.weforum.org/agenda/2016/01/the-fourth- industrial-revolution-what-it-means-and-how-to-respond/

Snell, S. A., Pedigo, P. R., Krawiec, G. M. (1995). Managing the impact of information technology on human resource management. In Ferris, G. R., Rosen, S. D., and Barnum, D.T. (Eds), *Handbook of Human Resource Management*. Cambridge: Blackwell Publishers, pp. 159–174.

Strohmeier S. (2007). Research in e-HRM: Review and implications. *Human Resource Management Review*, *17*(1): 19–37.

Strohmeier, S. (2020). Digital human resource management: A conceptual clarification. *German Journal of Human Resource Management: Zeitschrift Für Personalforschung*, *34*(3), 345–365. https://doi.org/10.1177/2397002220921131

Towers Perrin (2002) Health care cost survey: What consumers and employers are doing about the increases. USA: Towers Perrin.

Verlinden, N. (2018) Back to Basics: What Is Digital HR? AIHR Digital. URL: www.digitalhrtech.com/back-to-basics-what-is-digital-hr/.

Vrontis, D., Christofi, M., Pereira, V., Tarba, S., Makrides, A., & Trichina, E. (2021). Artificial intelligence, robotics, advanced technologies and human resource management: A systematic review. *The International Journal of Human Resource Management*, 1–30. https://doi.org/10.1080/09585192.2020.1871398

Wright, P. M., & Ulrich, M. D. (2017). A road well traveled: The past, present, and future journey of strategic human resource management. *Annual Review of Organizational Psychology and Organizational Behavior*, *4*(1), 45–65. https://doi.org/10.1146/annurev-orgpsych-032516-113052

3 Managing Acute Patient Flows in Hospitals

Svante Lifvergren and Axel Lifvergren

CONTENTS

3.1 Introduction 37
    3.1.1 Healthcare Challenges: Managing Acute Patient Flows 37
3.2 Process and Flow Management 39
    3.2.1 Principles and Challenges 39
3.3 Processes and Flow: A Theoretical Overview 39
    3.3.1 Process Versus Flow 39
    3.3.2 Frameworks on Flow Dynamics 40
3.4 Process and Flow: A Technical Management View 41
    3.4.1 Process Principles and Practices..41
    3.4.2 Flow Management Principles and Practices.....................................42
3.5 Process and Flow: A Social Management View 45
3.6 Context: The Hospital..49
    3.6.1 An Overview of the Case 49
    3.6.2 Acute Patient Flows: Challenges 49
    3.6.3 Some Words about the Method 50
3.7 Lessons Learned 51
    3.7.1 Clear Functions and Roles in Patient Flows.....................................51
    3.7.2 Apply Production and Capacity Planning to Level Out Flows 52
    3.7.3 Evaluate Acute Flows 53
    3.7.4 Improve Collaboration Between Professional Groups 54
    3.7.5 Manage and Support Coordination and Integration 55
    3.7.6 Involve Professional Groups in Managerial Decisions 55
3.8 Conclusions 56
References 56

3.1 INTRODUCTION

3.1.1 HEALTHCARE CHALLENGES: MANAGING ACUTE PATIENT FLOWS

A growing set of reports illuminates several recurrent trends that will result in huge challenges for European healthcare systems in general and for hospitals in particular. A growing population of elderly is already a fact in several European countries as well as in Sweden. As people are getting older, increasing numbers of patients are suffering from multiple illnesses that require extensive specialist care and more resources (Nergårdh, 2017; SALAR, 2014; SOU, 2009; Stiernstedt, Zetterberg & Ingmansson, 2016). At the same time, new and expensive drugs and treatments are being introduced at an accelerating rate, leading to increasing healthcare expenditures (Hallin & Siverbo, 2003). Subsequently, the number of elderly people with complex care needs will rise, resulting in increased patient flows to hospitals, partly due to exacerbations of conditions in a growing cohort of elderly patients with chronic diseases (Anell & Mattisson, 2009; SALAR, 2005; Stiernstedt, Zetterberg & Ingmansson, 2016). Thus, hospitals' capacities to manage acute patient flows will be of key concern to meet current and future challenges (Stiernstedt, Zetterberg & Ingmansson, 2016). However, several national assessments show that Swedish hospitals suffer from a low flow efficiency,[1] resulting in long waiting times at emergency wards. Unfortunately, the situation is not improving. On the contrary, waiting times have increased during recent years (Socialstyrelsen, 2015; Stiernstedt, Zetterberg & Ingmansson, 2016).

During the last years, slightly more than 10 percent of GDP has been spent on healthcare in Sweden (Norrbäck & Targama, 2009; OECD, 2013; Stiernstedt, Zetterberg & Ingmansson, 2016). From an international perspective, however, this proportion is quite low. Thus, an important question is whether more resources might solve the problem with long waiting times at emergency wards in Sweden. Several national evaluations, however, point out that the current problems are not resource related (Stiernstedt, Zetterberg & Ingmansson, 2016). Instead, more focus must be directed towards a continuous improvement of flow efficiency from a patient perspective, where changes come from within the system. In other words, sustainable improvement must entail co-workers and managers in the microsystems[2] along the acute patient care processes embracing a systemic view. Thus, improvement initiatives must involve all organizational levels in the hospital; long-term strategies must create long-term preconditions for successful and continuous improvement (ibid.).

Drawing from several empirical studies of the acute patient care process at a medium-sized Swedish hospital, barriers limiting an efficient acute patient flow has been identified. A literature overview shows that these findings seem to be of general concern and might be relevant to most European hospitals. Turning attention to theories and best practices from other sectors, such as industry, many lessons can be learned and translated to a healthcare context to improve flow efficiency in hospitals.

The chapter starts with a presentation of theoretical frameworks pertaining to technical as well as more social aspects of process and flow management. Drawing from the scientific literature, successful principles and practices to manage processes and flows have been identified. The chapter then continues to present a longitudinal case study where the theoretical frameworks have been applied analytically to illustrate how a hospital has tried to improve its capability to manage acute patient processes and flows. The chapter concludes with lessons learned, where evidence-based improvement suggestions are presented that, hopefully, will translate to other contexts and encourage other hospitals to embark on their own improvement journeys.

3.2 PROCESS AND FLOW MANAGEMENT

3.2.1 PRINCIPLES AND CHALLENGES

Understanding and managing organizational change that improves the quality of value-adding processes in an organization is a vast subject that constantly seems to increase in popularity. Improvement or change is often seen as something positive and manageable that can be accomplished by implementing strategies and programs following a linear step-by-step list (Collins, 2001; Inamdar & Kaplan, 2002; Kotter, 1995; Porter, 2010; 1996). Many improvement strategies have an origin in the quality and process management movement, for example business process re-engineering (BPR), just-in-time (JIT), total quality management (TQM), Lean, Six Sigma to mention a few, and address process improvements both in single organizations and in nationwide service sectors (De Feo & Barnard, 2004; Gupta, 2004; Magnusson, Kroslid & Bergman, 2003; Liker, 2004). The underpinning logic is the belief that fulfilling the needs of the customer through continuous improvement of value-adding processes is the pathway to competitive advantage (Knights & McCabe, 2002). This may not be surprising given the fact that process quality management and improvement as well as value-based management (Porter, 1996; 2010) is often stated to be the only valid strategy for especially non-profit organizations to reach their goals (Kaplan & Norton, 1992; 2001). According to this logic, organizations should have a business concept from which to formulate visions and goals to channelize managers and co-workers activities and efforts in a common direction. Vision and goals, in turn, help organizations to reach a future envisioned state (ibid.). The purpose of management systems, or rather, managerial practices, therefore, is to share, connect and translate vision and goals into improved processes entailing routines and procedures in everyday operational practices across divisions and units, eventually closing the gap between current and future state of the business (Bruzelius & Skärvad, 2004). Thus, a pronounced focus on value-adding processes is key to reaching the goals of the organization.

3.3 PROCESSES AND FLOW: A THEORETICAL OVERVIEW

3.3.1 PROCESS VERSUS FLOW

A process is commonly defined as a network of activities, repeated over time, the purpose of which is to create value for external as well as internal customers.[3] In a healthcare context, the patients and their near and dear are the most important external customers. In general, processes have a start and an end as well as a supplier, who delivers input for the process to start (Bergman & Klefsjö, 2020). Rentzhog (1998) uses a boat channel metaphor to illuminate key characteristics of the process concept. Continuing the metaphor, boats with their passengers can be seen as units that flow through the channel. Thus, the channel is equal to the process, that is, the horizontal organizational structure in which units (boats with passengers) flow and where value is co-created along the way. Flow units might be humans (patients), material or information. Most often, the flows in a process entail a combination of these three elements, where one element most often dominates the flow (Modig & Åhlström, 2015). Thus, processes can be seen as structures (channels) in which units flow. Subsequently, process and flow are complementary perspectives on the same phenomenon.

Translated to a healthcare context, patient journeys throughout the hospital are managed by interconnected functional units consisting of multi-professional teams, microsystems, along the patient process. Flow units are thus patients as well related information (for instance, medical information), where care activities are co-created together with the patient to continuously refine value in terms of improved health and well-being. Process models describe the structure of the process; for instance, how roles and responsibilities are distributed and how management and the continuous development of the process is organized and evaluated. Equally, process models also detail how resources in terms of vertical functional organizational units (for instance, hospital units, departments and clinics) are interconnected on a practical as well as a managerial level to avoid suboptimization and to ensure a smooth journey that fulfills the needs of the patient (Ljungberg & Larsson, 2012).

3.3.2 Frameworks on Flow Dynamics

Various frameworks define important preconditions regarding flow characteristic and are thus important pillars on which principles and practices for an improved flow management must rest (Goldratt, 1990; Modig & Åhlström, 2015). In industrial as well as in healthcare contexts, three preconditional concepts are often brought forward; Little's law, theory of constraints and the relationship between flow and resource efficiency (ibid.).

Little's law is a theorem in queuing theory, which states that the long-term average number of products in a stationary system equals the average arrival rate of products multiplied by the average time that a product spends in the system. In more popular terms, Little's law is often expressed as:

$$\text{Total throughput time} = \text{number of producs in the system} * \text{average cycle time}$$

where cycle time equals the time that passes between two products that leave the system.

A bottleneck is the activity/phase in a flow that consumes the most amount of time. Thus, the pace of the flow in a system is decided by the bottleneck and can never supersede that pace (Goldratt, 1990; Olhager, 2013). Queues always emerge before the bottleneck, whereas an overcapacity is present downstream, after the bottleneck. Bottlenecks are most often dynamic in healthcare systems. That is, they vary over time and in space.

Flow theory also puts focus on the difference between resource efficiency and flow efficiency. Resource efficiency is probably still the dominant approach in healthcare and focuses on maximizing the use of the most critical resource/process activity in the system (Graban, 2009; Haraden & Resar, 2004). Thus, according to this logic a higher utilization of the critical resource will thus improve resource efficiency.

Flow efficiency, on the other hand, focuses on the flow units in the process and how preconditions for an even and balanced flow can be continuously improved. Modig and Åhlström (2015) define flow efficiency as:

$$\frac{\text{The sum of value adding activities(time)}}{\text{Total throughput time}}$$

in a process. Maximizing utilization of the critical resource in a process impairs flow efficiency. When resource utilization in an activity reaches somewhere around 80 percent or more, flow efficiency drops dramatically, and the queue increases infinitely. However, less overall variation in the system makes it more resilient to this effect (ibid.). In sum, these theories entail some important implications for understanding flow characteristics:

- Many simultaneous products (patients) in a system impair flow efficiency and increase queues.
- There are always bottlenecks in a system or a process that eventually decides throughput time and that must be continuously managed.
- Balancing flow and resource efficiency is a crucial aspect of flow management, where more emphasis on evening out and balancing flows in the process is important.
- Reducing unwanted variation in a process makes it less sensitive to high resource utilization effects on flow efficiency.

3.4 PROCESS AND FLOW: A TECHNICAL MANAGEMENT VIEW

3.4.1 Process Principles and Practices

As pointed out earlier, organizations need to focus on their processes drawing from a customer perspective, customer needs can only be fulfilled by a well-functioning process. In processes, synchronized and integrated activities and microsystems interact with the customer to co-create value, thereby fulfilling the needs of the customer. In other words, it is essential for organizations to focus on managing and continuously improving their processes (Bergman & Klefsjö, 2020; Ljungberg & Larsson, 2012; Rentzhog, 1998).

There are many normative models that provide recommendations for organizational development of processes. These models overlap to a large extent and entail the same steps, albeit in various orders. Further, the models commonly make a distinction between *establishing* a process as a precondition for the further *development* of the process (Ljungberg & Larsson, 2012). Pivotal steps in common models for successful process development are outlined below and overlap with healthcare process models to a large extent (Alfalla-Luque, Medina-Lopez & Dey, 2013; Meiboom, Schmidt-Bakx & Westert, 2011; Ljungberg & Larsson, 2012; Nilsson & Mandoff, 2015; Nordenström, 2014; Rentzhog, 1998; SIQ, 2015; Wiger, 2013):

- Identify the most common and important customer processes, i.e., flows for which the organization must create value. Establish infrastructures for each

process in terms of roles and responsibilities; resource and process owners, process managers and process groups, which is an important prerequisite for the further management and improvement of the process. Make sure that all the functional units, including co-workers and managers, are represented along the process. Additionally, make sure to create sustainable support for process development. For instance, IT support, measurement systems, and process improvement facilitators providing analytical and improvement support.
- Understand the process and its characteristics, its current problems and improvement opportunities as well the demands on the process. Based on these assessments, initiate continuous improvement involving co-workers in the process to create an even deeper understanding of the process and to attend to the most urgent problems in the process.
- Establish a development plan for the process that includes goals from various perspectives and revise the plan regularly. Common perspectives on goals are customer availability, quality, cost, time (flow efficiency) and safety to mention a few. All in all, the goals should capture the efficiency and the appropriateness of the process. Various process maturity models are also available to evaluate the current state of the process and its capability. A balanced mix of process (how things are done) and outcome (the result of how things are done) goals is recommended. Moreover, both quantitative and qualitative evaluations of the process are important.
- Keep going with continuous improvement to reach goals and reduce the gap between the current state and a desired future state of the process. In addition, take on new and emergent opportunities for improvement.

3.4.2 Flow Management Principles and Practices

The flow of patients in need of acute care as well as the decision in the flow can be described as processes. Simplified, the patient is the customer in the acute flow. The process starts when the patient gets ill and eventually arrives at the emergency ward to be admitted to a ward. The process ends when the patient is discharged from the hospital. Value is created when the patient's health is improved, or at least, stabilized (Nordenström, 2014). The decision process not only entails medical and care decisions regarding the individual patient. From a flow management perspective, the process also involves recurrent information on inflow and outflow rate, occupancy rate at the emergence and common care wards and flow situation at the x-ray department. The information is continuously analyzed and refined to inform vital flow management decisions that aim to match overall care demand with available capacity over time (Haraden & Resar, 2004), commonly referred to as production and capacity planning (PCP).

Most PCP models are based on the notion of a hierarchy of planning levels, where decisions taken at a higher level frame the decisions at the next lower level. At higher planning levels, the planning horizon is longer, the planning object is more aggregated, and the planning frequency is higher. Required capacity is calculated from forecasts or customer orders, and then compared to available capacity for each period within the planning horizon (Brandt & Malmgren, 2015; Mattson & Jonsson, 2013; Plantin & Johansson, 2012; Rosenbäck, 2018). The key purpose is to match available resources (capacity) with current and forecasted production demands at various hierarchical levels (Walley, Silvester & Steyn, 2006). Translated to a healthcare context, PCP must be carried out at all levels in a hospital; unit, clinical and hospital levels, usually referred in an industrial context as operational, tactic and strategic levels. Further, these planning levels must be aligned along patient flows throughout the hospital. Each planning level must entail short-, medium- as well as long-term prognoses regarding production demands to match these against available forecasted capacities (Alfalla-Luque, Medina-Lopez & Dey, 2013; Olsson, 2014; Vissers, Bertrand & De Vries, 2001). In addition, PCP models focus on key common capacities in hospital systems, where recurring bottlenecks in hospitals flows inevitably emerge: Emergency departments, radiology departments, intensive care units, operation theatres, the number of available beds at care units or the outflow from the hospital (Haraden & Resar, 2004). The capacity in these functions determine the throughput rate for most patient flows, why production plans from the various clinics must be balanced to match forecasted hospital's common capacities to avoid suboptimizations between clinics or uneven flow rates upstream and downstream the hospital (Olsson & Aronsson; 2012; 2015).

Hierarchical PCP models have proved to be effective in many various contexts, although they are always necessary to adapt to the specific conditions of an operation, in terms of type of customer demand, the product, and the production system (Plantin & Johansson, 2012). A challenge for hospitals trying to establish PCP models is also patient variation. Although acute patient inflows to hospitals are remarkably stable over time, the individual patient variation is large. Hence, individual patient care plans forecasting the patient care journey are important planning perquisites in the healthcare context. The individual care plan must be established as soon as possible and continuously revised throughout the patient care journey (Ortiga et al., 2012).

Most assignable causes of variation in hospitals emerge due to internal variation when planning and managing internal flows (for instance, aligning the schedules of physicians and other healthcare professionals), whereas external factors affecting inflow variation are few (Haraden & Resar, 2004), the COVID-19 pandemic obviously being an important exception. Key principles to reduce unwanted variation are, for instance, to level out different flows *in the hospital* as far as possible; to balance demand and capacity over time; to separate acute and planned patient flows (Brandt & Malmgren, 2015; Walley, Silvester & Steyn, 2006) and to align the schedules of various professional groups to better match the capacity to handle patient flows (Brandt & Malmgren, 2015; Haraden & Resar, 2004). In addition, Utley and Worthington (2012) also point out that simple principles inspired by Lean approaches, such as pull rather than push, is an important mindset in flow management. Thus, improving and sharing updated real-time information regarding patient discharge processes makes room for (pulls) patients from emergency departments into the hospital (Allder, Silvester & Walley 2010; Haraden & Resar, 2004; Ortiga et al., 2012). Finally, other practices that might improve flow efficiency are, for instance, the establishment of multi-professional teams across units and clinics (Alfalla-Luque, Medina-Lopez & Dey, 2013; Meiboom, Schmidt-Bakx & Westert, 2011); well-designed communication processes that support integration and coordination of care activities along patient flows (Alfalla-Luque, Medina-Lopez & Dey, 2013); and process flow coordinators that dedicate their time to support flow management at critical decision points in the hospital. Research-based principles and practices are summarized in Table 3.1.

TABLE 3.1
Research-Based Recommended Principles and Practices

Principles	Practices	References
Strive for a flow-oriented design	Encourage relevant stakeholders to embrace process and flow perspectives instead of focusing on separate functions	Aronsson, Abrahamsson & Spens 2011
	Try to dissolve barriers between functions	Cao Baofeng, Yuan & Xiande, 2015
	Create teams that bridge various organizational functions	Alfalla-Luque, Medina-Lopez & Dey, 2013; Meiboom, Schmidt-Bakx & Westert 2011
	Create channels of communication to foster integration	Alfalla-Luque, Medina-Lopez & Dey, 2013
	Analyze how to redesign the organization to improve flow management	Meiboom, Schmidt-Bakx & Westert 2011
	Standardize discharge processes	Allder, Silvester & Walley 2010, Ortiga et al., 2012
Strive for a balanced allocation of resources in the flow	Use short-, medium- and long-term horizons for production and capacity planning	Walley, Silvester & Steyn, 2006
	Continuously enhance the understanding of mechanisms in the main flows of the organization	Allder, Silvester & Walley, 2010; Aronsson, Abrahamsson & Spens, 2011; Olsson & Aronsson, 2012; 2015; Walley, Silvester & Steyn, 2006
	Identify major patient flows in the organization	Walley, Silvester & Steyn, 20
	Identify resource utilization and resource needs in every part of the organization	Brandt & Palmgren, 2015; Mattson & Jonsson, 2012
	Coordinate resources on strategic, tactical, and operational levels	Brandt & Palmgren, 2015; Vissers, Bertrand & De Vries, 2001

TABLE 3.1 (Continued)
Research-Based Recommended Principles and Practices

Principles	Practices	References
	Allocate resources to minimize bottleneck effects on flow efficiency	Aronsson, Abrahamsson & Spens, 2011; Olsson & Aronsson, 2012; 2015
Strive for capable processes that are known, accepted, comprehensible and accessible	Appoint process owners	Ljungberg & Larsson, 2012; Nordenström, 2014; Rentzhog, 1998
	Support co-workers who work with process development	Nilsson & Mandoff, 2015; Wiger 2012
	Acknowledge hierarchies and action limitations when appointing process owners	Nilsson & Mandoff, 2015
Value is created by continuously defining, improving, and evaluating key processes of the organization	Analyze the current state of the process	Ljungberg & Larsson 2012; Nordenström, 2014
	Create long-term goals for every process	
	Incorporate process follow-up in the overall management system	
Identify and eliminate unwanted variation to improve flow	Separate acute and planned patient flows	Aronsson, Abrahamsson & Spens 2011; Brandt & Palmgren 2015; Olsson & Aronsson, 2012; 2015
	Separate patients with simpler care needs from patients with more complex care needs	Aronsson, Abrahamsson & Spens 2011; Olsson & Aronsson, 2012; 2015
	Plan for patient discharge when the patient is admitted to the hospital	Allder, Silvester & Walley, 2010; Haraden & Resar, 2004
	Identify and eliminate unwanted variation	Aronsson, Abrahamsson & Spens, 2011; Bergman, 2020; Elg, Palmberg & Kollberg, 2013; Olsson & Aronsson, 2012; 2015

3.5 PROCESS AND FLOW: A SOCIAL MANAGEMENT VIEW

Evidently, many evidence-based recipes for the technical management of acute flows have been around for a long time. However, recurring queues and high occupancy rates are still very common in many European countries and, especially, in Sweden. There are long patient queues to units, clinics and hospitals, at emergency departments and so on (Stiernstedt, Zetterberg & Ingmansson, 2016). Why? Subsequently, it seems important to investigate whether these practices are actually practiced. If not, why?

What are the obstacles? How can one move forward to improve flow efficiency? To seek answers to these questions, the attention is turned to a more critical view on the social aspects of management.

The presumed success of management principles and quality management practices has come into serious question during recent years (Alvesson & Sveningsson, 2007; Beer & Nohria, 2000; Beer, Eisenstat & Spector, 1990; Dawson, 2003; Duck, 1993; Schaffer & Thomson, 1992; Strebel, 1996). These scholars argue that organizations are not rational entities where people do as they are told and follow the latest strategic step-for-step model. For instance, organizations trying to follow a set of pre-planned and already determined steps when trying to improve operations in a new and more prosperous direction are rarely successful; A does not necessarily lead to B and successful results are rare (Stacey, 2003). Instead, a more complex view on organizational management is proposed (Alvesson & Sveningsson, 2007; Dawson, 2003; Stacey, 2003), which involves "applying an understanding of a complex and chaotic organizational reality. Unforeseen consequences of planned organizational change, resistance, political processes, negotiations, ambiguities, diverse interpretations and misunderstandings are part of this" (Alvesson & Sveningsson, 2007). In the same vein, Brunsson (2006) suggests that what is talked about, what is decided and what is enacted are three dimensions that are often disconnected and poorly synchronized in everyday organizational life. In fact, Brunsson suggests that organizations need to accept the fact that these phenomena are unavoidable characteristics that should be considered when managing complex organizations.

Subsequently, these scholars propose that viewing organizations or systems as complex adaptive or responsive processes (Stacey, 2003) might be a more fruitful alternative. This viewpoint involves the combination of relational psychology and complex systems theory: The individual consciousness is developed through a continuous conversation with the self and people close-by in one's environment (Mead, 1967). Voices, symbols, and emotions are central elements. The same pattern appears in an organization, where interactions between co-workers, within groups or between groups can be seen as ongoing conversations where relations are created and create one another. Conversational symbols can be fragments of texts and talk, body language or emotions, together creating new patterns, some of which will survive, the attractor[4], and organize experiences contributing to the emergence of new practices and activities. Thus, what is talked about in the organization becomes pivotal, especially if it results in jointly shared interpretations and integrated action (Stacey, 2003). However, the survival of new concepts is, to a high degree, emergent and dependent on the culture and history of the actual organization (Alvesson & Sveningsson, 2007; Dawson, 2003; Stacey, 2003) as well as on what conversations are allowed and whether these lead to integrated action or not.

Healthcare, and especially hospitals, are examples of complex organizations. Glouberman and Mintzberg (2001a, 2001b) argue that healthcare systems are among the most complex organizational systems to be encountered, mostly due to the interplay between four different worlds or mental models, the 4Cs. The 4Cs entail *cure, care, control* and *community* and represent four worlds inherent in the healthcare system. Specifically, *cure* signifies specialization and expertise together with a medical responsibility and a mission to diagnose, treat and cure the patient (a physician perspective). *Care*, in turn, represents a focus on integration and collaboration, emphasizing care for the patient as a whole individual as well as focusing on the system surrounding the patient to improve the integration of services (a nurse perspective). *Control* entails a focus on governance and management to allocate and distribute resources for the care provided (represented by administrators and managers). Finally, *community* signifies the political ownership of the care system, representing the community and its citizens, prioritizing, and setting the overarching goals of the system (politicians, owners, and steering committees).

Thus, it might not be surprising that traditional *n*-step managerial concepts, such as technical practices for flow management, have been contested more recently in the healthcare context (Mintzberg, 1997; 2011; Hallin & Siverbo, 2003, Morelli & Lecci, 2014; Röthlin, 2013; Yasin, Zimmerer, Miller & Zimmerer, 2002). This might also be explained by a high degree of professionalism and autonomy, especially in the Care and Cure worlds; vocational corps working directly with the patient (Hellström, Lifvergren & Quist, 2010). In general, autonomy as well as a high degree of professionalism among healthcare co-workers seem to foster resistance to external control and managerial attempts (Alvesson, 2013). Subsequently, professionals working directly in everyday care practices often feel that managerial efforts from the outside should be kept to a minimum (Glouberman & Mintzberg, 2001a; 2001b; Norbäck & Targama, 2015). Thus, goal conflicts between patient versus economic perspectives often emerge. For instance, if administrative or economic decisions (for instance, prioritization of patient groups and cost cuts) are non-compliant with everyday care practices, physicians almost always adhere to their own rules and routines. Physicians might even ignore new guidelines completely if these are seen to compete with their shared view of the phenomenon at hand (Hallin & Siverbo, 2003; Numerato, Salvatore & Fattore, 2012).

Subsequently, many scholars stress that there seems to be a tendency to focus too much on technical management aspects, especially when introducing new managerial practices, thereby underestimating the soft side of management in complex social systems. Hence, sustainable change in healthcare systems can only be accomplished through a fruitful collaboration between the four worlds, where cooperative networks tear down the barriers between the different worlds (Glouberman & Mintzberg, 2001b). Huzzard, Hellström, Lifvergren and Conradi (2014) argue that the patient perspective can be seen as a potent unifier in the development of these worlds and networks. Thus, a continuous collaboration between the four Cs from a patient perspective might create unanimity and a common direction (see also Ho, Chan & Kidwell, 1999; Mintzberg, 2011). Possible future management strategies should thus be outlined and evaluated together with various professional groups in iterative action–reflection cycles (Funck, 2016; Röthlin, 2013). Similarly, new guidelines, might they be economical, administrative, or clinical, have a much higher chance of being implemented if the professionals, co-workers in the microsystems, are involved early in the decision process (Funck, 2016). See Table 3.2 for an overview of this section.

TABLE 3.2
Research-based Inputs on Managerial Principles and Practices in Complex Social Systems

Principles	Practices	References
Management should focus on minimizing goal conflicts	Agree on goals among different vocational corps	Ho, Chan & Kidwell, 1999; Norbäck & Targama, 2015; Numerato, Salvatore & Fattore, 2012
Various managerial practices should be related to one another and, preferably, harmonized	Incorporate various types of competences in management teams	Morelli & Lecci, 2014
	Use several balanced perspectives in strategic management guidelines	Funck, 2016; Wiger, 2012
Management should strengthen coordination and integration in the organization	Encourage a profound understanding of various organizational units in the management team	Eriksson, Bergbrant & Mörck, 2011; Morelli & Lecci, 2014; Otley, 1999
	Avoid frequent swaps of management members	Morelli & Lecci, 2014
Acknowledge the structure, culture as well as competencies in key managerial decisions	Medical professionals must be directly involved in the design of strategies, not managed	Glouberman & Mintzberg, 2001a; 2001b; Hallin & Siverbo, 2003; Hellström, Lifvergren & Quist, 2010; Merchant; 1982; Mintzberg, 1997; 2011; Stacey, 2003; Yasin, Zimmerer, Miller & Zimmerer, 2002
Continuous evaluation should reflect gaps between present and intended state of performance	Use frequent measurements as a springboard for continuous improvement	Strome, 2013
	Create a common understanding on why measurements are important and explain the rationale behind them; clarify what they are supposed to do and how they can be used	Elg, Palmberg & Kollberg; 2013; Numerato, Salvatore & Fattore, 2012
	Feedback on results must be shared with co-workers that have a mandate to initiate improvements	Dey, Hariharan & Despic, 2008; Morelli & Lecci, 2014

3.6 CONTEXT: THE HOSPITAL

In this section, a longitudinal case study involving a medium-sized hospital in West Sweden serving a population of 140,000 people is presented. Focus has been on examining acute patient flow-related principles and practices, thereby identifying limiting factors regarding process and acute flow management. An analytical approach informed by evidence-based principles and practices outlined in the previous sections has been used.

3.6.1 AN OVERVIEW OF THE CASE

The hospital under study is part of a larger hospital group of three hospitals with a catchment area of 260,000 people. The hospital *group* consists of 30 medical specialties, two emergency departments, two intensive care units and three operation units. During a non-pandemic year, the hospital group manages somewhere around 50,000 inpatient episodes, 400,000 outpatient visits 20,000 surgical procedures out of which 5,000 procedures are acute. The actual hospital is the largest hospital in the group with a yearly inflow of slightly more than 50,000 patients to its emergency ward. The hospital also manages somewhere around 60–70 percent of the care production outlined above. The overall goals of the hospital are to focus on and continuously improve medical quality, patient safety, accessibility, patient experiences, evidence-based practices and working conditions, simultaneously maintaining a balanced economy.

The hospital has a matrix-inspired organizational design. Thus, there is a functional organizational structure with 12 clinics encompassing medical, surgical, psychiatric and pediatric specialties. Each clinic has a varied number of units/departments, where the number of units depends on the size of patient inflows. The management model is mainly hierarchical, consisting of a hospital director, a hospital staff, 12 clinical managers and 100 unit managers. Each clinic is led by a clinical manager together with their unit managers. In addition to the vertical organization, horizontal managerial structures are also in place. Each clinic has a designated patient process manager, process leaders and cross-professional process groups. Most often, the clinical manager is the process owner. Together, these co-workers are responsible for the continuous improvement of patient processes within each clinic (Hellström, Lifvergren, Gustavsson & Gremyr, 2015). The hospital process improvement guidelines are based on the four steps for process improvement outlined earlier (see also Ljungberg & Larsson, 2012). Thus, patient processes within each clinic are managed by the clinical process organization in collaboration with the clinical top management team, whereas patient flows crossing several clinics are managed by the hospital top management team entailing all the clinical managers. As of today, almost 50 patient processes have been identified and are continuously being developed according to process improvement guidelines.

3.6.2 ACUTE PATIENT FLOWS: CHALLENGES

During the last decade, the number of available beds at the hospital has decreased, primarily due to economic constraints, whereas patient inflow rates have increased. Thus, recurring episodes with bed occupancy rates above 100 percent have become more common. The situation has put a serious strain on working conditions as well as

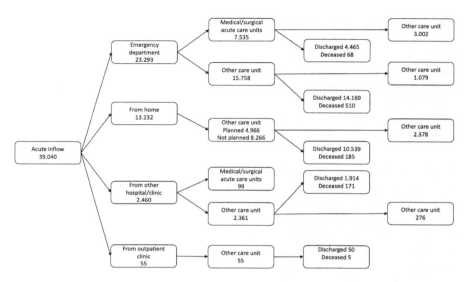

FIGURE 3.1 Visualization of patient flows.

making it more difficult to provide an accessible and safe patient care with high medical quality. As an example, the number of beds has been reduced by 8 percent since 2010, whereas patient inflows to the emergency department have increased by almost 10 percent. Subsequently, improving hospital flow efficiency has been one of several pivotal strategies to manage these challenges.

During the last five years, several hospital wide projects have been initiated to improve flow efficiency as well as production and capacity planning. Several quantitative assessments and analyses have been carried out. Data mining methods have been used to get an overview of the overall acute patient flow throughout the system (see Figure 3.1).

Based on data mining methods to visualize patient flows, clinical, process and unit managers as well as ward unit coordinators along the flow have collaborated regularly to identify clinical waste and bottlenecks to improve the overall flow efficiency. Areas of improvements are detailed in the concluding sections of the chapter and juxtaposed against best practices.

3.6.3 Some Words about the Method

This is an ongoing longitudinal case study that was initiated in 2016. Quantitative and qualitative methods have been used iteratively in an abductive approach to inform ongoing improvements initiatives (Gummesson, 2000). Thus, semi-structured interviews with coordinators, unit, process and clinical managers with key roles in the acute flow (see Figure 3.1; $n=20$) have been carried out. Additionally, key areas for improvement as well as improvement suggestions have been identified during numerous workshops with all the unit managers ($n=30$) from the departments outlined in Figure 3.1. Data from interviews and workshops have then been analysed thematically and compared with evidence-based best practices.

3.7 LESSONS LEARNED

Unfortunately, the COVID-19 pandemic has put an enormous stress on the hospital from the beginning of 2020 to early 2021, with peaks of more than 70 patients occupying hospital beds on certain days. Thus, repeated quantitative assessments of flow efficiency are difficult to interpret. Still, bed occupancy rates and length of stay at the emergency department have been maintained from 2019 to the spring of 2021 despite a huge additional inflow of COVID-19 patients, while resources in terms of healthcare staff and number of beds have remained the same, indicating an improved acute flow efficiency. Although improvement initiatives are still ongoing, extensive quantitative and qualitative data have been collected, analysed and compared with evidence-based best practices. Based on the case, key problem areas regarding flow efficiency are presented, and approaches to manage these problems are then suggested, drawing from local experience as well as from the scientific literature.

Six key areas of improvement have been identified, where social as well as more technical aspects of flow management are represented:

- Create distinct functions and roles pertaining to the main acute patient flows in the hospital
- Apply and continuously develop production and capacity planning at all levels to coordinate and level out flows
- Evaluate acute flows for continuous improvement
- Improve collaboration between professional groups
- Manage and support coordination and integration at all levels
- Involve professional groups in key managerial decisions

Each area will be presented in more detail, where examples on how problems might reveal themselves in everyday practises will be outlined. Further, tips for improvements based on evidence-based practices and local experiences will be presented.

3.7.1 CLEAR FUNCTIONS AND ROLES IN PATIENT FLOWS

There are no formal process functions or roles connected to the acute flow throughout the hospital. Roles and functions are only discernible on lower hierarchical levels, for instance in clinics and at units. Thus, current clinical and process managers, groups and leaders can only understand parts of the overall flow, which prevents a whole systems view of the flow (Aronsson, Abrahamsson & Spens et al., 2011; Bergman & Klefsjö, 2020). Clinical managers tend to only focus on their part of the acute flow:

> I do not have any detailed knowledge of the overall flow. And I do not think a manager at my level needs to have that
>
> (Interview, clinical manager)

Or, as put by another clinical manager:

> We would sincerely need one or two process managers responsible for the entire acute flow. They should have the top management support in making everyday decisions regarding the overall flow; that would actually mean something.

Evidently, the lack of a whole systems view of the entire flow impedes the possibilities to detect and manage critical bottlenecks in the overall system (Cao, Baofeng, Yuan & Xiande, 2015; Haraden & Resar, 2004; Plantin & Johansson, 2012). In other words, clinics tend to work on their own instead of collaborating to improve the throughput of the entire system. Since there are no distinct flow management roles at the hospital level, the management of cross-clinical flows is mainly dependent on informal relationships between co-workers at the various clinics:

> Some of us have known each other for a long time and we can usually come up with solutions when we have to manage patients that needs to be transferred from one clinic to another. But it is not the system that creates good prerequisites for the transfers, quite the contrary.
>
> (Interview, clinical manager)

Similarly, clinical and process managers feel that they do not know what mandate they have to make decisions regarding the overall management of the acute patient flow. This, in turn, limits their repertoire of possible actions (Nilsson & Mandoff, 2015). Collaboration between units at each clinic seems to work more smoothly. Within clinics, flow management meetings between units are usually in place, which enforces cooperation across units, decreases risks for conflicts and brings forward the patient perspective much more. Regular meetings across functions and units also create a deeper understanding of ones role in the overall system, thereby encouraging integration and coordination of care activities along the various patient journeys.

In summary, clear process roles and functions related to the overall flow across clinics and units must be in place. Good relations between co-workers and between units and clinics facilitate flow management. Creating multi-professional, cross-functional teams that repeatedly meet to manage patient flows from the patient's perspective are other important recommendations to improve flow efficiency (Alfalla-Luque, Medina-Lopez & Dey, 2013; Meiboom, Schmidt-Bakx & Westert, 2011).

3.7.2 Apply Production and Capacity Planning to Level Out Flows

The hospital has started to apply production and capacity planning at various units and clinics. However, the maturity of the plans varies greatly between clinics. In addition, the plans often lack medium- and long-term prognoses and, most often, they only entail planned care activities. Thus, medium- and long-term production and capacity plans regarding the acute flow are missing, and the clinical plans are not aggregated on a strategic hospital level. As put by a member of the top management team:

> We do not have information on how the clinics have planned their inflow to their departments or to surgical procedures. And I do not know the situation at the emergency ward from day to day. In short, it is very difficult to know the

current state of the system, how much resources we need, and how it might evolve next week, next month.

In addition, a large proportion of acute patients are admitted directly to the departments, without passing the emergency department. These patients are difficult to capture in information systems and are not acknowledged in daily flow management decisions on a higher system level. Subsequently, valid and relevant information regarding key aspects of current inflow versus available resources in terms of beds are very difficult to obtain. Thus, production and capacity planning must entail planned as well as acute flows from all clinics and include short-, medium- and long-term planning horizons (Haraden & Resar, 2004). Information systems for PCP must also be in place to inform decisions and the outcomes of these decisions at relevant places in the overall system.

Historically, the hospital has had a dominating economical management system, where resource efficiency and a balanced economy have been prioritized. As pointed out by several scholars, flow efficiency will be severely impaired by management decisions that encourages 100 percent resource utilization in various parts of the system (Brandt & Malmgren, 2015; Modig & Åhlström, 2015; Olsson & Aronsson, Abrahamsson & Spens, 2012). According to flow dynamics, queues will inevitably grow infinitely as the resource utilization approaches 100 percent. Paradoxically, a more pronounced managerial focus on quality and flow from a patient perspective will most probably save resources in the long run. However, quality and flow management require a long-term strategy that focuses on creating more flexible resources in terms of beds and personnel to manage variation in inflow without compromising flow efficiency. Other recommended approaches are to separate acute and planned flows as far as possible, as well as to create fast tracks for certain patient groups where standard treatments are possible (Olsson & Aronsson, Abrahamsson & Spens, 2015).

3.7.3 Evaluate Acute Flows

This improvement area is closely connected to the lack of formal process roles for the overall acute flows at the hospital. Put differently, the overall acute patient flow has, so far, not been recognized as a prioritized hospital process that requires continuous improvement according to evidence-based approaches. Although parts of the overall acute flow are identified as important *clinical* processes in need of continuous improvement, no roles or goals for the *overall* acute patient process are in place, which impedes a whole system view on the acute flow. In addition, there are no integrated measurements at the hospital level to continuously evaluate different perspectives of the acute flow:

> There are departments that can tell you how fantastic their process measurements are. But the hospital does not have it. And the hospital does not ask for it either.
> (Interview, clinical manager)

Thus, measuring key variables over time is a prerequisite to understand variation and to improve flow and production and capacity planning. The development of a process measurement system at hospital, clinical and unit-level is thus a pivotal improvement area in itself and requires special attention (Bergman & Klefsjö, 2020; Elg, Palmberg & Kollberg 2013; Ljungberg & Larsson, 2012).

There are other challenges pertaining to evaluation as well:

> At this hospital people find their own way of picking variables to follow. So it is not surprising that we cannot put the variables together to make sense of the system.
>
> (Interview, clinical manager)

Subsequently, the choice of variables must be thoroughly thought through and aligned across units, clinics and at the hospital level. Data analyses and visualization are other challenges. For instance, there is a tendency to focus too much on average and median values, thereby neglecting to follow how different variables vary over time and how they are interrelated. Monitoring and analyzing variation and co-variation are thus important preconditions to get a deeper understanding of cause-effect mechanisms in patient flows. Further, variation in variables upstream in the flow are often neglected. Thus, common measures such as average length of stay and average occupancy rates are only the result of decisions upstream, but do not help to manage flow variation upstream. Subsequently, these variables are not useful for upstream flow management. However, they are necessary as result variables to continuously refine flow management decisions in repeated improvement cycles (Bergman & Klefsjö, 2020).

There is also a tendency to neglect the relation between flow management guidelines and medical guidelines. Experiences from workshops at the hospital revealed that flow management guidelines are often overridden by decision that are medically rather than flow based, which is important to consider when developing flow management systems.

3.7.4 Improve Collaboration Between Professional Groups

As pointed out earlier, many problems in flow management are primarily related to social aspects. For instance, management groups and care professionals do not share the same vocabulary, partly because they represent different worlds (Gloubermann & Mintzberg, 2001a, 2001b). In the Cure and Care worlds, patient satisfaction, care quality and patient safety are common words in the discourse, whereas management prefer to talk about economy and resources:

> There is a certain language used by managers that the care professionals do not understand.
>
> (Interview, clinical manager)

Different discourses add to the already existing gaps between the 4Cs, especially between Care/Cure and Control. However, conflicts are not uncommon between Cure and Care either. For instance, unit managers (often representing the Cure world) point out that they cannot control the physicians although they play a key role in flow management at the unit. If pressed, some physicians even threaten to quit:

> Physician did not follow common guidelines at the unit. No one dared to confront him though, afraid that he would quit.
>
> (Interview, unit manager)

A key strategy to improve relations between the 4Cs is to use the patient as an attractor; a unifier for dialogue and collaboration at various levels in the organization (Huzzard, Hellström, Lifvergren & Conradi, 2014). Additionally, a pronounced focus on patient processes creates more arenas to discuss and balance patient, safety and medical perspectives with resource perspectives.

3.7.5 Manage and Support Coordination and Integration

The hospital has made many organizational changes and hospital directors have come and gone on numerous occasions. As pointed out repeatedly in the scientific literature, re-organizations and frequent changes of management team members hamper trust as well as confidence for the management (Mintzberg, 2011; Morelli & Lecci; 2014). On the contrary, a pronounced focus on integration and coordination in terms of process and flow development is dependent on a long-term stability in organizational and managerial structures. In addition, coordination and integration across units and clinics requires time for co-workers to meet in cross-professional improvement teams. However, time is a critical resource for development and is often lacking:

> We in the top management team demand too much information from clinical and unit managers. And that steals time from them and prevents them from being present at the units, close to their patients.
> (Interview, top management team member)

However, several studies point out that, embracing a longer timeframe, scheduling for development is indeed a possible as well as necessary strategy to encourage continuous process improvement (Kotter, 1995; Porter, 1996; 2010).

3.7.6 Involve Professional Groups in Managerial Decisions

The hospital has formulated common values as well as long-term goals that should permeate co-workers' activities. However, few organizations are successful when it comes to securing a common value platform and sharing common goals, especially if the goals are abstract and not related to everyday care practices (Alvesson, 2013). A similar situation prevails at the hospital:

> I must admit that, sometimes, we just do not care about the goals because the unit would not function if we do.
> (Interview, unit manager)

Large projects that aim at reaching long-term goals do not involve care professionals at the project start. Indeed, few of these projects reach their goals and they are soon forgotten among healthcare professionals. Instead, representatives from Care and Cure should be involved already when identifying relevant improvement areas that should primarily target patient processes. Healthcare professionals should also be involved in the prioritization and initiation of larger strategic projects (Röhlin, 2013). In addition, abstract strategic plans must be jointly translated by Cure, Care and Control worlds to concrete daily activities that are relevant as well as meaningful for the microsystems of the organization (Funck, 2016).

3.8 CONCLUSIONS

It is very tempting to glance at advanced technical and ICT-related solutions to meet future healthcare challenges and to manage acute flows in hospitals. However, as illustrated by this case, there are many low-hanging fruits that need immediate attention. The areas of improvement that have been illuminated in the chapter have been well known for decades in the scientific literature as well as from numerous case descriptions in many different contexts, and call for improved social as well as technical aspects of management that must evolve in a continuous dialogue with all the co-workers of the value-adding microsystems. Thus, the aim of the chapter has been to point to evidence-based improvement areas and examples of solutions that, hopefully, will help other hospitals in other contexts to embark on similar improvement journeys that acknowledge the technical as well the social aspects of managing change.

NOTES

1. In this context, flow efficiency is defined as the sum of value-adding activities in relation to the throughput time in a care process (Modig & Åhlström, 2015).
2. Microsystems are the smallest functional units in a healthcare system. They are complex adaptive systems that consist of cross-professional teams and include the patient. According to microsystem theory, the care quality of a healthcare system can never supersede the care quality delivered by integrated and coordinated microsystems along the patient's care journey (Nelson et al., 2008).
3. Customer is defined as a person for whom the organization should create value (Bergman & Klefsjö, 2020).
4. The term 'attractor' is commonly used in chaos mathematics to denote various centers of the chaotic system, around which agents in the system are organized in surprisingly regular patterns (Stacey, 2003).

REFERENCES

Alfalla-Luque, R., Medina-Lopez, C. & Dey, P.K. (2013). Supply chain integration framework using literature review. *Production Planning & Control, 24*(8–9), 800–817.

Allder, S., Silvester, K. & Walley, P. (2010). Understanding the current state of patient flow in a hospital. *Clinical Medicine, 10*(5), 441–444.

Alvesson, M. (2013). *Organisation och ledning: Ett något skeptiskt perspektiv.* ('Organization and management: A somewhat sceptical perspective'). Studentlitteratur AB.

Alvesson, M. & Sveningsson, S. (2007). *Changing organization culture: Cultural change in progress.* Routledge.

Anell, A. & Mattison, O. (2009). *Samverkan i kommuner och landsting.* ('Collaboration in communities and counties'). Studentlitteratur AB.

Aronsson, H., Abrahamsson, M. & Spens, K. (2011). Developing lean and agile health care supply chains. *Supply Chain Management: An International Journal, 16*(3), 176–183.

Beer, M., Eisenstat, R.A. & Spector, B. (1990). Why change programs don't produce change. Harvard Business Review, Nov–Dec, 158–166.

Beer, M. & Nohria, N. (2000). Cracking the code of change. Harvard Business Review, May–June, 133–139.

Bergman, B. & Klefsjö, B. (2020). *Kvalitet från behov till användning.* ('Quality from customer needs to customer satisfaction'). 6th ed., Studentlitteratur AB.
Brandt, J. & Palmgren, M. (2015). *Produktionsstyrning i sjukvård ('Production management in healthcare').* Lyxo.
Brunson, N. (2006). *The organization of hypocrisy.* Abstrakt.
Bruzelius, L.H. & Skärvad, P-H. (2004). *Integrerad organisationslära ('Integrated organizational studies').* Studentlitteratur AB.
Cao, Z., Baofeng, H., Yuan, L., & Xiande, Z. (2015). The impact of organizational culture on supply chain integration: A contingency and configuration approach. *Supply Chain Management: An International Journal, 20*(1), 24–41.
Collins, J. (2001). *Good to great: Why some companies make the leap ... and others don't.* Harper Collins.
Dawson, P. (2003). *Understanding organizational change: The contemporary experience of people at work.* Sage.
Dean, J.W. & Bowen, D.E. (1994). Management theory and total quality: Improving research and practice through theory development, *Academy of Management Review, 19*(3), 392–418.
De Feo, J. & Barnard, W. (2004). *Juran Institute's Six Sigma breakthrough and beyond: Quality performance breakthrough methods.* McGraw-Hill.
Dey, P.K., Hariharan, S. & Despic, O. (2008). Managing healthcare performance in analytical framework. *Benchmarking: An International Journal, 15*(4), 444–468.
Duck, J.D. (1993). Managing change: The art of balancing. *Harvard Business Review*, Nov–Dec, 109–118.
Elg, M., Palmberg, K. & Kollberg, B. (2013). Performance measurement to drive improvements in healthcare practice. *International Journal of Operations & Production Management, 33*(11/12), 1623–1651.
Eriksson, H., Bergbrant, I. & Mörck, I. (2011). Reducing queues: Demand and capacity variations. *International Journal of Health Care Quality Assurance, 24*(8), 592–600.
Forsberg Hvitfeldt, H., Aronsson, H., Keller, C. & Lindblad, S. (2011). Managing health care decisions and improvement through simulation modeling. *Quality Management in Healthcare, 20*(1), 15–29.
Funck, E. (2016). The balanced scorecard in healthcare organizations. In: Örtenblad, A., Abrahamson Löfström, C. & Sheaff, R. Thomson (Eds.), *Management innovations for healthcare organisations: Adopt, abandon or adapt?.* Routledge, pp. 61–79
Glouberman, S. & Mintzberg, H. (2001a). Managing the care of health and the cure of disease – part I: Differentiation. *Health Care Management Review, 26*(1), 56–69.
Glouberman, S. & Mintzberg, H. (2001b). Managing the care of health and the cure of disease – part II: Integration. *Health Care Management Review, 26*(1), 70–84.
Goldratt, E.M. (1999). *Theory of constraints.* North River Press.
Graban, M. (2009). *Lean hospitals.* CRP Press.
Gummesson, E. (2000). Qualitative methods in *management research.* Sage.
Gupta, P. (2004). *Six Sigma business scorecard.* McGraw-Hill.
Hallin, B. & Siverbo, S. (2003). *Styrning och organisering inom hälso- och sjukvård.* ('Managing and organizing in healthcare'). Studentlitteratur AB.
Haraden, C. & Resar, R. (2004). Patient flow in hospitals – Understanding and controlling it better. *Frontiers of Health Services Management, 20*(4), 3–15.
Hellström, A., Lifvergren, S., Gustavsson, S. & Gremyr, I. (2015). Adopting a management innovation in a professional organization – The case of improvement knowledge in healthcare. *Business Process Management Journal, 21*(5), 1186–1203. DOI 10.1108/BPMJ-05-2014-0041.

Hellström, A., Lifvergren, S. & Quist, J. (2010). Process management in healthcare: Investigating why it's easier said than done. *Journal of Manufacturing Technology Management*, *21*(4). 499–511.
Ho, S.J.K., Chan, L. & Kidwell, R.E. Jr. (1999). The implementation of business process reengineering in American and Canadian hospitals. *Health Care Management Review*, *24*(2), 19–31.
Huzzard, T., Hellström, A., Lifvergren, S. & Conradi, N. (2014). A physician-led and learning-driven approach to regional development of 23 cancer pathways on Sweden. In: S. Mohrman, S. and Shani, A.B. (Eds.). *Reconfiguring the eco-system for sustainable healthcare*, Volume 4. Emerald, pp. 101–131.
Inamdar, S.N. & Kaplan, R.S. (2002). Applying the balanced scorecard in healthcare provider organizations. *Journal of Healthcare Management*, *47*(3), 179–196.
Kaplan, R.S., & David Norton, D. (1992). The balanced scorecard: Measures that drive performance. *Harvard Business Review*, *70*(1), 71–79.
Kaplan, R.S. & Norton, D.P. (2001). *The strategy-focused organisation: How balanced scorecard companies thrive in the new business environment.* Harvard Business School Press.
Keegan, A. (2010). Hospital bed occupancy: More than queuing for a bed. *Medical Journal of Australia*, *193*, 291–293.
Knights, D. & McCabe, D. (2001). A road less travelled: Beyond managerialist, critical and processual approaches to total quality management. *Journal of Organizational Change Management*, *15*(3), 235–254.
Koestler, C.D., Ombao, H. & Bender, J. (2013). Ensemble-based methods for forecasting census in hospital units. *BMC Medical Research Methodology*, *13*(67). https://doi.org/10.1186/1471-2288-13-67
Kotter, J. (1995). Leading change: Why transformation efforts fail. *Harvard Business Review*, March–April, 59–67.
Lekwall, P. & Wahlbin, C. (2004). *Information för marknadsföringsbeslut* ('Information for marketing decisions'). Studentlitteratur AB.
Liker, J. (2004). *The Toyota Way*. McGraw-Hill.
Ljungberg, A. & Larsson, E. (2012) *Processbaserad verksamhetsutveckling: varför, vad, hur* (*Process-based operations development – why, what, how*)? Studentlitteratur AB.
Magnusson, K., Kroslid, D. & Bergman, B. (2003). *Six sigma: The pragmatic approach.* Studentlitteratur AB.
Mattson, S-A. & Jonsson, P. (2012). *Material- och produktionsstyrning ('Material and production management').* Studentlitteratur AB.
Mead, G. (1967). *Mind, self, and society: From the standpoint of a social behaviourist.* University of Chicago Press.
Meijboom, B., Schmidt-Bakx, S. & Westert, G. (2011). Supply chain management practices for improving patient-oriented care. *Supply Chain Management: An International Journal*, *16*(3), 166–175.
Merchant, K.A. (1982). The control function of management. *Sloan Management Review Summer*, *23*(4), 43–55.
Mintzberg, H. (1997). Toward healthier hospitals. *Health Care Management Review*, *22*(4), 9–18.
Mintzberg, H. (2011). To fix health care, ask the right questions. *Harvard Business Review*, *89*(10), 44.
Mintzberg, H. & Waters, J. (1985). Of strategies, deliberate and emergent. *Strategic Management Journal*, *6*(3), 257–272.
Modig, N. & Åhlström, P. (2015). *Detta är Lean* ('This is Lean'). Rheologica Publishing.

Morelli, M. & Lecci, F.J. (2014). Management control systems change and the impact of top management characteristics: The case of healthcare organisations. *Journal of Management Control*, 24(3), 267–298.
Nelson, E.C., Godfrey, M.M., Batalden, P.B., Berry, S.A., Bothe, A.E., McKinley, K.E., Melin, C.N., Muething, S.E., Moore, L.G., Wasson, J.H. & Nolan, T.W. (2008). Clinical microsystems, part I.: The building blocks of health systems. *The Joint Commission Journal on Quality and Patient Safety*, 34(7), 367–378.
Nergårdh, A. (2017). Samordnad utveckling för god och nära vård ('Coordinated devleopment of integrated good care'). Final report, Swedish official assessments 2017:01.
Nilsson, K. & Mandoff, M. (2015). Managing processes of inpatient care and treatment. *Journal of Health Organization and Management*, 29(7), 1029–1046.
Norbäck, L.E. & Targama, A. (2009). *Det komplexa sjukhuset. Att leda djupgående förändringar i en multiprofessionell verksamhet* ('The complex hospital'). Studentlitteratur AB.
Nordenström, J. (2014). Värdebaserad vård ('Value-based healthcare'). Karolinska Institutet University Press.
Numerato, D., Salvatore, D. & Fattore, G. (2012). The impact of management on professionalism: A review. *Sociology of Health & Illness*, 34(4), 626–644.
OECD. (2013). *OECD health data*. OECD.
Olhager, J. (2013). *Produktionsekonomi* ('Production economy'). Studentlitteratur AB.
Olsson, O. (2014). *Managing variable patient flow at hospitals*. Licenciate thesis, Linköping.
Olsson, O. & Aronsson, H. (2012). *Logistikhandbok för hälso- och sjukvården* ('Handbook of Healthcare Logistics'). Linköping University Electronic Press.
Olsson, O. & Aronsson, H. (2015). Managing a variable acute patient flow: Categorising the strategies, Supply Chain Management: *An International Journal*, 20(2), 113–127.
Ortiga, B., Salazar, A., Jovell, A., Escarrabill, J., Marca, G. & Corbella, X. (2012). Standardizing admission and discharge processes to improve patient flow: A cross sectional study. *BMC Health Services Research*, 12, 180.
Otley, D. (1999). Performance management: A framework for management control systems. *Management Accounting Research*, 10(4), 363–382.
Plantin, A. & Johansson, M. (2012). Implementing production planning processes in health care: A case study of a surgery clinic. Proceedings of PLAN Research and Application Conference, Lund, August.
Porter, M.E. (1996). What is strategy? *Harvard Business Review*, 74, 61–78.
Porter, M.E. (2010). What is value in health care? *New England Journal of Medicine*, 363, 2477–2481.
Rentzhog, O. (1998). *Processorientering – en grund för morgondagens organisationer* ('Process orientation – a foundation for the organizations of tomorrow'). Studentlitteratur AB.
Rosenbäck, R. (2018). *Produktionsstyrning i sjukvården – en väg framåt* ('Production management in healthcare – one way forward'). Studentlitteratur AB.
Röthlin, F. (2013). Managerial strategies to reorient hospitals towards health promotion: Lessons from organisational theory. *Journal of Health Organization and Management*, 27(6), 747–761.
SALAR (The Swedish Association of Local Authorities and Regions). (2005). *Hälso- och sjukvården till 2030*. ('Healthcare on to 2030'). SALAR.
SALAR (The Swedish Association of Local Authorities and Regions). (2007). *Strategier för effektivisering* ('Strategies for Efficiency'). SALAR.
SALAR (The Swedish Association of Local Authorities and Regions). (2014). *Skador i vården – en sammanställning av klinikvisa resultat* ('Healthcare care related injuires – an assessment of clinics'). SALAR.

Schaffer, R. & Thomson, H. (1992). Successful change programmes begin with results. *Harvard Business Review*, Jan–Feb, 80–89.

Stacey, R. (2003). *Strategic management and organizational dynamics – The challenge of complexity.* (Fourth Edition), Pearson Education Limited.

Statens offentliga utredningar, SOU. (2009). *En nationell cancerstrategi för framtiden* ('A national cancer strategy for the future'). SOU.

Stiernstedt, G., Zetterberg, D. & Ingmanson, A. (2016). *Effektiv vård* ('Efficient care'). Final report, Swedish official assessments 2016:2.

Strebel, P. (1996). Why do employees resist change? *Harvard Business Review*, May–June, 86–92.

Strome, T. (2013). *Healthcare analytics for quality and performance improvement.* Wiley.

Swedish Institute for Quality, SIQ (2015). *SIQ management model.* SIQ.

The National Board of Health and Wellness (Socialstyrelsen). (2015). *Tillståndet och utvecklingen inom hälso- och sjukvård och socialtjänst – lägesrapport* ('Report on state of affairs in healthcare and social services'). SoS.

Utley, M. & Worthington, D. (2012). Capacity planning. In: Hall, R (Ed.), *Handbook of healthcare system scheduling.* Springer, pp. 11–30.

Vissers, J.M.H., Bertrand, J.W.M. & De Vries G. (2001). A framework for production control in health care organizations. *Production Planning & Control*, *12*(6), 591–604.

Walley, P., Silvester, K. & Steyn, R.S. (2006). Managing variation in demand: Lessons from the UK national health service. *Journal of Healthcare management/American College of Healthcare Executives*, *51*(5), 309–320.

Wiger, M. (2012). *Logistics management in a healthcare context: Methodological development for describing and evaluating a healthcare organisation as a logistics system.* Licentiate thesis, Linköpings Universitet, Linköping Studies in Science and Technology.

Yasin, M.M., Zimmerer, L.W., Miller, P. & Zimmerer, T.W. (2002). An empirical investigation of the effectiveness of contemporary managerial philosophies in a hospital operational setting. *International Journal of Health Care Quality Assurance*, *15*(6), 268–227.

4 Influence of Commercial Excellence and Digital in Healthcare Professionals Relationships Management
A Pharmaceutical Focus Group Study

António Pesqueira

CONTENTS

4.1 Introduction	61
4.2 Literature Review	63
4.2.1 Pharma Commercial Excellence	63
4.2.2 Digital in Commercial Pharmaceuticals	66
4.3 Methodological Approach	68
4.3.1 Focus Group Questions	71
4.3.2 Focus Group Key Takeaways and Raised Points	72
4.3.3 Key Observations from the Focus Group Interviews	76
4.4 Conclusions	79
4.4.1 Limitations and Future Recommendations	79
4.4.2 Conclusions	80
References	84

4.1 INTRODUCTION

Currently, the pharmaceutical industry is increasing all the necessary efforts to optimize and leverage the models of engagement and interactions with patients, healthcare professionals (HCPs), payers, or key opinion leaders (KOLs) to activate modern and advanced business models by leveraging all existing opportunities and gain competitive advantage.

DOI: 10.1201/9781003227892-4

The go-to-market models from the industry are being forced to maximize customer engagement activities while human and financial resources are being optimized (Graham and Ariza, 2003).

Some of the main reasons for large investments in customer engagement models and commercial strategies during 2021 are of course connected to the COVID-19 pandemic crisis, which brought into the organizations different challenges and pitfalls of how digital and commercial excellence (CommEx) are being used.

But other reasons are more connected with the sequential and evolutive transformation of behaviors, needs, and ways of interacting with all the society members but most of the time with healthcare professionals (HCPs) as their workloads and capacity to engage with the pharmaceutical industry has suffered major changes.

HCPs' contact time was significantly reduced where physicians, during COVID-19, increasingly preferred digital contacts where major behavioral changes shifted demand and preference points toward more digital contacts with the pharmaceutical industry. HCPs from a general point of view have been increasingly open to digital communications from the healthcare and life sciences industries, where major trends like digital-savvy professionals and digital immigrants are being representative of some of the growing behavioral patterns of the HCPs and mainly as a need to cope with all patient visits pressure and provide as more care services as possible (Lee and Lee, 2021).

Many pharmaceutical companies have objectives to create effective customer journeys and effective engagement models that can bring agility and value to the customer engagement process, but as well to meet the opportunity to increase or being more efficient in reaching hard-to-see or sometimes even no-see customers.

In nowadays, many pharma companies are investing actively in financial and human resources systems to bring innovation and business value to customer journeys and needs analysis and optimization, while communicational and engagement channels and content are being redesigned, depending on the HCPs' preferences and needs (Kaplan, Norton, and Davenport, 2004).

Previously, the industry was very much used to traditional commercial models like share-of-voice or face-to-face interactions visits, with HCPs interacting with the industry in a single channel or with in-person events like clinical sessions, congresses, speaker programs, or many other examples.

But that reality changed tremendously where currently the healthcare management systems and specifically the healthcare professionals are under massive scrutiny and pressure with the COVID-19 pandemic crisis forcing the interactions and engagement models with the pharmaceutical industry to redesign and rethink all the business models and strategies.

Also applicable to all the industries, there is a strong barrier of replacing old systems or outdated processes, where those tasks will always result in complex execution plans or simply create disruptions of how sensitive organizational functions in pharma like regulatory, legal, and data privacy and compliance will undertake all the necessary scrutiny (Volberda, Van Den Bosch, and Heij, 2018).

In this chapter, we will focus our attention on all implications, use cases, and value propositions from CommEx and digital operational and strategic models to the pharmaceutical industry and assessing the way HCPs are being engaged and their relationships with the industry being managed.

A new set of capabilities must be acquired by commercial professionals to keep pace with the advancing times, such as analytics, digital, and data acumen. The purpose of this review is also to explore all challenges and opportunities arising from the transformation of the CommEx and digital teams and the skills that are needed for the successful implementation of CommEx programs.

Thus, throughout this research effort, we will proceed with a conceptual and practical understanding of applications, opportunities, and challenges derived from CommEx and digital capabilities and processes to the pharmaceutical industry with the resource of a focus group method with 32 industry professionals who were engaged between January and April 2021 and still during the pandemic crisis of COVID-19.

We will analyze the different applications in the pharma context of CommEx and digital but understanding as well throughout the focus group all dependencies with the major challenges, opportunities, use cases, future trends, and current knowledge.

As we aimed to focus not only on the conceptual spectrum but as on a more practical understanding of the defined research questions, this research went deeper in terms of understanding all different nuances and technical and business details.

This chapter is organized as follows: we will start to lay the major questions and concepts during the introductory section and then move to a second section where will detail the relevant literature review of key topics and all connected background and related academic work with the defined research questions and defined problem. The third section will then explain the focus group review and provide all necessary details of that focus group discussions. This section will also describe the planning and implementation of the focus group methodology and showing the discussions and interview results. In the last section of this chapter, we will present the fourth section where the conclusions and future work recommendations will be explained.

4.2 LITERATURE REVIEW

4.2.1 PHARMA COMMERCIAL EXCELLENCE

The COVID-19 pandemic crisis brought several changes to all pharmaceutical industries, from research and development to commercial activities. As several companies are struggling to control the dispersed changes and impacts from COVID-19 in this unprecedented period of uncertainty, currently the way that sales and marketing activities are being conducted needed to change paradigms and strategies.

The COVID-19 pandemic and its challenges on pharma commercial activities have been extremely impactful the way the industry interacts with all stakeholders, including the way products and services are delivered and all dependencies from sales, marketing, medical issues, and much more.

During the last years, we have been witnessing a clear biosimilars uptake that has been a major market trend and especially during pandemic times, where small molecules, biologics, have been driving several innovations in terms of product launches techniques and models when we consider the rapid and efficient business models coming from those major trends.

In the pharmaceutical industry, commercial excellence (CommEx) is seen as one of the most important functions of scientific expertise. For years, the CommEx function has served the needs of business development and strategy teams, as well as sales, marketing, and medical affairs departments and leadership teams (Melnychuk, Schultz, and Wirsich, 2021).

On the sales marketing front, CommEx has been involved in proposing and assessing key analytical and strategic business components but as well overall organizational commercial capabilities to drive key pharmaceutical processes like key account management (KAM), salesforce effectiveness (SFE), digital, customer relationship management (CRM), competitive intelligence (CI), tenders and supply chain management, and many other processes or guidelines and regulations (Ghiraldelli, 2020).

Other responsibilities include training commercial teams to improve their understanding of multichannel and virtual selling, support the excellent achievement in customer engagement tactics or sales efficiency, preparing content for sales or marketing programs, facilitating events design, and supporting decision-making needs from all organizational tiers (Chaffey, Smith, 2017).

Drug development, supply chain management, and digital transformation brought a greater emphasis on the growing demands for customer engagement strategies, stricter guidelines, and regulations, increased scrutiny on interactions between pharma companies and healthcare providers, and a rapid shift in stakeholders' expectations.

CommEx is experiencing challenging times as data-driven businesses and increasing digital business paradigms have made this function a critical strategic partner to a variety of internal functions. Since the pandemic, CommEx has become one of the five strategic pillars in the pharmaceutical organization along with research and development, commercial, medical affairs, and market access (Ghiraldelli, 2020).

Digital engagement models for HCPs and the industry as a whole have been disrupted by the ongoing situation, and many of these changes will persist and continue to define HCPs' experiences after the pandemic.

Global lockdowns and the immediate impact of COVID-19 on global markets caused numerous challenges in adapting to the changing environment and addressing huge demands for clinically meaningful information.

Some of the most immediate results of the pandemic and enforcement of national lockdowns, travel restrictions, and physical contact were as follows:

- Decreased access to HCPs and institutions.
- Face-to-face interactions with HCPs were significantly impacted.
- Reduction in exchange of scientific information and insights gathering.
- Physical access to health care providers is decreasing, resulting in an increase in virtual engagements.
- A shift from face-to-face channels by HCPs.
- Inadequate network connectivity from HCPs and increasing appetite for multichannel platforms and digital capabilities from the pharmaceutical companies.
- Refocus of HCPs priority areas like the COVID-19 pandemic crisis.

Commercial Excellence and Digital in Healthcare   65

- Aiming to raise awareness among healthcare professionals, dispel misinformation, and provide them with enhanced knowledge.
- A shift from a focus on brands to a focus on diseases and therapeutic areas of specialization for the HCPs.
- By going completely virtual, we will be able to ensure faster dissemination and a wider reach to HCPs.

Competitive advantage and operational performance are referenced and connected with CommEx and digital management in several different ways and with a growing need from pharmaceutical organizations to leverage and activate frameworks and connected processes that can drive innovation, business agility, and value (Yanik and Kiliç, 2018).

The duration and impact of this pandemic crisis will dramatically change as CommEx plays a role within the industry, with major impacts in core commercial processes like business development, supply chain management, market access, and key account management. Post-pandemic, CommEx and digital will play a major role for commercial organizations in supporting the understanding of HCPs' opinions, preferences, and the best way of engagement (Hung, Huang, and Lin, 2005).

For most of the pharma companies, the relevance and importance of engagement activities with the HCPs represents the efficiency and performance levels from market penetration levels but as well how much those companies can place and promote their products across retail, hospitals, or other sales channel available (Matricardi, Dramburg, Alvarez-Perea, et al. 2020).

Among the above-mentioned changes and especially during the COVID-19 pandemic crisis are also the rising of HCPs' expectations and needs, regulation pressure, technological forces including marketing automation, virtual engagement activities or events, and digital sales channels.

Many critical areas have been playing a decisive role in CommEx executing programs, including the following: long-term relationships building with HCPs, segmenting and profiling customer value, driving business opportunities, creating and optimizing sales organizational structures, gaining competitive insights, and business outcomes generation.

In today's competitive market landscape, the way that pharmaceutical companies are engaging with key HCPs and healthcare-relevant stakeholders are more from a long-term relationships perspective and by using a wider range of digital channels to effectively exchange products' key messages, clinical data, and information and as well interact from a promotional and medical perspective (Ulrich, Jick, and Von Glinow, 1993).

An effective CommEx program should have as main pillars the vision and strategy, capabilities, tools, and platforms. The three main pillars should run as a business value chain sequence where it is not possible to have a proper set of capabilities without having in place and design the vision and strategy for the CommEx operations. The last step of the sequence should be the full alignment of the platforms and tools to the developed capabilities with the main support of the initial sequence step of the vision and strategy.

Connecting and managing the three pillars of the program, we then have all activities and initiatives of the change management processes and the effective project management strategy. The fourth dimension of a CommEx program acts almost like the glue of all the connecting sequences of the main three pillars, having as main outcomes the full and effective business integrations along with the quality control setting.

Influencing the main operating block of the program, we have the organizational culture and external landscape playing an effective role in all existing dynamics and adoption of the program by all the stakeholders and involved teams. It's the absorption of the organizational culture with the external landscapes that allow the full effectiveness of the change and project management control processes.

As part of the organization, we have the most complex and sensitive factors that strongly influence the running mode of any CommEx in pharma, where the values or leadership or even the business model can dramatically change the output of any CommEx program.

To put in place an effective CommEx program, we cannot discard the governance and monitoring phase, which is the key element for the success of all operations and activities. During this phase, the monitoring of the business and systems performance with clear milestones, planning, timelines, and metrics will strongly influence all producing outcomes and results.

For CommEx, most of the time is requested for a strong alliance with all existing departments and teams to execute most of the programs. Good examples of that collaboration are the patient support programs (PSPs), sales territories alignments, brand plan activations, strengthening reimbursement plan or incentive compensation programs. The typical touchpoints that any CommEx department needs to do are to allow the holistic view of several different processes and projects with distinct and systemic connecting points with a large number of internal departments (e.g. IT, compliance, medical affairs, regulatory, commercial, marketing, etc.) and external groups (patient support groups, KOLs, competitors, payers, etc.).

4.2.2 Digital in Commercial Pharmaceuticals

The pharmaceutical space has been implementing digital programs for some time now in mainly creating more effective customer journey maps and assuming a high business relevance in terms of understanding critical areas like tele-detailing, multichannel marketing engagement, continuous medical education, events management, and media content dissemination.

The pharmaceutical commercial models will largely shift from a pre-pandemic low digital engagement model to a more intensive and agile setup. A more commercial capability will bring disruptions to what might be considered organizational changes, more favorable customer engagement models, wherefrom in one side of the spectrum those cultural changes will require more capacity in terms of digital learning and change management to allow digital sales and marketing processes to play a decisive role (Cutcher-Gershenfeld, Baker, Berente, et al. 2020).

Commercial digital in the pharma industry has many applications and its contribution to strategic pillars in any commercial organization is vast. Some examples of how commercial digital plays a decisive role in pharma commercialization processes are:

- Connect channels to provide a holistic customer experience.
- Omnichannel.
- Digital marketing.
- Measure digital sales and marketing activities with evidence-based marketing.
- Customer insights.
- Predicting what the customer wants.
- More targeted HCPs communications.
- Deliver compelling health-economic arguments and demonstrate long-term benefits.
- Overcome demands for quality information and with effective HCPs virtual engagement.
- Virtual engagement technologies enable your field force to easily bring product and medical experts into the conversation.

Currently and already with a post-pandemic understanding, many pharma companies are redesigning business processes across different functions such as finance, supply chain management, sales, marketing, and many others for improved inventory optimization, demand planning, sales force performance, and product supply productivity.

The organizations that were already using the advantages of digital transformation programs and leveraging new business models with key principles like multichannel, digital engagement, key account management, CommEx, and sales digital transformation were the organizations that better survived the initial and ongoing challenges of COVID-19 organizational and social changes.

The massive quantities of data currently generated digitally and the new dynamics between the healthcare system and the pharmaceutical manufacturers have been presenting new challenges to all kinds of pharma companies, no matter the size, dimensions, and market positioning (Stolpe, 2016).

Younger generations of HCPs started to express bigger needs for medical and scientific content delivery and an increase in preference for online sources and on-demand information to be available as self-serving content from non-commercial HCP portals or websites.

An effective commercialization strategy requests from the organization a clear understanding of the approach to engage HCPs to ensure a solid awareness of the disease and medical or clinical solution.

Not all pharma companies have been able to excel in digital programs but the ones that do are more capable of achieving a high level of productivity and performance, mainly because the time-to-market business agility and competitive reach of digital technologies allow them to achieve higher profits and business excellence.

HCPs are already accessing their medical records online and interacting with external and internal stakeholders through online portals. Furthermore, HCPs are using mobile applications more often to complete online forms and interact with other peers in online communities.

In pharmaceutical and healthcare marketing, where data is becoming more available, healthcare professionals are conducting telemedicine visits, or interacting with pharmaceutical manufacturers, are undergoing rapid transformations that require a significant digital shift.

Several companies are already being able to implement multiple real-time data sources across sales and marketing to provide a consolidated and competitive view of different parts of the business and provide deeper insights into operations and activities of global businesses. The way digital marketing can support business innovations and help businesses to embrace new customer-centricity models to the customer communities at a faster pace is tremendous and crisis like COVID-19 and healthcare management system redesign initiatives are clear examples of this model (Davenport, Harris, Morison, 2010).

Digital is the active transformation of the pharmaceutical industry and building stronger digital marketing capabilities and business acumen to unlock untapped and hidden value within the organization and accelerate real-time decision making and improve digital product innovation lifecycles.

One of the biggest hurdles to a pharma company during COVID was the unexpected risks and market changes derived from several different sources and the complete incapacity for any organization to control or monitor any future event (Schuster, Allen, Brock, 2007).

Only in the last five years, we have been witnessing clear advancements in digital programs in mainly big pharmaceutical companies around mobile communications, learning, and the Internet of Things (IoT). COVID-19 brought growing disruptive potential from several different digital initiatives to pharma, some of which are blockchain, IoT, omnichannel frameworks, and advanced analytics (Uvarova and Pobol, 2020).

Many pharma companies needed to be more efficient in demonstrating their value and the value of their drugs in new complex times and growing demands for real-world evidence and value-based treatments and outcomes.

In the digital age and especially during COVID-19, HCPs are less open to spending physical time in knowing new products, treatments, or pharmaceutical innovations. The industry is at a pivotal moment to transform the methods to interact and engage with the HCPs with an unprecedented opportunity to converge knowledge, experiences on digital innovations, cloud technology, and data science solutions.

The HCPs are much less dependent on a sales representative to understand pharmacological, scientific, or medical innovations as they are increasingly able and willing to take control of their continuous medical education and disease awareness through the vast amount of information available online and on mobile applications.

4.3 METHODOLOGICAL APPROACH

In this section, we present the methodology we used in addition to other key metrics that helped us draw meaningful conclusions and better understand the relationship between key variables during our questionnaire data analysis.

In this section, we will provide an overview of the selected methodology including the key discussion topics and highlights from all the research selections.

The main purpose of this research work is to investigate how CommEx and digital can be perceived as an effective approach and model to current commercial pharma challenges with efficient utilization of resources, data, decision-making capacity, and agility gain in commercial procedures.

An analysis of the study consists of a focus group consisting of 32 pharma professionals ranging in level from analysts to executives.

As part of this study, systematic integration of CommEx and digital will be developed through analyzing the following research question: which CommEx and digital opportunities, challenges, and use cases are most relevant to pharmaceutical commercialization processes and capabilities?

Our research goals aim to understand how pharmaceutical organizations' commercial departments and functions are integrating CommEx and digital into internal processes.

By identifying the CommEx and digital projects that most impact commercialization processes and capabilities within the pharmaceutical industry, the selected research question aims to identify the most influential CommEx and digital projects. A secondary objective under discussion here is to better understand which CommEx and digital techniques and models can be used to influence commercial pharma projects.

The focus group was conducted during the first four months of 2021, where several different questions were made to understand all the participants' knowledge and experiences and to provide all the necessary backbone knowledge and foundation to better understand all answers and nuances connected with the already presented research questions.

For reasons of anonymity, we will not be able to disclose the focus group participant names as all the participants are leaders of the industry and with leadership positions that require special treatment in terms of data privacy.

All focus group interviews were conducted in digital channels via Zoom or Microsoft Teams videoconference technologies during January, February, March, and April 2021.

Participants represented a broad range of positions levels, such as vice presidents, global and local heads, senior directors, managers, and specialists (see Figure 4.1).

From January 4, 2021, to April 15, 2021, we conducted four virtual sessions through Zoom and Microsoft Teams, dividing participants into small groups of eight participants with similar titles and responsibilities.

Participants were guided and facilitated through PowerPoint slides. The presentation lasted 60–90 minutes.

The defined participants' population had a solid dispersion of therapeutic areas and business unit representation from the pharmaceutical industry, where central nervous system (CNS), rare diseases, hematology, oncology, dermatology, and respiratory units were mainly represented during the focus group discussions and giving a good overview of different practices and processes across different business functions, clinical and medical scope, and also use cases across different components of the industry.

The focus group participants were selected according to pre-defined criteria like the participant role, expertise and knowledge in the presented topics, industry experience, and availability to be part of the study. One of the key objectives was to have a good representation of the pharmaceutical biotechnology industry, with practical experience in the selected topics and domains (Koshy and Waterman, 2010).

FIGURE 4.1 Focus group participants job titles and positions.

The focus group study consisted of a total of nine related focus group questions that were grouped into three key topic areas.

The focus group study design was developed through a collective set of observations and conclusions from the previously presented literature review, and the process was validated with a subset of the selected participants and as experts in the selected area, where the overall design was discussed, including samples of the questions and topics collected from the literature review (Bell, Morse, and Shah, 2012).

To simplify the data collection process and maintain consistency in the design and implementation phase of the focus group study, the authors produced a simple roadmap and interview processual 'script' to better facilitate and support the management of all the interviews with the participants.

This focus group study presented some risks of discovering sensitive and confidential information, for instance concerning commercial confidential practices, or specific metrics and key performance indicators that might pose specific competition and confidential information concerns.

Commercial Excellence and Digital in Healthcare   71

In the focus group introduction and debriefing, the focus group facilitators emphasized that participants will have all the necessary time to review all collected notes and observations from the focus group interviews, and in case of any correction or inconveniences, the facilitator and moderator will remove all potential risks and avoid any potential risks.

Some of those situations and risks happened during the study and to protect the identity of the participants and working organizations, it was decided previously in the focus group sessions that all organizations, diseases, products, and participants' names will not be referred and will therefore be retracted. The participants were also informed about these matters in the informed consent form whereby signing the form, the participants agreed to maintain the confidentiality of the information discussed by all participants and researchers during the focus group session.

4.3.1 Focus Group Questions

As previously presented, a list of nine questions was developed to guide the discussions from the focus group sessions. The participants received the questions in advance of their scheduled session to have a better preparation and to support the participants in terms of reflecting as pre-read the major considerations and opinions regarding different topics. Since the composition of each group varied by job duties and titles, several perspectives were represented in their responses (see Table 4.1).

TABLE 4.1
Focus Group Questions

Area	Questions
People	What are the most relevant organizational competencies and skills in performing CommEx and digital programs and activities?
	What was the investment made in people skills and competencies to improve the quality of CommEx and digital and increase all involved project members' capabilities?
Processes and strategy	In your department/organization, what strategies will you use to raise awareness of the benefits of communications and digital? How have both concepts been used to create value?
	Is your organization currently using CommEx and digital for specific digital and business development metrics?
	Can you describe examples of your organization and department's implemented and is currently managing CommEx and digital processes?
	Detail all existing standards for CommEx and digital and examples of its applications
	Describe all procedures and regulations, compliance, legal or privacy considerations for any existing CommEx and digital processes and projects
Technology and data	What are the systems and tools being used for CommEx and digital?
	Can you please describe the data sets, data points, and sources for your CommEx and digital processes?

In the following subsection, we explore how the answers to the above questions were presented, key insights from the focus group interviews, and key discussion topics.

4.3.2 Focus Group Key Takeaways and Raised Points

In general, the participants assigned high relevance and a good commitment and dedication to all interviews, where they also highlighted that reading questions beforehand was extremely useful and important for the interview preparation.

On a more abstract level, several participants referred to the idea of having a more well-developed digital organizational culture and overall data governance capacity from all involved functions, where the different challenges of this industry in making the most from CommEx and digital have to be deeply understood.

A widespread opinion across all participants was that a framework, best practices guidelines, and training need to be further discussed; there were two concrete suggestions on how to tackle this issue. The first one was to involve consultancy companies in shaping and providing discipline-specific pieces of digital training, while the second was to create a generic framework between the academic and professional world with details of how CommEx can be taken to the next level and generically adopted across the life sciences industry.

Since the very beginning of the focus group discussions, we understood that most of the participant's companies use CommEx and digital to understand the HCPs' landscape with pillar deliverables like targets identification, gain HCP insights, execute effective virtual engagement events, and access relevant stakeholders' maps. Throughout the focus group discussions, we were able to understand the most relevant HCPs engagement tactics:

- Scientific platforms commercialization
- Congresses and publications or guidelines awareness
- Speaker bureaus
- Disease education sessions
- Key account management strategy plans
- Field force engagement plans
- Awareness campaigns
- Advisory boards

The focus group interviews and related answers and observations provided by the participants also highlighted a range of other important issues that are pertinent to be explained in this section.

Table 4.2 describes and highlights the key topics of each of the presented questions with an analysis in terms of the most used and frequent words in the provided answers.

The table is a summary of all collected answers from the focus group discussions and is grouped by four major columns: questions, main categories, most relevant discussed topics, and keyword frequency with more than five recurrent words analysis (see Table 4.2).

TABLE 4.2
Focus Group Interview Sessions Topics/Subtopics, Key Takeaways, and Additional Raised Points

Topic/Subtopics	Key Takeaways from All Sessions
Question 1: What are the most relevant organizational competencies and skills in performing CommEx and digital programs and activities?	It's important to implement a rigorous and creative process to consistently understand what relevant healthcare professionals, institutions, patients, and KOLs are discussing about the company, competitors, and products. Critical digital or multichannel leaders core competencies: customer experience, building relationships between customers and front-end technologies, brand and content marketing, performance marketing (SEO, PPC, SEM), P&L management, product management, CRM, data insights and analytics, and net promoting score. Important skills to CommEx leaders: 360 degree and high-level strategy view, commercial DNA, hybrid understanding of technology and commercial processes, P&L mindset/management, technology agility for digital needs, advanced analytics, competitive intelligence, platforms, and technology architecture foundational knowledge.
Question 2: What was the investment made in people skills and competencies to improve the quality of CommEx and digital and increase all involved project members' capabilities?	Currently, biopharma needs to improve the capacity to estimate and understand the social media reach and digital content. Still, not a lot of companies are examining social and non-social media consistently and better understand the content performance. From most of the participants' comments, we understood that almost all of the represented companies needed to invest in more employee skills in virtual selling, usage of digital technology, and remote engagement techniques, which were the main goals to increase the organizational capacity around remote interactions and interactive contact time between sales representatives and HCPs (e.g. nurses, physicians, etc.). Some additional impacting factors for skills and competencies development was the impact of COVID-19 on mental health and symptoms from some ranges of the organizational structure of anxiety, digital fatigue behaviors, and depressive disorders. To overcome these factors, the companies needed to invest in HR resilience and emotional balance and well-being programs with different functions of the organization.

(continued)

TABLE 4.2 (Continued)
Focus Group Interview Sessions Topics/Subtopics, Key Takeaways, and Additional Raised Points

Topic/Subtopics	Key Takeaways from All Sessions
Question 3: In your department/ organization, what strategies will you use to raise awareness of the benefits of communications and digital? How have both concepts been used to create value?	Customer journeys and digital relationship creation with HCPs across channels were mentioned as important success factors when pharma organizations are rolling out digital projects. HCPs and patient journeys using CommEx principles allow commercial organizations to target the correct HCPs at the right time by using Commercial Excellence principles around segmentation, profiling, and targeting and also use precise segmentation information, behavioral data, and CRM insights. Commercial salesforce teams are strongly dependent on promotional and brand awareness-related budgets where pharma companies' visibility across the market and reputation are key foundations to a commercial strategy success. By defining the strategy and coordinating insights gathering through pre- and post-product launches, CommEx has been able to produce valuable insights and business outcomes to critical decision-making. CommEx has been able to add concrete advantages to commercial functions and able to execute and refine insights and gather critical information pieces to tactical plans.
Question 4: Is your organization currently using CommEx and digital for specific digital and business development metrics?	Digital agility has a strong connection with satisfaction levels from HCP engagement activities and the way pharma companies develop loyalty and production adoption programs are strongly connected with digital. Digital assets are being deployed to target HCPs that are exposed to appropriate content development by brand teams and marketing operations. Some of the benefits of applying digital strategies in HCP engagement activities are the possibility to deliver brand messages across multiple channels with personalized content and digital assets performance being captured in real-time. New marketing automation solutions are being considered to support the business in growing with automating marketing and sales processes, tracking customer engagement activities, and delivering personalized experiences to each HCP across marketing, sales, and medical affairs. There are growing needs from commercial functions to automate and personalize HCP journeys with multiple and effective touchpoints. Additional referred technical capabilities: automate online marketing campaigns, minimize the time and effort spend on creating marketing campaigns, align the brand strategy with campaign execution, and creation of valuable and actionable metrics that can track the return on investment from marketing campaigns.

TABLE 4.2 (Continued)
Focus Group Interview Sessions Topics/Subtopics, Key Takeaways, and Additional Raised Points

Topic/Subtopics	Key Takeaways from All Sessions
Question 5: Can you describe examples of your organization and department's implemented CommEx and digital processes that are currently managed?	Fine-tuning campaigns are being designed to provide advanced behavioral analytics to inform the performance of the brand strategy throughout marketing campaigns and allowing to close the loop between brand strategy and marketing digital campaigns. CommEx is a key function when multichannel and brand strategies need to be assessed and evaluated from a return on investment and value generation perspective and when deep engagement strategies and relationships need to be built with HCPs across different channels. Digital is being used to drive the gain of market share and market positioning through digital activities around disease awareness and speaker programs. In product launches, some commercial activities are benefiting from active CommEx processes around market data overview and landscape assessments, operating plan readiness, tenders, and contracts management.
Question 6: Detail all existing standards for CommEx and digital and examples of its applications	For digital projects standards and key considerations around tailored and promotional brand messaging capabilities to support marketing, operations were mentioned and allowing to enable multichannel capabilities across standard channels like email, web, social media, SMS. The pharma industry needs more capacity and business standards for effective and agile customer-centric digital transformation programs with more solutions in: • Decision-making support • Multichannel strategy • Logistics • Digital supply chain management • eProcurement
Question 7: Describe all procedures and regulations, compliance, legal or privacy considerations for any existing CommEx and digital processes and projects	Measurements of master data usage connected with digital and commercial systems like CRM or analytics were mentioned as being critical to quantitative insights generation and to support legal and compliance validations from transactional and activity-based data. Some qualitative research activities are being conducted to understand from the HCPs the following points: Understand their values, beliefs, motivations, and behaviors relating to the patient's self-care journey. Understand what they perceive (belief) and value about their role concerning self-care and the patient relationship.

(continued)

TABLE 4.2 (Continued)
Focus Group Interview Sessions Topics/Subtopics, Key Takeaways, and Additional Raised Points

Topic/Subtopics	Key Takeaways from All Sessions
	To explore the potential design of HCP archetypes founded in insight. Understand how approaches to HCPs need to be tailored based on the insights and the range of emerging archetypes. We need to improve our understanding of how they interpret their role and how this manifests itself through HCP, patient, and career relationships. HCPs have an integral role in patient health outcomes and therefore the self-care agenda.
Question 8: What are the systems and tools being used for CommEx and digital?	The time-to-value acceleration capacity was mentioned by several participants as a core principle to gain business agility, innovation, and support digital programs to business development activities. Digital was always mentioned to this point as a major propeller to activate digital services and enable innovation. One example was presented of an advanced artificial intelligence (AI) solution being implemented within marketing for HCPs' remote communications automation for audience segmentation, real-time interactions, modular content creation, and AI analytics. This solution is being presented to commercial leaders and salesforce teams' HCP journeys and social media insights but as well more capacity for digital engagement strategy.
Question 9: Can you please describe the data sets, data points, and sources for your CommEx and digital processes?	Essential and structural governance and design principles that can enable integrated digital commercial capabilities are key considerations for all mentioned organizations. Additional considerations are important to be evaluated concerning consent management, data privacy regulations, and master data management principles.

4.3.3 Key Observations from the Focus Group Interviews

The following section highlights the major conclusions and takeaways from the collected information and knowledge from the interview sessions. Several observations and comments were made during the focus group interview sessions in terms of key areas connected with CommEx and digital. The key concepts listed below were presented as the key success areas for both CommEx and digital in commercial pharmaceutical companies:

- Insights and analytics: decision-making process, information and insights management, and advanced analytics.

Commercial Excellence and Digital in Healthcare

- Processes: market and sales data analysis, promotional content management, multichannel marketing, strategic account analysis and planning, strategy and tactics management, key account management, segmentation and targeting, territory planning, and many others.
- Systems and tools: Customer relationship management (CRM), analytics, tenders, and contracts management.
- Talent management: competencies, skills, talent, and training planning.
- Governance: roles, decision rights, and policies.

The focus group interviews also highlighted clear examples of applied use cases in the industry and offered a better understanding of the strengths and weaknesses of CommEx and digital activities and projects to different pharma companies. Many observations were made in terms of decisive CommEx and digital intervention areas like the ones we described and detail in Table 4.3.

TABLE 4.3
Focus Group Topic Areas and Key Observations Summary

Topic Areas	Comments and Observations
Strategy	A thorough analysis of the company portfolio is typically conducted by CommEx
	P&L and forecasts analysis by brand with growth objectives perspectives can be considered as one of the outcomes from CommEx activities
	It's fundamental to be active in competitive intelligence, customers journeys, and product launches scenarios when requesting support from CommEx departments
	CommEx key responsibilities also include long-term portfolio vision and best model predictions in terms of complementary and agile approaches
	Normal expectable CommEx services important to sales strategy: customer reach analysis like coverage, responses curves, key customers identification, field force distribution by territory
	High relevance for common engagement models to operate in a cross-channel environment more efficiently
Analytics	Market share, sales volumes and values analysis, and investment decisions assessments
	Predictive analysis is being used to propose the next best offer and channel to the right customers
Operational	Digital and CommEx are fundamental to upgrade sales processes to dimensions of value in the pharmaceutical space
	Digital disruption is not only a possibility for a pharma company's future but in most cases and especially in a post-pandemic era it's the only possibility. Once business decision-makers can accept that premise, decisions that previously seemed courageous or outrageous will instead appear to be rational and inevitable
	Tailored messaging and engagement approaches produced the biggest impact from a customer-centric perspective, where centered actions around key customers are most likely to succeed

(continued)

TABLE 4.3 (Continued)
Focus Group Topic Areas and Key Observations Summary

Topic Areas	Comments and Observations
	Still existing different approaches and language for same commercial processes across countries instead harmonized, streamlined and common processes for all countries
Digital Sales	In 2021, some companies were able to take a strategic digital strategy to embed digital into a more multichannel brand planning
	Digital is being an enabler of commercial functions, supporting both delivery and value proposition with the major outcome to optimize commercial performance
	There are key three principles for a successful multichannel approach: knowledge of commercial processes, value proposition models, and effective operational tools
	New projects are being implemented in omnichannel and cross-channel strategies to evolve channel capabilities in line with communication trends
Segmentation and targeting	To improve the knowledge of key and relevant HCPs, it's important to adopt an agile and 360-degree approach of target selections, messages definition, and channel mix selection
	Extremely relevant to understand the customer preferences in terms of contact channels and specifically measure the usage level of a given product
	Part of an effective segmentation strategy is important to understand the potential, influencing levels, channel preference, accessibility of an HCP toward a specific product brand
Systems and tools	Tools are crucial for the process execution and an effective operational and analytical CRM is a key component for CommEx
	For companies with pharmacy channels having a system like a CRM to track activities that need to be linked and integrated into any existing data, the warehouse is key
	Most relevant systems for commercial organizations: Customer continuous communication, events management, sales incentives and rewards, content management, and analytical systems capable of performing data mining and patients pattern identification
	The integration of all sales channels and company data to determine channel effectiveness and best practices are key objectives of a multichannel digital strategy
	Important to have an integrated view with all the channel execution tools and the system that supports the company digital solution landscape
	Digital channels integration to boost traditional marketing mix is a key component and to enhance the impact of traditional channels to optimize contact frequency and customer reach
	Currently, several digital solutions lack the flexibility to adapt to quick changes in the organization

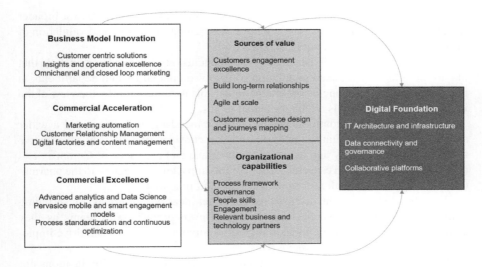

FIGURE 4.2 Discussion topics high-level overview.

Additionally, during the sessions, we were able to understand commonalities and similarities in several areas that allowed us to frame an overview of areas involved in digital foundations, organizational capabilities, and sources of value from digital and CommEx processes and models.

To help some of the discussions, we had this necessity to draw a figure that allowed in some of the focus group follow-ups where additional questions and clarifications were presented and to better understand some of the connections and dependencies between terms and concepts.

The discussions were extended even after the completion of the focus group analysis and almost in an informal way with some of the most interested participants. One of the recommendations received from the participants was to highlight and summarize all the key conclusions into one single figure.

Therefore, we present what was the result of some of the conclusions and observations concepts and high-level overview and executive summary (Figure 4.2) of what might be considered in the future as a possible commercial digital framework for any digital program implementation connected with commercial business models. Although we didn't have the objective to produce one framework or single image overview of digital and CommEx impact into pharmaceutical commercial business models, we didn't want to remove this useful piece of content and result from the discussions.

4.4 CONCLUSIONS

4.4.1 Limitations and Future Recommendations

The research study was created and designed to better understand two specific areas in a pharmaceutical context where the presented conclusions and observations provide already solid pieces of evidence but are recommendable to exist a broader research activity in terms of a possible large-scale empirical testing and validation of the discussed techniques, models, and processes.

From the results from all data analysis and conclusions, we can also conclude that due to the relatively reduced population of involved companies and departments in this study, some of the conclusions might not be able to represent fully the view of the industry and professionals on the presented topics.

For future recommendations, it will be recommendable to be designed a wider scope in terms of qualitative analysis and the consideration of new quantitative research to better qualify and assess the importance and relevance of CommEx and digital to the industry and especially after the full impact of COVID-19 to the pharmaceutical companies and all related departments and functions.

We witnessed during the discussions that novel ideas and solutions are springing within smaller commercial business units and even product sales teams as they attempt to bring a more agile logic of digital prototyping and pilots and drive implementation of digital solutions themselves.

The research study in this chapter provides several practical implications for decision-makers and managers in organizations. Based on the study, management must support the creation of a culture that tracks innovative digital and CommEx models, technologies, processes, and employee skills to better empower their commercial departments and structures and develop better outcomes frameworks and insight models.

The ability to test and pilot new digital and CommEx technologies and processes will typically require new capabilities in terms of concrete and practical implementation of use cases and techniques as well as how to align them with organizational structures and processes, which was also noted in discussions with focus group participants.

4.4.2 Conclusions

In this study, all stakeholders involved in the blockchain-based life sciences ecosystem provided very clear examples and details.

The COVID-19 pandemic crisis has been changing any historical precedents of the way medicines are prescribed, distributed, and administered with major changes to business operations and the existing drivers of growth for several pharma companies.

During the interview sessions, the feedback from the participants was not always consistent in terms of the key CommEx and digital areas and major points of observations but overall, all participants highlighted the importance of CommEx and digital programs to commercial strategy models and delivery.

Some of the raised topics during the interview sessions showed the major shifts to new market trends where the prescribing flexibility allowed physicians and policymakers to act more quickly to constant changes and the way pharmaceuticals interacted with the HCPs have undergone major changes.

We start this subsection topic with an overview of the major conclusions from the focus group discussions, where some of the presented benefits of changes resulting from CommEx and digital programs that we were able to understand from the interview are described as following:

- Augmented impact of new digital sales channels to commercial activities of several pharma companies and they are already part of the business commercial strategy to incorporate major digital concepts and principles.
- Increase number and frequency of contacts performed remotely with a higher distribution of efforts by digital channels via CRM systems or external video conferencing tools.
- Digital reach beyond and more effective than face-to-face interactions, where specific events like preceptorships, congresses, clinical sessions, symposiums, and advisory boards are already assuming digital importance.
- Support physician and nurses' interactions with value-based discussions and with more scientific and clinical exchange expertise from the industry.
- Optimization of marketing budgets and offer additional marketing mix options with higher importance and influence from digital assets management and digital marketing.
- Higher need to promote the right channel mix for the right HCP to enable effective customer segmentation, behaviors, preference understanding, where CommEx has been assuming critical importance during the pandemic crisis times.

From an analysis of all data, we can also conclude that all participants involved in the study represented with some accuracy levels the industry views and perspectives; also the spectrum of all the involved participants' functions and departments offer the research a good representation of the presented research questions.

From the focus group discussions, we were able to understand that modern customer journeys assume a high business relevance in terms of understanding critical areas like:

- HCPs interests and product or disease awareness.
- Most relevant engagement channels for the future: email, text messaging, loyalty programs, online continuous medical education, clinical/scientific newspaper, and word of mouth.
- Advocacy and brand experience: expectations management versus reality, social posts, speaker programs, or bureaus.
- Academic and medical research: blogs and articles, medical publications, social media, and word of mouth.

Several participants highlighted that currently, the organizational digital leaders are more able and capable during the COVID-19 times to operationally and structurally support the long-term changes of foundational levels and framework structures connected with data governance and digital readiness to be more prepared in the future for massive transformations from customers and the healthcare management landscape.

During the focus group discussions, some major trends were also presented in terms of new ways of working around telemedicine, patient empowerment to use more digital solutions, and facilitation of remote offering around remote medical monitoring and remote patient visits to hospitals and clinics. These ongoing trends drove new methods from pharmaceutical companies in stakeholder management (e.g., HCPs) but impacted as never before the volume of remote sales representative's contacts due to the lockdown restrictions and the way remote digital engagement activities were conducted.

During the last years, several pharma companies have been launching CommEx programs with a different scope of services and processes coverage. With the increase of market challenges and competition pressure within the sector, most organizational leaders have been redefining their commercial business models and strategies. But not only the models and strategies needed to change, but also the efficacy in sales and marketing operations suffered several changes as the demand from organizational leaders increased in terms of performance, agility, results, and value-driven outcomes.

In most of the cases, these companies also understood that currently the market is transforming quickly into a more digital and cost-effective setting and all commercial departments need now to focus their attention and focus on more digital customer-oriented activities and a more multichannel engagement strategy.

CommEx developed professionally can play a decisive role in optimizing sales and marketing operations, leverage and increase the results of commercial initiatives, and drive new relationships levels with key opinion leaders (KOLs), HCPs, medical communities, payers, or different public institutions.

Based on the focus groups' findings, both models, CommEx and digital, can be used to predict the best tactics and actions to support sales and marketing.

A consensus was reached that pharma companies have, more than ever, a responsibility to enable analytics, data governance, digital factoring, and data standards so that they can support the decision-making processes. CommEx will work with different departments to ensure that they have the capability and capacity to meet the needs for decision-making processing and insights generation.

Nine out of the 32 participants reported the affiliated companies have elevated digital and increased investments in commercial digital solutions during the COVID times. Also, 60 percent of all participants shared that digital is a dedicated, standalone function and there isn't a formal digital representation in the executive management teams. Throughout the focus group discussions, it was also clear that most of the participants felt that customer data are essential to launching new products or manage the sales model effectively, but most of the organizations are not satisfied with the quality and service they receive from their legacy customer data provider or have the right customer data foundation to fully support the digital transformation.

Data from the focus group discussions show that content management, data governance, organizational culture, and skills are vital skills for communications and digital in pharmaceutical. In order to continue answering our primary research question, we also concluded from our focus group discussions that most opportunities to CommEx and digital pharma pertain to the following:

- Automated processes: the ability to change how commercial departments manage business needs and application uses to drive real business value.

- Focus group participants discussed how to manage and control data quality, security, and integration.
- The majority of digital projects fail because of a lack of effective processes and governance that are incapable of providing systems integration consistency and a high level of data quality.
- Several participating companies do not yet have all the confidence and solidity needed to scale their digital services and have the appropriate resources due to the current knowledge and technical architectures.

The focus group, therefore, helped us to understand the following general benefits and constraints of CommEx and digital projects in the pharmaceutical industry that most influence commercialization processes and capabilities.

The focus group provided the following conclusions to both research questions and we better understood how CommEx and digital can be applied to both research questions:

- Digital programs based on HCPs classification and profiling models.
- HCP-centric engagement models with active usage of CRM medical surveys, interactions reporting, eDetailing, approved emails, and account plans.
- Determining disease awareness across social media platforms.
- Health-related services and information from a variety of channels and sources.
- Wearable devices enable patients with certain diseases like diabetes to monitor physiological parameters and biometrics with the ultimate capacity to use digital in continuously adapting insulin dosage accordingly and in real-time to optimize treatment times and adherence.
- Data management, progressive data management, and mining of untapped data sources in creating elevated agile digital solutions and prescriptive insights.
- Digital customer journeys and integration of digital tactics with multichannel marketing.
- Events management like remote speaker education and programs, video interviews with HCPs and KOLs, clinical case discussions, congresses updates, virtual engagement calls with GPs, webinars, and remote educational meetings.
- HCP portal/website development and implementation with possibilities for updates on web seminars and congresses.
- Online communities tracking and literature/guidelines review.

The major conclusions collected from the focus group discussions were that organizations are engaged in major efforts in data collection to support digital and advanced analytics that relate to CommEx operations. The investment in master data management, strong data governance, democratization and speed of data analysis, and external collaboration with technology and healthcare partners (including HCPs) are transforming the way pharma companies are capable of processing and understanding data and providing access to a digital ecosystem in the use and deployment of data and can make good usage of core organizational capabilities like CommEx.

REFERENCES

Agarwal, S., Punn, N. S., Sonbhadra, S. K., Tanveer, M., Nagabhushan, P., Pandian, K. K., & Saxena, P. (2020). Unleashing the power of disruptive and emerging technologies amid COVID-19: A detailed review. ArXiv, abs/2005.11507.

Bell, S., Morse, S., & Shah, R. A. (2012). Understanding stakeholder participation in research as part of sustainable development. *Journal of Environmental Management, 101*, 13–22.

Chaffey, D., & Smith, P. R. (2017). *Digital marketing excellence: Planning, optimizing and integrating online marketing*. Taylor & Francis.

Cutcher, G., J., Baker, K. S., Berente, N., Berkman, P. A., Canavan, P., Feltus, F. A., ... & Veazey, P. (2020). Negotiated sharing of pandemic data, models, and resources. *Negotiation Journal, 36*(4), 497–534.

Davenport, T. H., Harris, J. G., & Morison, R. (2010). *Analytics at work: Smarter decisions, better results*. Harvard Business Press.

Ghiraldelli, F. (2020). Business continuity plan: How to keep your company continuous: case studies: how companies have implemented and activated the BCP during the Covid-19 emergency. Master Thesis. Libera Università Internazionale degli Studi Sociali. http://tesi.luiss.it/29134/1/704341_GHIRALDELLI_FLAVIO.pdf

Graham, A. K., & Ariza, C. A. (2003). Dynamic, hard, and strategic questions: Using optimization to answer a marketing resource allocation question. *System Dynamics Review, 19*(1), 27–46.

Hung, Y. C., Huang, S. M., & Lin, Q. P. (2005). Critical factors in adopting a knowledge management system for the pharmaceutical industry. *Industrial Management & Data Systems, 105*, 164–183.

Kaplan, R. S., Kaplan, R. E., Norton, D. P., Davenport, T. H., & Norton, D. P. (2004). *Strategy maps: Converting intangible assets into tangible outcomes*. Harvard Business Press.

Koshy, E., Koshy, V., & Waterman, H. (2010). *Action research in healthcare*. Sage

Lee, S. M., & Lee, D. (2021). Opportunities and challenges for contactless healthcare services in the post-COVID-19 Era. *Technological Forecasting and Social Change, 167*, 120712.

Matricardi, P. M., Dramburg, S., Alvarez Perea, A., Antolín-Amérigo, D., Apfelbacher, C., Atanaskovic-Markovic, M., ... & Agache, I. (2020). The role of mobile health technologies in allergy care: An EAACI position paper. *Allergy, 75*(2), 259–272.

Melnychuk, T., Schultz, C., & Wirsich, A. (2021). The effects of university-industry collaboration in preclinical research on pharmaceutical firms' R&D performance: Absorptive capacity's role. *Journal of Product Innovation Management, 38*, 355–378.

Schuster, E. W., Allen, S. J., & Brock, D. L. (2007). *Global RFID: The value of the EPC global network for supply chain management*. Springer Science & Business Media.

Stolpe, M. (2016). The internet of things: Opportunities and challenges for distributed data analysis. *Acm Sigkdd Explorations Newsletter, 18*(1), 15–34.

Ulrich, D., Jick, T., & Von Glinow, M. A. (1993). High-impact learning: Building and diffusing learning capability. *Organizational Dynamics, 22*(2), 52–66.

Uvarova, O., & Pobol, A. (2020). *SMEs digital transformation in the EaP countries in COVID-19 time: Challenges and digital solutions*. Eastern Partnership Civil Society Forum.

Volberda, H. W., Van Den Bosch, F. A., & Heij, K. (2018). *Reinventing business models: How firms cope with disruption*. Oxford University Press.

Yanik, S., & Kiliç, A. S. (2018). A framework for the performance evaluation of an energy blockchain. In *Energy management – Collective and computational intelligence with theory and applications* (pp. 521–543). Springer.

5 Psychology on the Edge of Health Tech Challenges

Lourdes Caraça and Vera Proença

CONTENTS

5.1 Introduction 85
5.2 Psychology/Psychotherapy and the Paradigm of Relationship 87
5.3 Psychology and Online Intervention 90
5.4 Clinical and Health Psychology 92
References..94

5.1 INTRODUCTION

After more than 21 centuries of humanity that has incessantly sought to dominate nature and its processes, we can currently say that we have amazing technologies that result directly from the human capacity to act, to initiate new and unprecedented processes.

At the same time, while this is happening, fertility rates are decreasing – especially in rich countries – the rate of aging is growing and the full renewal of the population is not taking place. Problems related to the renewal of the natural environment arises and climate changes (as a result of global warming) also pose serious challenges for public health: (a) diseases usually limited to tropical countries currently raise concerns for the health of humanity, as it is the case of malaria or dengue, because they spread to other places; and (b) the major meteorological catastrophes, hurricanes, floods, fires, extreme droughts, and the problems inherent to the migratory movements of people, who seek out rich countries, because they present themselves as having a better quality of life, also bring health challenges to world population (Fielding, 2014).

In fact, technological development has undeniably converged toward the increase in people's life expectancy due to the achievements: (a) in medical diagnoses and treatments; (b) in the control of chronic diseases; and (c) recently even as a strong ally in the control of the SARS pandemic/COVID-19. In a technological world and in the aftermath of a pandemic, we look at the decisive role that technologies have brought and may come to bring in health.

As it is known and as Neville Symington (1996) mentioned, moments of crisis brings key changes to our lives. Insecurity and vulnerability are dimensions inherent to the human condition, with which we have always lived and that the virus has

DOI: 10.1201/9781003227892-5

highlighted. Such consequences were felt in different areas, namely in health, physical, but above all in mental health. As this area of health is already so vulnerable, it is, in our view, extremely important to look at the dilemmas that we are faced with and grow from there.

Scientifically, the advantages of interdisciplinarity in solving problems are also increasingly recognized, and as experts in human behavior and mental processes, as well as promoters of healthy and sustainable behavior, the involvement of psychology and clinical and health psychologists is essential (Gilford, 2014). Other authors even add that, on essential themes for public health and on issues relevant to the organization of health responses, specialists in psychological sciences should be consultants to policymakers and health management bodies (Clayton et al., 2015; Fielding, Honsey, & Swim, 2014; Van Lange, Joireman & Millinski, 2018).

Thus, clinical and health psychology, which was first concerned with the response of mental health, started to respond to general levels of health and currently also contributes to solutions in the context of the disproportionate pressure and demand for health services, avoiding overload and collapse, failing to align with the expected functioning (Jamoulle, 2015). This integration of knowledge resulted from the evolution of the concept of health itself, which expanded in recent decades, took off from the traditional biomedical model, and began to include broader, dynamic, and comprehensive health perspectives, which include physical and mental well-being, spiritual, and social complete and not just the absence of disease (Caraça, 2021). It was also assumed that it is possible to live healthily with or without illness by changing behavior and lifestyles (Gilford, 2014).

Psychological and behavioral dimensions were included in health as a way to empower people to take care of their own well-being (Glanz, Rimer & Viswanath, 2008). And, progressively, conceptual structures that explain the behavior implied in health literacy and in health self-management (McAndrew, Mora, Quigley, Leventhal, & Leventhal 2014) have been affirmed, with the invaluable help of models that analyze what determines motivation for change, with regard to disruptive or unhealthy behaviors (Armitage & Conner, 2000).

Based on the conviction that almost all human behavior or activities have an impact on our health or well-being condition, there has been an attempt to boost the participation of people in the adoption and maintenance of healthy lifestyles (OMS, 1986, 2019), empowering them for a responsible and competent participation, with strong and positive repercussions in their lives.

In view of the current challenges, clinical and health psychology, as well as psychotherapy, are at a unique and crucial moment, which highlights their need for renewal. This change has already started a few decades ago; however, it is necessary to analyze if the new lines of work correspond to the real challenges (Eiguer, 2005), and in what way they can be decisive for the revitalization of healthcare in general (Kaplan, 2009).

Online care in clinical practice is a reality that, before the pandemic, was becoming a possibility, but whose use has grown exponentially since the beginning of the pandemic period, mainly due to the receptivity of all stakeholders, professionals, and clients. The benefits that come from such assistance are undeniable; however, we cannot help but think also about the disadvantages, challenges and difficulties.

In fact, the benefits in accessibility and economic impacts bring an excellent cost–benefit ratio to the use of technology (Donker et al., 2015) and has even made it possible to reach populations and people with greater mobility or movement difficulties, bringing gains in follow-up and monitoring of health in general. The use of eHealth in digital health programs and platforms inevitably entails challenges, the first of which stems from the certainty that the use of these technologies cannot be generalized; in other words, for some people a more traditional model of intervention is needed, for which it is essential to ensure continuity. On the other hand, not all professionals will be able to adapt and they may be the ones to provide these answers. We cannot forget about all the adaptations that will be necessary and how this interferes with the relationship established between different professionals and clients, as well as between different professionals among themselves, who will have to rediscover new ways of working.

5.2 PSYCHOLOGY/PSYCHOTHERAPY AND THE PARADIGM OF RELATIONSHIP

The relationship between psychologists and clients has historically occupied a prominent role in theories of therapeutic processes (Horvath, 2005). The human being is not an island. They form, live and progress in bond with others (Eiguer, 2005). Roudinesco (2005) also emphasizes that in psychology and psychotherapy the "healing" process is closely linked to the influence that the psychologist can have on the client, as well as in the latter's belief on the therapeutic power of the clinician. We can then affirm that the establishment of a relationship based on trust, between the client and the psychologist, becomes fundamental for the success of therapy. This is a concept that dates back to Freud; however, it has undergone important changes in its definition and only later did specific terms begin to be used to refer to it.

Zimerman (2007) states that the first term used – therapeutic alliance – belongs to E. Zetzel, an American psychoanalyst who, in a 1956 work, conceived an important aspect related to the transference process; in other words, the fact that the client has a mental condition, both consciously and unconsciously, that allows him to remain truly allied to the therapeutic task. This concept appears in different texts over time, with different names, but with equivalent meanings.

However, in this chapter, we seek to look more deeply at the movement that has been emphasizing the individual's relational experience with others and in particular with the psychologist himself (Fonagy & Target, 2003). The therapeutic relationship established as a result of these theories takes on a new configuration, meeting the client's needs. Thus, we see the concept of intersubjectivity introduced, where the established bond is assumed as a reciprocal relationship between two subjects whose unconscious influence each other, creating a new relationship space. We are facing not only a developmental model, but also a therapeutic model, where client and psychologist are dyad partners. As Daniel Stern states, there is not a one-person psychology (Eiguer, 2005).

The intersubjectivity between the client and the psychologist is increasingly considered fundamental and a facilitator of change. For most psychologists, the quality of the relationship between them and clients has an important impact on the success of therapy (Horvath, 2000; Horvath, Del Re, Fluckiger, Symonds, 2011). Neutrality, anonymity, and abstinence are essentially negative principles, as they describe what not to do. When in doubt, don't answer, don't speak, don't express yourself, don't reveal. Silence and emotional indifference are safe. By contrast, the postclassical view allows the psychologist to alternate restraint with expressiveness, making his feelings an essential part of the process (Mitchell, 2000).

We can then consider that intersubjectivity is a central part of psychotherapeutic intervention, where it becomes essential, as has been referred to, the establishment of an alliance, characterized by a strong client–psychologist relationship, as well as a good affective bond, characterized by trust, mutual respect, feelings of liking and caring for each other, consensus on treatment goals, commitment and responsibility on the part of the client, and the same involvement on the part of the psychologist (Horvath & Bedi, 2002).

In therapy we try to create a space where the absence of conscious intentions allow feelings to arise. These transference feelings are not devoid of unconscious intentionality. Thus, the client is asked to love and hate irresponsibly, allowing feelings to arise without awareness or concern about their implications and usefulness. For the psychologist, feelings inside and outside the relationship are very important. Because he is responsible for maintaining, at all times, the established relationship. The psychologist is asked to love and hate responsibly by allowing feelings to emerge, but never failing to look at their implications in the process of which he is the guardian. His personality is present in analytic work as in any other activity, although with different versions (Mitchell, 2000). Thus, establishing a good alliance depends on its ability to adapt to the client's needs, expectations, and capabilities. However, fluctuations in this relation within or between sessions will be normal and when resolved are associated with good results. According to Eiguer (2005), the field of intersubjectivity and its bond must be extended to all human relationships.

But what is at stake in establishing these relationships? In the twenty-first century, we are faced with a turn to virtual reality. Something that has been happening, but that was certainly driven by the pandemic. Where are we and our body in the virtual reality? Because a relationship is established by different types and levels of communication.

Reality and the body are formed in relation to each other, both constituted by stories, myths, desires, and representations, which are related to the dialogue that gives rise to them. We and our body will influence the other and his body, forming its own communication system. Thus, it becomes imperative to think about the importance of the body in the process of communication and socialization, whether in person or virtual. Although in face-to-face contact, the real body is a fundamental part of the relationship, in the virtual world this is not true; we see emerging representations of the self and its emotions as a way to make up for this lack of expressions and real emotions (Baldanza, 2006).

A conversation in real space and time exposes minds to the process of mutual discovery of thoughts and feelings. Our presence exposes us to silences or murmurs and to our evolution, coming from children's ideas into more mature ones (Farber, 2016).

How is the therapeutic relationship compromised by a deterioration of nonverbal communication, which is considered an essential element in the interaction between psychologist and client? What questions arise when the body is not physically present? The entry of the internet into our daily lives enables dialogues between people who only know each other through the screen. Enabling new relational forms, including in therapy and other health fields. We must then think about the presence or absence of the body in relationships, the level of depth to which they occur, and all their advantages and disadvantages, in a world ruled by social networks, by people or avatars, which may or may not be real. Stephen Porges (2011) emphasizes the importance of the psychologist providing a safe experience to the client during all therapeutic process. The author states that physical proximity leads to the activation of a mechanism that checks whether it is safe to be close to another person. Because this is our reality, psychologists can and should be involved in the assessment of constraints and adaptations to be implemented.

In a 2016 article in which Farber reviews Turkle, he underlines a sentence that makes perfect sense to us, and which we consider important to reflect on – regardless of the indispensable things that computers do for us, it is essential to think about what computers do to us, as individuals and as society. We cannot deny that there is a visible facilitation of interaction and sociability through the internet, where the emphasis is placed on communication without social barriers or prejudices. The internet has become a very attractive place, enabling interactions without the presence of the body, therefore without geographic contingencies. Without the body, we can be everywhere, at any moment, without leaving where we are.

We are facing a visible evolution toward relationships where physical presence is less and less necessary and mandatory. However, the internet keeps people in touch, but it doesn't necessarily keep them close. Little is revealed, little is involved; instead of deep communication, we have communications coming from superficial portions of the mind (Bollas, 2015). Our corporeality is particularly challenged. For psychologists, the incorporation of many physical aspects is a key point in the dynamics of the process, from the rhythm of breathing to posture, speech, smell, the clothes that are worn, the tics, the look. Face-to-face communication does not only involve verbal communication, but also there is a set of bodily expressions and behaviors that often communicate much more than words. In turn, in the internet intervention, all these interactions are absent. The real body is transformed into other emotional representations not mediated by the body.

Castells (2003, p. 108) underlines that the relationships built through the internet are rarely lasting, "people connect and disconnect from the internet, they change their interests, they do not necessarily reveal their identity". When this reality is transported to therapeutic relationships or other relations with health professionals, it can enhance the lack of adherence to treatments. The involvement of the clinical and health psychologist is essential to consider the phases and psychological determinants of behavioral change (Teixeira et al., 2020).

It is important for us, psychologists, to think about these contingencies and how we can get around them and adapt. The decay of the "talking cure", consequence of a culture whose goals aim financial results, is a sad reality to which we have been submitted. Our current challenge focuses on making ourselves, health professionals, relevant in such a changed world (Farber, 2016). The author also mentions Tuckle underlining that the ties we form on the internet are not the ties that connect us, but the ties that concern us, as we are looking for intimacy with machines, or through machines. And where is this intimacy? What does it become? Since intimacy as we know it is based on personal contact, being with others, smelling them, hearing them, see them, learning to read everything that is not verbal, everything that passes through the body.

Chaos and discontinuity, disruption and change are essential features of the internet (Farber, 2016). We cannot overlook the new addictions resulting from technology-induced mental habits: quick excitements, sudden disillusionments, mute when bored or understimulated. The distance provided by technology can hide people with difficulties in tolerating delays in gratification, which is fundamental for us to lose the illusory omnipotence and consequently allow us to feel, think, symbolically represent and accept the separation (Farber, 2016). Bollas (2015) draws attention to the work of Danah Boyd in 2014, which takes us to a stark reality – we are adults, who, due to our fears and our perception of the surrounding society, do not allow children to play with each other. The countless activities, with which we fill our children's daily lives, do not allow them to socialize unless it is through the internet. By occupying the children, we end up preventing them from feeling the absence of reference figures, in a strategy with socially accepted contours; however, in this discontinuity we perpetuate the difficulties of the connection. According to the author, the psychic values of the contemporary client are based on indirect perceptions generated by the information revolution; in other words, they live at various degrees of distance from involvement in reality, taking refuge in technology. We are increasingly programmed for a world of fast networks, giving priority to speed over reflection (Bollas, 2015).

5.3 PSYCHOLOGY AND ONLINE INTERVENTION

When thinking about online psychology and psychotherapy appointments, there is no doubt that one of the first challenges is to know if this is psychotherapy (in the original sense of the term) and, consequently, if we can be sure that a therapeutic relationship will be formed without any type of face-to-face contact, since as we have seen, that is a fundamental condition of the process and to treatment adherence.

Numerous questions have been raised about the effectiveness of online therapy, which have been answered by researchers (Barak, Hen, Boniel-Nissim & Shapira, 2008). However, the topic has not achieved consensus. In recent years, Barak, Klein and Proudfoot, (2009) have proven the benefits (effectiveness and efficiency) of psychological interventions using information and communication technologies. However, other studies have demonstrated the existence of greater difficulties for psychologists to interpret clients body language, make eye-to-eye contact, and establish an empathic relationship in online consultations (Mitchell, Meyers, Swan-Kremeier & Wonderlich, 2003). In turn, clients reported that, in online therapy, the relationship with the psychologist is different from face-to-face, but not necessarily worse or better (Simpson, Bell, Knox & Mitchell, 2005).

What differences are there and what changes we face from traditional psychology and online appointments? There are several factors to take into account. Morón and Aguayo, (2018) underline the American Psychology Association guidelines for the practice of online psychological intervention, which addresses the following aspects:

1. *Competence of the psychologist*: The psychologist who provides services must guarantee their competence both with the technologies used and with the potential impact of the technologies on their clients, supervisees, or other professionals.
2. *Comply with the rules of protection in the provision of online psychology services*: The psychologist must ensure that the ethical and professional standards of care are guaranteed and respected throughout the entire process and duration of the telepsychology services offered.
3. *Informed consent*: Psychologists must provide and document informed consent that specifically addresses issues unique to the online psychology services they provide. In doing so, psychologists are aware of applicable laws and regulations.
4. *Confidentiality of data and information*: Psychologists who provide online psychology services must protect and maintain the confidentiality of their clients/patients' data and information and inform them of the confidentiality risks inherent in the use of technologies, if any.
5. *Security and transmission of data and information*: Psychologists who provide online psychology services should take steps to ensure that there are security measures in place to protect the data and should assess what risks exist in each case before entering into such relationships to avoid the disclosure of confidential data.
6. *Elimination of data, information, and technologies*: Psychologists have to get rid of data and information used. According to the APA Record Keeping Guidelines (2007), psychologists are encouraged to create policies and procedures for the safe destruction of data.
7. *Tests and evaluation*: Psychologists must adapt the instruments to the online mode without losing the psychometric properties (reliability and validity). They must also be aware of the specific issues that may affect the assessment.
8. *Interjurisdictional practice*: Psychologists must familiarize themselves with and comply with all relevant laws and regulations when providing telepsychology services. Laws and regulations vary by state, province, territory, and country, so they must provide services within the jurisdictions in which they are located.

These guidelines are similar to those given by the Portuguese Psychological Association (OPP, 2018). However, it raises new and important aspects to take into account, namely the aspects of the setting, especially those psychologists cannot guarantee. It is important to inform clients of the risks and limitations of internet therapy and the technological requirements necessary for online sessions, paying special attention to the clarity of the messages (Barnett, 2011). Issues related to the lack of internet security, privacy and confidentiality, the fragile dependence on technology, cultural differences or even forms of payment.

Legal issues regarding jurisdiction and regulation of practice are also significant, as well as considerations about the need to train psychologists in this new psychotherapeutic modality (Barak, et al., 2008). Another aspect to have in mind is the management of crisis situations, which, together with insufficient training on the part of professionals, can trigger situations of risk for those requesting online help (Morón & Aguayo, 2018).

There is also an urgent need to reflect on some issues, essentially anonymity, facilitated by this malleability of internet movements, which may contribute to ethical abuses. Any professional or client, from both the private professional practice and from public health service, can carry out and experience ethical abuses. And health platforms require the ethical attention of professionals and teams, especially if these technologies facilitate anonymity, for either clients or professionals. Here, we understand the challenge of anonymity as much more than not knowing who the other is. The vulnerability to which we are exposed is accentuated when contacting a professional via online, where computer secrecy is not guaranteed, nor can the client be sure of who is serving him on the other side of the computer (Morón & Aguayo, 2018).

There are also questions regarding the environment that present themselves as challenges. For example, in most online appointments, psychologists and clients sit too much closer to the camera, making the body not visible. But we cannot forget the experience lived during the pandemic, in face-to-face care, where, due to the imposed limitation of the situation, we have to wear masks, putting us in front of the difficulty in observing the other's faces, but allows us to communicate through the body.

However, not everything is negative and the use of technology also brings several benefits. The adaptation of therapy to the internet allows the use of profiles and avatars as an experimentation of benign and age-appropriate identity. We can be our real selves without fear of judgment. We must think in terms of new forms of human subjectivity (Farber, 2016). It reduces stigma; it is an easy and attractive method that allows us to reach some minorities or persons, who for some reason, are prevented from using face-to-face models (OPP, 2018).

It is essential to meet the limitations of practices, reserving the use of technologies for situations in which it is even essential and decisive for the maintenance of therapeutic help, whether provided by the psychologist or another health professional.

5.4 CLINICAL AND HEALTH PSYCHOLOGY

Relationships are vital in health as in any other area, and as an important tool in psychology and psychotherapy, on which we constantly think about and work with. Given the scenario we are living nowadays and what is foreseen in the future, it seems pertinent to emphasize the importance of clinical and health psychologists in the training of various health professionals, whether within interdisciplinary teams or in the academic programs, with potential for a more comprehensive and ambitious growth. Increasingly, it is important to find in the academic curriculum of many professionals, disciplines that prepare health professionals to work with people and for the complex challenges we face (Ramos, 1988).

We believe that this professional training, carried out by clinical and health psychology, will focus on three levels: (1) valuing and empowering the individual in managing their health; (2) preparation of all professionals for the full integration of the various knowledge that make up the teams; and (3) facilitating the design and implementation of innovative responses suited to the characteristics and needs of the people to whom they are dedicated.

Intrinsically, there are also the difficulties of identifying, or becoming aware, immediately or in due course of all the implications of the behavior (in its entirety), the acceptance of responsibility for the consequences (any isolated act, or lack of action, has consequences, for the self, for others and for the natural environment). In the health field, the challenge of using technologies is sometimes foreshadowed in the decision to suspend treatments, or in the decision to persist, which imposes the need to guarantee, in addition to training, a specific supervision and intervision for professionals.

As we understand, there are several aspects to take into account when designing future scenarios, where the use of technology, either in the area of mental health as in other health areas, allows easy access to health care, but always in strict and sustainable practice. In Portugal, in 2019, the Portuguese Psychologists Association recognized the potential in psychoeducation, clinical and health psychology – with the promotion of general and mental health, prevention, treatment, relapse prevention and rehabilitation, the benefits of screening, evaluation, and monitoring (Carvalho et al., 2019).

In conclusion, we believe the future certainly reserves us an auspicious path at a technical level, where incredible advances and discoveries will continue to exist; the challenge will be in the wisdom of making this path of decision-making, strengthened by the virtue of the integration of knowledge, people, teams, health systems, with the awareness that "the process of a single act can literally extend to the end of time, until humanity itself has come to an end" (Arendt, 2001 p. 285). Fortunately, science has revealed the virtues of interdisciplinarity and transdisciplinarity around an epistemology of the convergence of the person. And the pandemic we are experiencing has underlined what genetics told us and we were not always able to recognize – in the human genome, uncoded DNA condenses 98 percent of the total storage capacity. And it is this DNA, formerly called "garbage", that is dedicated to managing interactions, and the expression of its information depends on our behavior, emotions, and lifestyles. In other words, we are far from being "pre-determined" everything happens as a result of interactions.

Finally, it is fundamental to value the person as a whole, including motivation, the power of hope and spirituality, the quality of affective interactions and social cohesion itself, which is only possible with the enhancement and full development of new and (inter) cultural communication skills in health. In the pandemic and post-pandemic period, we have the opportunity to value the decisive role that technologies have given and can continue to add to health, knowing that only in an articulated way, we can improve the adaptation of various professionals to this new reality, without losing the bonds between people, processes, and information systems, in order to respond to the real needs of individuals. The specificity of the therapeutic relationship established in a clinical context is an advantage in favor of the involvement of clinical and health psychologists, who can work in teams and with people in the search for solutions.

Online psychology and psychotherapy is a fact, but there is still a long way to go and investigate. Increasing studies on the effectiveness of teletherapy must be a number one priority for all psychological associations. We still live the dilemma between the physical and the virtual, but only knowing the advantages and disadvantages in one way and another, we can look for a solution that includes a balance, and from what we perceive, it can rarely be based only on contact at a distance. Certainly, in these new health intervention scenarios, we will move toward mixed models, with the necessary ethical care in the development and use of technologies, and integrating the skills developed over time by specialists who work specifically with mental processes and behavior associated with health.

REFERENCES

American Psychological Association. (2007). Record keeping guidelines. *American Psychologist.* 62(9), 993–1004. doi: 10.1037/0003-066X.62.9.993

Arendt, A. (2001). *A Condição Humana.* Lisboa. Relógio d'água.

Armitage, C.J., & Conner, M. (2000). Social cognition models and health behaviour: A structured review. *Psychology and Health, 15,* 173–189. doi: 10.1080/08870440008400299

Baldanza, R.F. (2006). *A Comunicação no Ciberespaço: Reflexões Sobre a Relação do Corpo na Interação e Sociabilidade em Espaço Virtual.* Trabalho apresentado ao NP Tecnologias da Informação e da Comunicação, do VI Encontro dos Núcleos de Pesquisa da Intercom

Barak, A., Hen, L., Boniel-Nissim, M, & Shapira, N. (2008). A comprehensive review and a meta-analysis of the effectiveness of internet-based psychotherapeutic interventions. *Journal of Technology in Human Services, 26*(2/4), 109–160. DOI: 10.1080/15228830802094429

Barak. A, Klein, B., & Proudfoot, J.G. (2009). Defining Internet-supported therapeutic interventions. *Annals of Behavioral Medicine, 38*(1), 4–17. DOI: 10.1007/s12160-009-9130-7

Barnett, J.E. (2011). Utilizing technological innovations to enhance psychotherapy supervision, training, and outcomes. *Psychotherapy, 48*(2), 103–108. DOI: 10.1037/a0023381

Bollas, C. (2015). Psicanálise na era da desorientação: do retorno do oprimido. *Revista Brasileira de Psicanálise, 49*(1), 47–66.

Caraça, L. (2021). A Interdisciplinaridade na Saúde. In J. Fialho (Coord.). *Manual para a Intervenção Social - Da teoria à ação,* 307–317. Lisboa, Portugal: Edições Sílabo.

Carvalho, R.G., Dias da Fonseca, A., Dores, A.R., Santos, C.M., Batista, J., Salgado, J., & Marlene Sousa, M. (2019). *Linhas de Orientação para a prestação de serviços de Psicologia Mediados por Tecnologias da Informação e da Comunicação (TIC).* Proposta do Grupo de Trabalho em Psicologia e eHealth da Ordem dos Psicólogos Portugueses. Lisboa: OPP.

Castells, M. (2003). A *galáxia da internet: reflexões sobre a internet, os negócios e a sociedade.* Rio de Janeiro: Jorge Zahar Editor.

Clayton, S., Devine-Wright, P., Stern, P.C., Whitmarsh, L., Carrico, A., Steg L., (…) Bonnes, M. (2015). Psychological research and global climate change. *Nature Climate Change, 5*(7), 640–646. Doi:10.1038/nclimate2622.

Donker, T., Blankers, M., Hedman, E., Ljotsson, B., Petrie, K., & Christensen, H. (2015). Economic evaluations of internet interventions for mental health: A systematic review. *Psychological Medicine, 45*(16), 3357–3376. doi: 10.1017/S0033291715001427

Eiguer, A. (2005). *Nunca eu sem ti.* Portugal: Edições Parsifal.

Farber, D. (2016). Reviews: Book Review Essay: "Alone Together: Why We Expect More from Technology and Less from Each Other, Sherry Turkle New York: Basic Books; 2011; 360 pp. Reclaiming Conversation: The Power of Talk in a Digital Age Sherry Turkle New York: Penguin; 2015; 436 pp.". *Fort Da, Published by Northern California Society for Psychoanalytic Psychology*, *22B(2)*, 91–104. https://pep-web.org/browse/FD/volumes/22B?preview=FD.022B.0091A

Fielding, K.S., Hornsey, M.J., & Swim, J.K. (2014). Developing a social psychology of climate change. *European Journal of Social Psychology*, *44*, 413–420. doi:10.1002/ejsp.2058.

Fonagy, P., & Target, M. (2003). *Psychoanalytic Theories: Perspectives from Developmental Psychopathology*. London: Whurr Publishers Ltd.

Gilford, R. (2014). *Environmental Psychology: Principles and Practice*, 5th ed. Colville, WA: Optimal Books.

Glanz, K., Rimer, B., & Viswanath, K. (2008). *Health Behavior and Health Education: Theory Research, and Practice*. San Francisco, CA: John Wiley.

Horvath, A.O. (2005). The therapeutic relationship: Research and theory. An introduction to the special issue. *Psychotherapy Research*, *15*(1–2), 3–7. doi: 10.1080/10503300512331339143

Horvath, A.O., & Bedi, R.P. (2002). The Alliance. In J. Norcross (Ed.), *Psychotherapy Relationships That Work: Therapist Contributions and Responsiveness to Patients* (pp. 37–70). New York: Oxford University Press.

Horvath, A.O., Del Re, A.C., Fluckiger, C., & Symonds, D. (2011). Alliance in individual psychotherapy. *Psychotherapy*, *48*(1), 9–16. doi:10.1037/a0022186

Jamoulle, M. (2015). Quaternary prevention, an answer of family doctors to overmedicalization. *Internacional Journal of Health Policy and Management*, *4*, 1–4. doi:10.15171/ijhpm.2015.24

Kaplan, R.M. (2009). Health psychology: Where are we and where do we go from here? *Mens Sana Monographs*, *7*(1), 3–9. doi:10.4103/0973-1229.43584.

McAndrew, L.M., Mora, P.A., Quigley, K.S., Leventhal, E.A., & Leventhal, H. (2014). Using the common sense model of self-regulation to understand the relationship between symptom reporting and trait negative affect. *International Journal of Behavioral Medicine*, 21, 989–994. doi: 10.1007/s12529-013-9372-4

Mitchel, S.A. (2000). *Relationality: From Attachment to Intersubjectivity*. Hillsdale, NJ: Analytic Press.

Mitchell, J.E., Meyers, T., Swan-Kremeier, L., & Wonderlich, S. (2003). Psychotherapy for bulimia nervosa delivered via telemedicine. *European Eating Disorders Review*, *11*(3), 222-230. Doi:10.1002/erv.517

Morón, J.J.M. & Aguayo, L.V. (2018). La psicoterapia on-line ante los retos y peligros de la intervención psicológica a distancia. *Apuntes de Psicologia*, *36*(1–2), 107–113. Retrived from www.researchgate.net/publication/328307157_La_psicoterapia_on-line_ante_los _retos_ y_ peligros_de_la_intervencion_psicologica_a_distancia.

OMS (2019). *Cuidados de Saúde Primários*. OMS, 2 de Janeiro, 2020, em www.who.int/world-health-day/world-health-day-2019/fact-sheets/details/primary-health-care.

OMS (1986). *As metas de Saúde para Todos: metas da Estratégia Regional Europeia de Saúde para Todos*. Lisboa. PT: Ministério da Saúde, Departamento de estudos e planeamento.

OPP (2018). *Utilização das TIC na Intervenção Psicológica*. Lisboa.

Porges, S.W. (2011). *The Polyvagal Theory: Neurophysiological Foundations of Emotions, Attachment, Communication, and Self-regulation*. New York: W.W.Norton.

Ramos, V. (1988). Prever a medicina das próximas décadas: que implicações para o planeamento da educação médica? *Acta Médica Portuguesa*, *2*, 171–179. https://scholar.google.com/scholar_lookup?title=Prever+a+Medicina+das+Pr%C3%B3ximas+D%C3%A9cadas:+que+Implica%C3%A7%C3%B5es+para+o+Planejamento+da+Educa%C3%A7%C3%A3o+M%C3%A9dica?&author=Ramos+V.&publication_year=1988&journal=Acta+M%C3%A9dica+Portuguesa&volume=2&pages=171-179

Roudinesco, E. (2005). *O paciente, o psicólogo e o Estado*. (A. Telles, Trans.). Rio de Janeiro: Jorge Zahar Ed.

Simpson, S., Bell, L., Knox, J., & Mitchell, D. (2005). Therapy via videoconferencing: A route to client empowerment? *Clinical Psychology & Psychotherapy, 12*(2), 156–165. Doi:10.1002/cpp.436

Symington, N. (1996). *The Making of a Psychotherapist*. London: Routledge.

Teixeira, P.J., Marques, M.M., Silva, M.N., Brunet, J., Duda, J., Haerens, L., (…) Hagger, M.S. (2020). A classification of motivation and behavior change techniques used in self-determination theory-based interventions in health contexts. *Motivation Science, 6*(4), 438–455. Doi: 10.1037/mot0000172

Van Lange, P., Joireman, J., & Milinski, M. (2018). Climate change: what psychology can offer in terms of insights and solutions. *Current Directions in Psychological Science, 27*(4) 269–274. doi:10.1177/0963721417753945.

Zimerman, D.E. (2007). *Fundamentos psicanalíticos: teoria, técnica e clínica*. Porto Alegre: Artmed.

6 Perceptions of Clients on Quality of Health Services

*Maria Carolina Martins Rodrigues,
Luciana Aparecida Barbieri da Rosa,
Maria José Sousa and Waleska Yone Yamakawa
Zavatti Campos*

CONTENTS

6.1 Introduction 97
6.2 Theoretical Reference..99
    6.2.1 Knowledge Transition 99
    6.2.2 Quality Services 99
    6.2.3 Quality Management in Health Care Services 101
6.3 Methodology 102
6.4 Results and Discussion...103
    6.4.1 Effects on Customers' Reliability 103
6.5 Final Considerations...108
References..109

6.1 INTRODUCTION

The search for quality services has been growing in recent years. Quality is a management system that analyses the most appropriate strategies to improve the provided services. Quality management is a necessary part of organizations for measuring the quality of services provided.

Over time, in the search for competitiveness, quality systems have been adopted for effectiveness and efficiency of processes and high-performance rates with successful results. Quality management is of paramount importance for service organizations, and quality is no longer an exclusive feature of industries (Nogueira, 2008; Gholami, Kavosi & Khojastefar, 2016).

Among the service provider companies, the ones focused on health care are included, and they started to be careful with the issue of managing the quality of health services. It is noticed that this was one of the last sectors to be concerned with quality management, and this was due to the fact that service providers understand that quality in their operations was not questionable (Souza & Lacerda, 2009). Today, the quality of services provided in health care is increasingly emphasized within the growing competitive market (Feldman; Gatto, & Cunha, 2005; Melão et al., 2017).

For Faria (2006), the health care market has been seeking to meet existing needs, as quality management, characterized by any coordinated activity aiming to assure customer satisfaction with the offered services, is exceeding expectations. So, to satisfy their needs, clients also started to seek for qualified health care assistance services. Health quality management is run by a philosophy of search to get user satisfaction, through the motivation of the professionals and employees' involvement in the working process (Duarte & Silvino, 2010).

Programs' intervention is not applied directly upon the clinical act, but it is fundamental in the administrative processes of the health care organization, in order to reach the qualities of the services provided (Berwick, 1994). The choice of citizens regarding health care services, namely, in public hospitals is legally conditioned by their address/area of residence. However, these services can be provided by public and private hospitals, which allows users/patients to widen their alternatives as long as they have financial affordability to make that choice.

Hospitals, in turn, seek to invest in a higher quality of services provided, which are related to the need in creating value and sustainability for health organizations (Bosi & Uchimura, 2007). These authors consider that the evaluation of health actions as one of the fundamental elements of the strategic management of health units. Fernandes and Lourenço (2006) state five reasons for the importance of quality in health: professionals, through performance improvement; ethical, by users' trust in professionals; policies, for the strategic concern of quality; by the increased demand from users; and lastly, economical, due to the need of producing income resources but responding to the needs of the population.

Thus, improving health care quality is of great importance for the citizen, but also for the health professionals themselves, considering aspects such as their own health and safety at work.

Therefore, it is important that health quality management is measured, as it will favor decision-making based on evidence, due to the results found, improvements in health care management are trustworthy (Vituri, & Évora, 2015). Thus, in this context, the general objective of the study was to analyze the relationships between the perceived reliability of clients of organizations providing health care services with the perception of tangible aspects and interaction with staff leading to knowledge transition. Therefore, its specific objectives are as follows: (1) Check if the tangible aspects positively influence the reliability of clients in the health care service; (2) Identify whether the interaction leading to knowledge transition between service providers and clients positively influences clients' reliability in the health care service.

To explore all the presented issues here, this chapter is organized as follows, to fit in with the identified objectives: it presents a theoretical framework in relation to the quality of service, followed by a conceptualization of health services; then, the methodology and the analysis of the questionnaire data on the quality of health care service provision are presented; and finally, the main conclusions of the study are identified.

6.2 THEORETICAL REFERENCE

6.2.1 KNOWLEDGE TRANSITION

Knowledge "is a fluid mix of framed experience, values, contextual information, and expert insight that provides a framework for evaluating and incorporating new experiences and information," state Davenport and Prusak (2000). Other authors, such as Polanyi (1958), link knowledge to action. "Knowledge is the ability to act," he argues. Knowledge, according to Nonaka and Takeuchi (1997), is created by the flow of information combined with the beliefs and commitment of those who possess it.

The combination of resources that organizations utilize to supply their products and services is a function of knowledge management systems, according to the perspective of a knowledge-based framework in organizations (Alavi & Leidner, 2001). This viewpoint highlights the question of how to best manage the knowledge resource, particularly how to make knowledge transition behaviors easier. This necessitates a broader understanding of knowledge than the traditional one – an object that can be codified and distributed outside of the human who developed it (Fahey & Prusak, 1998) – knowledge also lives within the thoughts and experiences of the employees (Sousa et al., 2021; Grant, 1996). The knowledge transition encompasses both types of knowledge that are incorporated in individual behaviors and communications (Dal Mas et al., 2020; Fahey & Prusak, 1998; Spender, 1996; Swap, Leonard, Shields, & Abrams, 2001). Individual or workgroup experiences, as well as interpretations and routinization of work processes, may all contribute to embedding knowledge (Shujahat et al., 2019; Alavi & Leidner, 2000). Effective knowledge transfer is critical for competitiveness, but the challenge is that knowledge is difficult to capture, even if we all recognize that tacit knowledge is ingrained in an organization's routines and people (Frappaolo and Wilson Todd, 2000).

6.2.2 QUALITY SERVICES

Services are rendered everywhere; for example, a trip to a beauty spa, markets, stores, purchase for a cinema ticket, all these are done by service providers. Thus, in order to improve their performance, the quality of the service provided needs to be measured.

It is noted that the services are intangible, that is to say, that they do not result in the priority of something (Kahtalian, 2002; Gulc, 2017). These services, provided efficiently, generate value for the consumer through the experience in providing the service. For Grönroos (2009), services consist of several types of resources, which are used several times with direct intentions on consumers, in order to find solutions to a problem and thus offer satisfaction to those who consumed it.

According to Lovelock and Wright (2002, p. 5), services are resumed into "economic activities which create value and provide benefits to consumers in a specific time and places, as a result of a desired change in – or in the name of – recipient of the service". Therefore, services are important for organizations, because the composing elements facilitate the creation of value for the final product (Lovelock & Wright, 2002).

Consumers are keen that organizations are able to offer for quality services, as this is one of the main factors for decision making when looking for services providers. Quality cannot be considered as an accessory that organizations should have as an additional cost in their services; they need to understand that consumers require and need that the provided services are of good quality (Deming, 1990, Meidutė-Kavaliauskienė, Aranskis & Litvinenko, 2013).

According to Moura (2013), the quality of service may be one of the secrets to satisfy and value the consumer, and consequently increased the organization's profitability. When consumer acknowledges the quality of the service provided, the dimension of result or technique exist, as well as the dimension related to the process (Grönroos, 2009).

The fundamental issue is related to the quality of service; while delivering it, what satisfies consumers can be determined (Paladini 2002). Thus, from consumer's perspective, it will be possible to adopt decisions that will be able to meet their demands and consequently decide on continuous improvements of services provided (Paladini 2004, Sharabi, 2013).

The quality of services perceived by the consumer is a comparison between the dimension of the service perceived and what was expected (Grönroos, 1984). The perception of the end-user in a service provision is considered satisfactory when comparing the performance of a service with previous expectations with the ways it should be performed (Bitner, 1990). Services are developed and consumed simultaneously (Fitzsimmons & Fitzsimmons, 2010; Tan et al., 2014).

The quality of the services provided is an essential factor for the competition and success of an organization (Ladhari, 2009). For Angelova & Zekiri (2011), the customer's perception is a judgment about something perceived and evaluated, varying from consumer to consumer and their different perceptions, regarding specific services that play an important role in determining pleasure.

Fulfillment of users' wishes aim that all issues are, in fact, answered and translated into actions (Emmi & Barroso, 2008, Dias et al., 2017). Therefore, the feeling of satisfaction brings a positive judgment about the characteristics and quality of the services provided (Favaro & Ferris, 1991; Singh, Sharma & Garg, 2016).

The surveys carried out related to the consumer's satisfaction with services provided should render improvements in daily life, in addition to significant advances in service management (Esperidião & Trad 2005). According to Machado and Nogueira (2008), measuring consumer's satisfaction is a personal assessment of each one, and its importance for assessment should not be ignored.

Quality assessment and the availability of quality results may, in a way, contribute to adjusting expectations regarding the particular choices of each consumer (Lourenço & Knop, 2011; Rayle et al., 2016). For Souza et al. (2009), the importance of evaluating the performance of quality in the provision of the service is of paramount importance for monitoring, verification, and evaluation of the management process. Therefore, the quality of the services provided must be evaluated and measured in such a way that they deliver positive results to the organization and user satisfaction.

Therefore, the quality of services is crucial for improving processes and providing quick responses to consumers (Galvão; Corrêa & Alves, 2011). Positive attitudes in service provision can contribute favorably to a considerable increase in consumer satisfaction and retention (Milan; Toni & Maioli, 2014; Edward & Sahadev, 2011; Sultan

& Wong, 2013). Good service attracts new consumers and consequently improves the organization's image.

6.2.3 QUALITY MANAGEMENT IN HEALTH CARE SERVICES

In the current scenario of increasing competitiveness, companies seek constant innovations in their processes, services, and products. To cope with that they use appropriate tools to assess their performance. The purpose is to achieve the quality of services provided to customers, aiming to achieve excellence in quality service for their organization.

Thus, quality needs to be measured and improved so that the company continues to compete in the market. This issue has become essential for organizations that are part of the market competition, as well as search for quality turned to be an important variable for consumers (Ferreira; Cabral & Saraiva, 2010; Zafiropoulos & Vrana, 2008).

When consumers choose a specific service, the quality of an organization's services becomes one of the fundamental factors. Therefore, quality can be one of the secrets to satisfy and create value for consumer and also increases the company's profitability (Moura, 2013; Kanté et al., 2016). When a consumer perceives the quality of the services provided, we get a result or technical dimension, and a related dimension to the process (Grönroos, 2009; Karkee et al., 2015). Therefore, the customer is influenced by the way his experience was conducted and how he perceived the service provided.

Proving quality health care is an essential concern in managing the sector; its basis is not only on improving operational capacity, but also the center of all processes in the life of society (Kern, 2019). Therefore, quality in health is defined as "the provision of accessible and equitable health care, with an optimal professional level, which takes into account the available resources and achieves adherence and citizen satisfaction" (DGS, 2013). It can be said that quality in health is the provision of equitable and accessible health care, with an excellent professional level that has available resources and achieves consumer approval and satisfaction (PNS, 2012–2016).

So, in order to achieve satisfactory results, institutions providing health care services should have adequate resources and structures and apply and adopt effective management techniques and performance standards that allow offering quality services to consumers (Kern, 2019). Thus, the continuity of care through an adequate structure is one of the perspectives which allows assessing quality in health, confirming the importance of this dimension in the entire dynamics of the services provided (Mota et al., 2014). Thus, the study's first hypothesis emerges:

H1 – The tangible aspects have positive impact on clients' reliability in the health care services.

Quality in health care sector is a main concern, that's why it is considered to be of most importance that its operation goes through the intervention of systematic techniques and tools so that service shows a quality strategy and a rational perception, with decisions aimed for continuous progress in assistance (Oliveira et al., 2016). It is important that the activities developed by health institutions are consolidated in methodological assumptions, which are capable of ensuring a high standard in quality (Pertence et al., 2010, Gordon et al., 2014).

The improvement of quality in health care needs to be part of the daily routine of each professional in performing their tasks (Ascenção, 2010, Milne, 2016). For this reason, the evaluation of quality in health services must include all elements of the multidisciplinary teams involved in this consumer satisfaction process. Therefore, the second hypothesis of the study arises:

H2 – The interaction leading to knowledge transition among services providers staff and clients positively influences clients' reliability in health care services.

For Burmester (2013), health quality management should not only be concerned with product or service quality, as these are not final, but it should be understood as a consequence of the practice of systematic and coherent management models, which allow planning, control and continuous quality improvement in the provision of health services. Thus, the techniques used to measure the quality of health services must be transformed into a continuous stimulus for improvement (Kern, 2019).

Service organizations need to ensure improved performance, quality and safety assessment of the assistance provided by health care services, whether public or private, for the management system conducts an analysis of the services provided, reflecting on the quality of services (Bonato, 2011). It is of utmost importance to know how users evaluate the service provided by the services that are being offered, so that they can reflect on the actions of professionals or change the way services are provided, aiming at their improvement. (Ramos et al., 2003).

Good quality and access to health services are fundamental characteristics when aiming to improve health care for individuals and the community. Thus, it is important that the activities developed by health institutions are based on consolidated methodological and philosophical assumptions, capable of ensuring a higher standard of quality (Pertence et al., 2010).

It is a great challenge for the health service provider, consumer's safety and the quality of the services so that the services used reach all care practices (Santiago et al., 2015). Taking this into consideration, using indicators as instruments to measure the quality of the services provided is a great alternative, allowing services to keep high performance. Measuring the quality of the health services provided can act as alerts for aspects that need to be restructured (Vituri & Matsuda, 2009).

Therefore, with technological advances and the growth of quality management, the complexity of health services has also become more visible (Bueno & Fossarella, 2012); therefore, measuring the quality of health services promotes an increase in the efficiency of health services. It is necessary to evaluate, monitor and verify the performance of health care services, so that the services provided have an efficient management process (Souza et al., 2009), generating positive results for organizations.

Next, the methodology of the study will be presented.

6.3 METHODOLOGY

The study is quantitative in nature through a survey of health care service providers.

For the study, the instrument used by Gomes (2013) was employed, which is adapted to the situations of the companies, consisting of 16 variables, as shown

TABLE 6.1
Dimensions of the Study

Dimensions	Items	Perception/Expectation
Tangible aspects	1	Hospital has visually attractive facilities/buildings.
	2	Used equipment and furniture have modern lines.
	3	Used materials at the hospital have an attractive look (leaflets/brochures).
	4	The hospital keeps clean facilities.
	5	The layout of hospital facilities is easy to move around.
Reliability	6	The hospital has all its sectors duly identified.
	7	The hospital keeps the schedule of services within the previously indicated time.
	8	Staff keep their services accordingly and without errors.
	09	There are no errors in logging of provided services.
Staff interaction	10	Staff clarify patients doubts/questions adequately.
	11	Staff behavior inspires trust in them.
	12	Employees are never too busy to answer your requests/questions.
	13	Staff always perform in time/quick services to patients.
	14	Employees are nice and caring.
	15	The hospital shows interest in solving their patients' problems.
	16	Staff is able to deal adequately with patients claims.

Source: Adapted from Gomes (2013).

in Table 6.1. The sample was for convenience, obtaining the participation of 102 customers and elicited in the months of January to March 2020. The variables were measured using the seven-point Likert scale, characterized by the anchors from Totally Disagree to Totally Agree.

Data analysis of this paper, through linear regression, was performed by using SAS 9.4 software, and the application SAS Enterprise Guide, version 7.13.

The results of the study are presented below.

6.4 RESULTS AND DISCUSSION

6.4.1 Effects on Customers' Reliability

Descriptive statistics on the study variables are shown in Table 6.2.

From the analysis of Table 6.2, we can see that the number of responses in the survey questionnaire was of 102 responses with highest average applied to the variable Tangible Aspects (5.74902) and the highest standard deviation related to the Reliability variable (0.73725).

Table 6.3, on the other hand, shows Pearson' correlation between the study variables.

TABLE 6.2
Descriptive Statistics

Variable	Number of Responses	Average	Standard Deviation
Tangible aspects	102	5.74902	0.60472
Reliability	102	5.44118	0.73725
Staff interaction	102	5.17087	0.73699

Source: Authors' elaboration.

TABLE 6.3
Correlation Analysis between the Model Variables

	Tangible Aspects	Reliability	Staff Interaction
Tangible aspects	1.00000	0.63835***	0.51229**
Reliability	0.63835***	1.00000	0.68314***
Staff interaction	0.51229**	0.68314***	1.00000

*** p-value <0.0001.

The correlation analysis shows that all variables are correlated to each other in at least 51 percent, which indicates that the model variables may be suitable for the elaboration of regression. In other words, all variables are positively and significantly correlated, with a p-value <0.01. The highest coefficient is 0.63835, between "Intangible aspects" and "reliability". The variables interaction with personnel and reliability are positively related to a value of 0.68314. Finally, the variables "Intangible aspects" and "interaction with personnel" are positively related to a value of 0.51229. The model has a standardized Cronbach's alpha of 0.825, a value considered to be excellent (HAIR JR. et al., 2005).

In the multiple regression analysis, customer reliability was modeled as a dependent variable (y), while tangible aspects (x_1) and interaction with staff (x_2) were modeled as predictor variables. According to the ANOVA test, shown in Table 6.4, before the p-value <0.0001, H0 is rejected, and therefore, it is assumed that some $\beta \neq 0$. In other words, when rejecting the null hypothesis, it shows it is assumed that the regression equation is possible to be outlined. According to the multiple regression, the model explains 57.1 percent of the variation in customer reliability, with an adjusted R^2 value of 0.571.

Table 6.5 shows the levels of freedom of the estimated parameters for multiple regression.

From the analysis of Table 6.4, it appears that there are problems in the reliable gap of the intercept, which can cover the zero value, and further, its high p-value (0.6969) does not allow to reject the null hypothesis, and therefore its β can be equal to zero. Therefore, the intercept was not included in the regression.

TABLE 6.4
Table ANOVA

Analysis of Variance

Source	DF	Sum of Parcels	Average Parcel	F Value	P-value
Model	2	31.80944	15.90472	68.20	<0.0001
Error	99	23.08762	0.23321		
Total revised	101	54.89706			

Note: DF, degrees of freedom.

Source: Authors' elaboration.

TABLE 6.5
Appraisal of Multiple Regression Parameters

Independent variable	DF	PE	T value	p-value	VIF	RG 95%
Intercept	1	0.20318	0.43	0.6696	0	−0.73883/0.14519
Tangible aspects (x_1)	1	0.47667	5.15	<0.0001	1.35583	0.29309/0.66026
Staff Interaction (x_2)	1	0.48301	6.36	<0.0001	1.35583	0.33237/0.63365

Note: DF, degrees of freedom; EP, parameter estimate; VIF, variation inflation factor; IC 95%, reliable gap of 95%.

Continuous act, the independent variables "Tangible aspects" (x_1) and interaction with personnel (x_2) have *p*-value <0.0001, significant at α = 0.05%, which means that the two variables can be included in the model of multiple regression. Such inclusion was endorsed by the very low value of variation inflation factor (VIF), whose value of 1.35553 indicates that there are no problems of collinearity between the variables. Regarding the reliable gap, variables x_1 and x_2 do not present zero value problems within the range, which allows us to conclude that the two variables can integrate the model. From the above, the multiple regression equation was outlined:

$$Regression\ equation := 0{,}47667x_1 + 0{,}48301x_2$$

Subsequently, the four premises of regression models were verified from the analysis of the residues: (1) The residues are normally distributed; (2) The model is linear; (3) Residues have constant variations (homoscedastic); (4) The residues are independent (Black, 2010, p. 479). Figures 6.1 and 6.2 check the first premise about the normality of the waste distribution.

From the analysis of Figure 6.1, it is possible to infer that the waste has a normal distribution. The QQ-plot graph is also used to check the premise of normality of the residues, as shown in Figure 6.2. In view of the analysis in Figure 6.2, it is possible

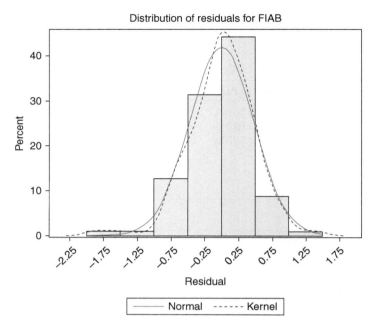

FIGURE 6.1 Premise on the normality of waste.

Source: Authors' elaboration.

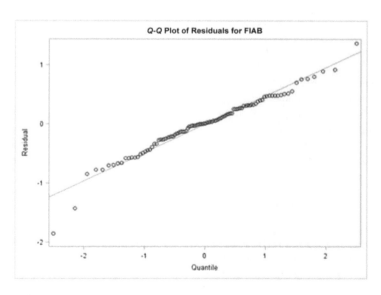

FIGURE 6.2 Premise on the normality.

Source: Authors' elaboration.

TABLE 6.6
Normality Tests

Normality tests	
Test statistic	*p*-value
Shapiro–Wilk	0.0182
Kolmogorov–Smirnov	0.1495
Cramer–von Mises	0.1374
Anderson–Darling	0.1115

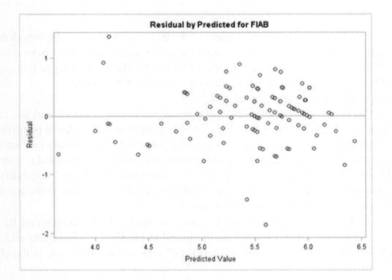

FIGURE 6.3 Assumptions about homoscedasticity and independence of residues.
Source: Authors' elaboration.

to state that the distribution is normal since the residues are relatively aligned to the line (Black, 2010).

In addition, to confirm the sample's normality, the Shapiro–Wilk, Kolmogorov–Smirnov, Cramer–von Mises and Anderson–Darling tests were performed, as shown in Table 6.6.

In these tests, the null hypothesis (H0) means a normal distribution. So according to the results of the tests performed and shown in Table 6.5, it is evident that the *p*-values of all tests are greater than 0.05, and with this, H0 is rejected, meaning that, the distribution is, therefore, normal.

Finally, assumptions 3 and 4 were checked regarding homoscedasticity and independence of residues, as shown in Figure 6.3.

TABLE 6.7
Meeting the Study Hypotheses

Hypotheses	Situation
H1 – Tangible aspects (x_1) positively influence the client's reliability on health care services (y).	Supported
H2 – Interaction leading to knowledge transition among service providers staff (x_2) and clients positively influence clients' reliability in health care services (y).	Supported

Source: Authors' elaboration.

In view of the analysis of Figure 6.3, it is evident that the residues are homoscedastic, that is, they have constant variation since they do not have a cone or parabola shape. It is also possible to infer that the residues are independent, as they do not show upward or downward behavior in the form of a line.

The hypotheses are summarized in Table 6.7.

From the above, the hypotheses of the study were supported, due to the regression equation $y = 0.47667x_1 + 0.48301x_2$, it is highlighted that: x_1 has a positive sign, demonstrating that the tangible aspects positively impact the reliability of customers; x_2 has a positive sign, indicating that the interaction leading to knowledge transition with the person providing health care services increases the clients' reliability; and finally, the sum of x_1 and x_2 is able to increase the reliability of customers in a model that explains 57.1 percent of the variability of y.

These findings are in line with the literature, as studies by Wakefield and Blodgett (1999) and Ladhari (2009) show that the tangible physical environment and interaction leading to knowledge transition with staff play an important role in the client's enthusiasm.

The final considerations of the study will be explained below.

6.5 FINAL CONSIDERATIONS

The general objective of the study was to analyze the relationships between the perceived reliability of clients of organizations providing health care services with the perception of tangible aspects and interaction leading to knowledge transition with staff. The objective was achieved as positive relationships were found between reliability with tangible and relational aspects in the health sector.

In relation to specific objectives, the first specific objective was achieved as it was found that there was a relationship between the tangible aspects and the reliability of clients in the health service. Concerning the second specific objective related to the existence of a relationship between the interaction leading to knowledge transition between service providers and clients in promoting the reliability of clients in the health service, this was met, so that it was found that a good relational service can increase customers' perception of trust.

The results shed light on the practical and managerial implications of tangible and relational aspects in the general perception of trust in the health service provided, which ultimately can foster organizational performance and customer loyalty. Therefore, managers must pay attention to the need of organizing tangible aspects such as physical spaces, material resources and structure, while encouraging good relations between employees and customers.

In other words, considering the context of the COVID-19 pandemic in which we live, a disease caused by the SARS-COV2 virus, health care must take into account the quality of interpersonal relationships and the use of adequate tangible aspects, with aiming to foster the perception of trust in clients/patients.

The limitations of the study involve the application of only one survey instrument. It is suggested for future studies to expand studies to other countries with the intersection of the instrument applied to other constructs.

REFERENCES

Alavi, M., & Leidner, D. (2001). Review: knowledge management and knowledge management systems: Conceptual foundations and research issues. *Mis Quarterly*, *1*, 107. https://doi.org/10.2307/3250961

Angelova, B., & Zekiri, J. (2011). Measuring customer satisfaction with service quality using American Customer Satisfaction Model (ACSI Model). *International Journal of Academic Research in Business and Social Sciences*, 1(3), 232–258.

Ascenção, H. (2010). *Da qualidade dos cuidados de enfermagem à satisfação das necessidades do utente*. Tese de Mestrado em Ciências de Enfermagem. Instituto de Ciências Biomédicas Abel Salazar, Faculdade de Medicina da Universidade do Porto.

Berwick, D.M. (1994). *Aplicando o gerenciamento da qualidade nos serviços de saúde*, São Paulo: Makron Books, 18–27.

Bitner, M. (1990). Evaluating service encounters: The effects of physical surroundings and employee responses. *Journal of Marketing*, *54*(2), 69–82.

Black, K. (2010). *Business statistics for contemporary decision making*. Sixth Edited. University of Houston—Clear Lake: John Wiley & Sons, Inc.

Bosi, M., & Uchimura, K. (2007). Avaliação da qualidade ou avaliação qualitativa do cuidado em saúde? *Revista De Saude Publica - REV SAUDE PUBL*, *41*. https://doi.org/10.1590/S0034-89102007000100020

Bueno, A.A.B., & Fassarella, C.S. (2012). Segurança do Paciente: uma reflexão sobre sua trajetória histórica. *Revista Rede de Cuidados em Saúde*; 6(1), 1–9.

Dal Mas, F., Biancuzzi, H., Massaro, M., & Miceli, L. (2020). Adopting a knowledge translation approach in healthcare co-production: A case study. *Management Decision*, *58*(9), 1841–1862.

DGS (2013). Departamento da Qualidade na Saúde – Uma cultura de melhoria contínua da qualidade. [on-line] Disponível em: www.dgs.pt/ms/8/default.aspx?id=5521. Consulta: 22/05/2020.

Dias, F.F., Lavieri, P.S., Garikapati, V.M., Astroza, S., Pendyala, R.M., & Bhat, C.R. (2017). A behavioral choice model of the use of car-sharing and ride-sourcing services. *Transportation*, *44*(6), 1307–1323.

Duarte, M.S. da M. & Silvino, Z.R. (2010). Acreditação hospitalar: qualidade dos serviços de saúde. Revista de Pesquisa: cuidado é fundamental, São Paulo, pp. 182–185.

Edward, M., & Sahadev, S. (2011). Role of switching costs in the service quality, perceived value, customer satisfaction and customer retention linkage. *Asia Pacific Journal of Marketing and Logistics*, 22(3), 327–345.

Emmi, D.T., & Barroso, R.F.F. (2008). Avaliação das ações de saúde bucal no Programa Saúde da Família no distrito de Mosqueiro, Pará. *Ciência Saúde* Colet, 13(1), 35–41.

Esperidião, M., & Trad, L.A.B. (2005). Avaliação da satisfação de usuários. *Ciência Saúde Coletiva*, 10(Supl.), 303–312.

Fahey, L., & Prusak, L. (1998). The eleven deadliest sins of knowledge management. *California Management Review*, 40(3), 265–276.

Faria, C. (2006). *Princípio da gestão hospitalar*. Info-Escola: navegando e aprendendo, São Paulo, pp. 1–4.

Favaro, P., & Ferris, L.E. (1991). Program evaluation with limited fiscal and human resources. *Cad Saúde Pública*, 11(3), 425–438.

Feldman, L.B., Gatto, M.A.F., & Cunha, I.C.K.O. (2005). História da evolução da qualidade hospitalar: dos padrões a acreditação. *Acta Paul Enferm*, 18(2), 213–219.

Fernandes, A.A.C.M., & Lourenço, L.A.N. (2006). O modelo da EFQM na melhoria da qualidade: O estudo das relações entre os critérios do modelo no Hospital Amato Lusitano. Covilhã: Universidade da Beira Interior.

Ferreira, I., Cabral, J., & Saraiva, P. (2010). An integrated framework based on the ECSI approach to link mould customers' satisfaction and product design. *Total Quality Management & Business Excellence*, 21, 1383–1401.

Fitzsimmons, J.A., & Fitzsimmons, M.J. (2010). *Administração de serviços: operações, estratégia e tecnologia da informação* (6th ed.). Porto Alegre: Bookman.

Galvão, H.M., Corrêa, H.L., & Alves, J.L. (2011). Modelo de avaliação de desempenho global para IES. ReA – Revista de Administração. *UFSM*, 4(3), 425–441.

Gholami, M., Kavosi, Z., & Khojastefar, M. (2016). Services quality in the emergency department of Nemazee Hospital: Using SERVQUAL model. *Journal of Health Management & Information*, 3(4), 120–126.

Gomes, C. (2013). *Avaliação da qualidade de serviços no comércio retalho: o caso Pingo Doce*. Tese de Mestrado, Lisboa

Gordon, A.L., Franklin, M., Bradshaw, L., Logan, P., Elliott, R., & Gladman, J.R. (2014). Health status of UK cares home residents: A cohort study. *Age and Ageing*, 43(1), 97–103.

Grant R. M. (1996), Toward a knowledge-based theory of the firm. *Strategic Management Journal*, 17, 109–122.

Grönroos, C. (1984). A service quality model and its marketing implications. *European Journal of Marketing*, 18(4), 36–44.

Grönroos, C. (2009). *Marketing gerenciamento e serviços*. 3ªed. Rio janeiro: Elsevier.

Gulc, A. (2017). Models and methods of measuring the quality of logistic service. *Procedia Engineering*, 182, 255–264.

Hair Jr., J. F., Babin, B., Money, A.H., & Samouel, P. (2005). *Fundamentos de métodos de pesquisa em administração*. Porto Alegre: Bookman.

Kahtalian, M. (2002). *Marketing de serviços: Coleção Gestão Empresarial*. Curitiba: Gazeta do povo.

Kanté, A.M., Exavery, A., Phillips, J.F., & Jackson, E.F. (2016). Why women bypass front-line health facility services in pursuit of obstetric care provided elsewhere: A case study in three rural districts of Tanzania. *Tropical Medicine and International Health*, 21, 504–514.

Karkee, R., Lee, A.H., & Binns, C.W. (2015). Bypassing birth centres for childbirth: An analysis of data from a community-based prospective cohort study in Nepal. *Health Policy and Planning*, 30, 1–7.

Kern, A.E. (2019). *Gestão de qualidade riscos e segurança do paciente.* Ed. Senac, São Paulo, SP.
Ladhari, R. (2009). A review of twenty years of SERVQUAL research. *International Journal of Quality and Service Sciences, 1*(2), 172–198.
Lourenço, C.D.S., & Knop, M.F.T. (2011). Ensino superior em administração e percepção da qualidade de serviços: uma aplicação da escala SERVQUAL. RBGN – Revista Brasileira de Gestão de Negócios, *13*(39), 219–233.
Lovelock, C., Wirtz, J., & Hemzo, M.A. (2011). *Marketing de serviços: pessoas, tecnologia e estratégia* (7th ed.). São Paulo: Pearson Prentice Hall.
Machado, N.P., & Nogueira, L.T. (2008). Avaliação da satisfação dos usuários de serviços de Fisioterapia. *Rev Bras Fisioter., 12*(5), 401–408.
Meidutė-Kavaliauskienė, I., Aranskis, A., & Litvinenko, M. (2014). Customer satisfaction with the quality of logistics services. *Procedia – Social and Behavior Sciences, 110*, 330–341.
Melão, N., Guia, S., & Amorim, M. (2017). Quality management and excellence in the third sector: Examining European Quality in Social Services (EQUASS) is non-profit social services. *Total Quality Management & Business Excellence,* 28 (7–8), 840–857.
Milan, G.S., De Toni, D., & Maioli, F.C. (2013). Atributos e dimensões relacionadas aos serviços prestados por uma instituição de ensino superior e a satisfação de alunos. *Gestão e Planejamento, 13*(2), 199–214.
Milne, A. (2016). Depression in care homes. In *mental health and older people: A guide for primary care practitioners,* eBook, Chew-Graham C, and Ray M (Eds.). Cham: Springer International Publishing, 145–160.
Mota, L., Pereira, F., & Sousa, P. (2014). SI de enfermagem: Exploração da informação partilhada com os médicos. *Revista Enfermagem Referência, 4*(1), 85–91.
Moura, P.A.S.C. (2013). *Percepção da qualidade do serviço como ferramenta para satisfação do cliente: estudo de caso da Escola Tia Marisa, na cidade de São Sebastião.* São Sebastião: FATEC São Sebastião.
Oliveira, J.L.C., Gabriel, C.S., Fertonani, H.P., & Matsuda, L.M. (2017). Management changes resulting from hospital accreditation. *Revista Latino-Americana* de Enfermagem, *25*, e2851.
Paladini, E.P. (2002). *Avaliação estratégica da qualidade.* São Paulo: Ed. Atlas.
Paladini, E.P. (2004). *Gestão da qualidade: teoria e pratica.* 2. ed. São Paulo: Atlas.
Pertence, P.P., & Melleiro, M.M. (2010). Implantação de ferramenta de gestão de qualidade em Hospital Universitário. Revista da Escola de Enfermagem da *USP, 44*(4), 1024–1031.
PNS – Plano Nacional de Saúde (2012–2016). Eixo Estratégico – Qualidade em Saúde. Disponível em http://pns.dgs.pt/files/2012/02/0024_-_Qualidade_em_Sa%C3%BAde_2013-01-17_.pdf (acedido em 24/05/2020).
Polanyi, M. (1958). *Personal knowledge: Towards a post-critical philosophy* (Vol. 20). University of Chicago Press.
Prusak, L., & Davenport, T.H. (2013). Knowledge after the knowledge creating company: A practitioner perspective. In G. von Krogh, H. Takeuchi, K. Kase, & C. G. Cantón (Eds.), *Towards Organizational Knowledge: The Pioneering Work of Ikujiro Nonaka.* UK: Palgrave Macmillan, pp. 255–262. https://doi.org/10.1057/9781137024961_15
Ramos, D.D., & Lima M.A.D.S. (2003). Acesso e acolhimento aos usuários em uma unidade de saúde de Porto Alegre, Rio Grande do Sul, Brasil. *Cad Saúde* Pública, *19*(1), 27–34.
Rayle, L., Dai, D., Chan, N., Cervero, R., & Shaheen, S. (2016). Just a better taxi? A survey-based comparison of taxis, transit, and ride-sourcing services in San Francisco. *Transport Policy, 45,* 168–178.

SAS Institute Inc. (2016). SAS 9.4 – SAS Enterprise Guide 7.13. Copyright © 2016 SAS Institute Inc. SAS e todos os outros nomes de produtos ou serviços do SAS Institute Inc. são marcas comerciais ou marcas registradas da SAS Institute Inc., Cary, NC, EUA.

Sharabi, M. (2013). Managing and improving the service quality. *International Journal of Quality and Service Sciences*, 5(3), 309–320.

Shujahat, M., Sousa, M., Rahim, S., Nawaz, F., Wang, M., & Umer, M. (2017). Translating the impact of knowledge management processes into knowledge-based innovation: The neglected and mediating role of knowledge-worker productivity. *Journal of Business Research*, 94. https://doi.org/10.1016/j.jbusres.2017.11.001

Singh, R.K., Sharma, H.O., & Garg, S.K. (2016). Study on supply chain issues in auto component manufacturing organization: Case study. *Global Business Review*, 17(5), 1196–1210.

Souza, A.A. et al. (2009). Controle de Gestão em Organizações Hospitalares. *Revista de gestão USP*, 16(3), 15–29.

Souza, T.C. de R., & Lacerda, P.T. (2009). Planejamento estratégico e qualidade: acreditação hospitalar – um estudo de caso no Hospital Vita Volta Redonda. V Congresso Nacional de Excelência em Gestão: gestão do conhecimento para a sustentabilidade, Niterói, Rio de Janeiro, Brasil, pp. 2–22, July.

Spender, J.C. (1996). Making knowledge the basis of a dynamic theory of the firm. *Strategic Management Journal*, 17, 45–62. https://doi.org/10.1002/smj.4250171106

Sultan, P., & Wong, H.Y. (2013). Antecedents and consequences of service quality in a higher education context: A qualitative research approach. *Quality Assurance in Education*, 21(1), 70–95.

Swap, W., Leonard, D., Shields, M., & Abrams, L. (2001). Using mentoring and storytelling to transfer knowledge in the workplace. *Journal of Management Information Systems*, 18, 95–114. https://doi.org/10.1142/9789814295505_0006

Tan, A.W.K., Yifei, Z., Zhang, D., & Hilmola, O.P. (2014). State of third-party logistics providers in China. *Industrial Management & Data Systems*, 114(9), 1322–1343.

Vituri, D. & Matsuda, L. (2009). Validação de conteúdo de indicadores de qualidade para avaliação do cuidado de enfermagem. *Revista da Escola de Enfermagem da USP*, 43(2), 429–437.

Vituri, D.W., & Évora, Y.D.M. (2015). Gestão da Qualidade Total e enfermagem hospitalar: uma revisão integrativa de literatura. *Rev Bras Enferm. Set-out*, 68(5), 945–952.

Wakefield, K.L., & Blodgett, J.G. (1999). Customer response to intangible and tangible service factors. *Psychology and Marketing*, 16(1), 51–68.

Zafiropoulos, C., & Vrana, V. (2008). Service quality assessment in a Greek higher education institute. *Journal of Business Economics and Management*, 9(1), 33–45.

7 Innovation Management Applied to Primary Care
An Integrative Review

*Patricia Gesser da Costa, Guilherme Agnolin,
João Paulo da Silveira, Andreia de Bem Machado,
Gertrudes Aparecida Dandolini,
João Artur de Souza and Maria José Sousa*

CONTENTS

7.1 Introduction	113
7.2 Methodology	114
7.3 Analysis and Discussion of Results	115
7.3.1 Analysis of Article 1	119
7.3.2 Analysis of Article 2	122
7.3.3 Analysis of Article 3	123
7.3.4 Analysis of Article 4	124
7.3.5 Analysis of Article 5	124
7.4 Innovation Management and Primary Health Care	126
7.5 Final Conclusion	126
References	127

7.1 INTRODUCTION

Primary health care (PHC) corresponds to clinical services offered at the community level and integrated into a properly structured health care network. PHC is proposed to be the most appropriate place to deal with complaints, demands, and health problems presented by a population, except in emergency situations where there is another specialized service available to meet them (Starfield, 1998).

In conceptual terms, PHC is the first contact with patients for the production of a continuous and person-centered care, in order to meet their health needs, which only refers to other services in very unusual cases that require other specialized action. PHC also coordinates care when people receive assistance at other levels of care (Macinko, 2009).

From a contemporary perspective, PHC represents the most cost-effective model for nation-states to offer and sustain health services for the entire population, to adequately resolve 90 percent of the complaints and demands presented, and to adequately direct the others within the complex network of services that make up the health sector in these societies (Macinko, 2009).

In the last three decades, several health reforms have been introduced not only in Brazil, with the Unified Health System (SUS), but in most countries in the Americas, aiming at structuring primary care as the main element of health systems. The reforms were implemented for a variety of reasons, including rising expenses, inefficient and low-quality services, limited government budgets, new technical breakthroughs, and the state's changing role.

Despite the fact that investment, in terms of better health and equity, most reforms appear to have minimal, mixed, or even negative results. Studies conducted with health professionals in the Americas confirm that disagreements and misconceptions about PC exist in abundance, even within the same country.

In the day-to-day administration of health care in Brazil, coordinated efforts are verified in the qualification and consolidation of a comprehensive PC model, which satisfies the needs and expectations of Brazilians in terms of outpatient care provided by the state, in its capacity as funder, regulator, and provider of SUS (Silva, 2015). With the advent of the COVID-19 pandemic in 2020, PHC services were challenged as to the response that was needed to respond to the demand for services, both for patients with respiratory symptoms and suspected COVID-19 involvement, and to maintain the usual care that PHC provides to the population in a pandemic context (Silveira, 2020).

In this context of implementing a health care model in order to reorganize the supply of services nationwide, the relevance of innovation management in the planning and execution of care production by the different spheres of SUS is evident.

From the perspective of innovations applied to the public sector, these are justified not only by government failures, but mainly by the growth of increasingly complex and wicked problems, by the demands for better quality services, and by the desire for greater participation of society in public policymaking.

The fundamental principles (Calvalcante, 2017) that guide innovation processes in the public sector are: interaction, collaboration, trial-and-error, focus on knowledge, and long-term perspective in the maturation of innovations. Innovation in governmental organization can also be understood under three formats: (i) improvement of something that is already underway to expand the government's impact on people's lives; (ii) adaptation of an already tested idea to a new context in order to scale it up; and (iii) development of something entirely new to meet or exceed governmental goals.

Such concepts are applicable to the management, planning, and execution of PHC services, and in this sense, it is of interest to know how innovation management is applied and correlated by the literature to the PHC field.

7.2 METHODOLOGY

In order to understand how innovation management applied in primary care, the present study applied the integrative literature review method. According to Whittemore and Knafl (2005), integrative review has the potential to build science by informing research, practice, and policy initiatives through the analysis of diverse data sources, which allows for increased holistic understanding on a given topic.

For the development of this review, we adapted the method proposed by Kitchenham and Charters (2007), presented in Table 7.1. Once the research protocol was defined, the systematic search began, following each of the defined steps.

As for the timeline of activities, the research began on September 10, 2021. The selection of articles was performed on September 14, 2021, and the analysis of the articles and their discussion was completed on September 21, 2021. The first draft was completed on September 30, 2021, and the final version of the article was completed on November 26, 2021.

The research has primary health care and innovation management as its central aspect. In this sense, the following research question was established: *How can innovation management be applied in primary care in health?*

For the development of the study, the databases Scopus (www.scopus.com) and Scielo (www.scielo.br/) were consulted. The inclusion criteria determined for the research were: studies in the format of full articles available freely via access from the Federal University of Santa Catarina, published from 2012 and in Portuguese, English, and Spanish languages. The exclusion criteria were theses, dissertations, and duplicate articles in the databases. The Mendeley program was used to identify duplicates. The process of studies selection is demonstrated in Figure 7.1.

The search strategies were defined as follows. By collecting data in the Scopus research base, using a construct consistent with the theme addressed, 43 studies were identified, and in Scielo, base only two articles were found, as seen in Table 7.2.

The first stage of the analysis consisted of applying the inclusion and exclusion criteria. No duplicate articles referring to the theme were found in the Scielo database. Open access publications from 2012 to 2021 were considered. Thus, at the end of this stage, 22 articles were obtained from Scopus and two articles from Scielo. The titles, abstracts and keywords of the respective studies were read. Then, considering the most relevant, Table 7.3 with 16 studies was generated.

In the second stage, the 16 studies were read in full to detect those that matched the scope of the research and the quality criteria for selecting articles, namely: choosing the most cited articles or articles published in journals with a high impact factor, among those that responded to research question. Five articles were selected and analyzed as shown in Table 7.4.

After reading the articles in full, the data extraction strategy used was the reading of the studies' results and discussions, where the characteristics that permeate innovation management applied to primary care in public health were identified. Summaries of the articles were prepared highlighting the themes defined for data extraction, in order to facilitate their analysis and interpretation.

7.3 ANALYSIS AND DISCUSSION OF RESULTS

This topic presents the results obtained from the articles selected in the integrative literature review. The themes analyzed are related to the research question and

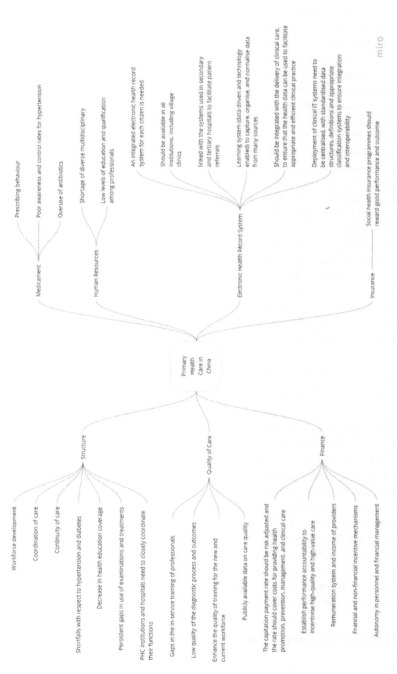

FIGURE 7.1 Criteria for inclusion and exclusion of articles.
Source: The authors (2021).

TABLE 7.1
Protocol of the Integrative Literature Review

1. Date
2. Background of the research
3. Research question
4. Databases consulted
5. Inclusion and exclusion criteria
6. Search strategies
7. Quality criteria for article selection
8. Data extraction strategies
9. Data analysis strategies
10. Knowledge dissemination strategies
11. Timeline of activities

Source: Adapted from Kitchenham and Charters (2007).

TABLE 7.2
Search Strategy Applied to the Databases without Considering Inclusion and Exclusion Criteria

Base	Expression	Quantity
Scopus	("innovation management" OR "service innovation" OR "innovation practices" AND "health" AND "primary care" OR "basic care")	43
Scielo	("innovation management" OR "service innovation" OR "innovation practices" AND "health" AND "primary care" OR "basic care")	02

Source: Elaborated by the authors.

TABLE 7.3
Articles Pre-selected for the Integrative Review

Title	Authors	Year
€ Function First': How to promote physical activity and physical function in people with long-term conditions managed in primary care? A study combining realist and co-design methods	Law, R.-J., Langley, J. et. al.	2021
Evaluating the feasibility and acceptability of a co-design approach to developing an integrated model of care for people with multi-morbid COPD in rural Nepal: A qualitative study	Yadav, U.N., Lloyd, J. et. Al.	2021
Quality of Primary Health Care in China: Challenges and Recommendations	Li, X., Krumholz, H.M., et. al.	2020

(*continued*)

TABLE 7.3 (Continued)
Articles Pre-selected for the Integrative Review

Title	Authors	Year
A pilot study of an integrated mental health, social and medical model for diabetes care in an inner-city setting: Three Dimensions for Diabetes (3DFD)	Ismail, K., Stewart, K. et. al.	2020
A € Function First – Be Active, Stay Independent' – Promoting physical activity and physical function in people with long-term conditions by primary care: A protocol for a realist synthesis with embedded co-production and co-design	Law, R.-J., Williams, L. et. al.	2020
Reshaping public hospitals: An agenda for reform in Asia and the Pacific	Gauld, R., Asgari-Jirhandeh, N.	2018
Tending to innovate in primary health care in Sweden: a qualitative study	Avby, G., Kjellström, S., Andersson Bäck, M.	2019
'Doing more with less': a qualitative investigation of South African health service managers' perceptions of implementing health innovations	Brooke-Sumner, C., Petersen-Williams, P., Kruger, J., Mahomed, H., Myers, B.	2019
The same but different: Online patients' perceptions of access to electronic health records among health care professionals	Wass, S., Vimarlund, V.	2019
Strengthening Integrated Care Through Population-Focused Primary Care Services: International Experiences Outside the United States	Loewenson, R., Simpson, S. et. al.	2017
Primary care interventions and current service innovations in modifying long-term outcomes after stroke: A protocol for a scoping review	Pindus, D.M., Lim, L. et. al.	2016
Pro-PET-Health facing the challenges of the professional training process in health	Da Costa, M.V., Borges, F.A.	2015
Configuration of innovation practices in primary health care: case study	Nodari, C.H., Camargo, M.E., et.al.	2015
A method to determine the impact of patient-centered care interventions in primary care	Daaleman, T.P., Shea, C.M. et. al.	2014
HOBE +, a case study: a virtual community of practice to support innovation in primary care in the Basque Public Health Service	Abos Mendizabal, G., Nuño Solinís, R., Zaballa González, I.	2013
Use of strategic scenarios for health human resources planning: Community pharmacists case in Portugal 2010–2020	[Uso de cenários estratégicos para planeamento de recursos humanos em saúde: o caso dos farmacêuticos comunitários em Portugal 2010–2020]	Gregório, J., Velez Lapão, L.	2012

Source: Elaborated by the authors.

TABLE 7.4
Articles Selected for the Integrative Review

Title	Authors	Year
Quality of Primary Health Care in China: Challenges and Recommendations	Li, X., Krumholz, H.M., et. al.	2020
Tending to innovate in primary health care in Sweden: a qualitative study	Avby, G., Kjellström, S., Andersson Bäck, M.	2019
'Doing more with less': a qualitative investigation of South African health service managers' perceptions of implementing health innovations	Brooke-Sumner, C., Petersen-Williams, P., Kruger, J., Mahomed, H., Myers, B.	2019
The same but different: Online patients' perceptions of access to electronic health records among health care professionals	Wass, S., Vimarlund, V.	2019
HOBE +, a case study: a virtual community of practice to support innovation in primary care in the Basque Public Health Service	Abos Mendizabal, G., Nuño Solinís, R., Zaballa González, I.	2013

Source: Elaborated by the authors.

objectives of this study, offering theoretical support for the interpretation of the results.

Although there were few articles, and all of them were empirical studies – one from Spain, two from Sweden, one from South Africa, and one from China brought significant contributions.

The oldest, by Mendizabal, Solinís, González (2013), describes the adoption of communities of practice as supporting innovation in PC in Spain. The second and third are developed in Sweden, with Avby, Kjellström, and Anderson Bäck (2019) investigating what promotes innovations in primary care; and Wass and Vimarlund (2019) the perception of health professionals (ambulatory and primary care) about an already implemented innovation: patient accessible electronic health records (PAEHRs). The fourth, by Brooke-Sumner, Petersen-Williams, Kruger, Mahomed and Myers (2019), surveys managers' perceived experience of adopting an innovation in the Western Cape with respect to new medical practices. Finally, in the Chinese context, Xi et al. (2020) address the challenges to innovate in the quest for higher quality primary care.

In the sequence, then, each of these researches is described, seeking to identify elements of innovation management.

7.3.1 ANALYSIS OF ARTICLE 1

Li et al. (2020) present an evaluation of the quality of primary health care in China, as well as the country's significant concerns and recommendations. China has prioritized the development of its PHC system over the last decade of health care reform;

nevertheless, the current system makes it difficult to provide high-quality and useful care to the public. The study's recommendations are measures that have transformed health care from a hospital-centric system to an integrated system anchored in primary care and facilitated by cutting-edge technology and data.

A literature review in PubMed, MEDLINE, and China National Knowledge Infrastructure (CNKI) was utilized as a research approach to find relevant studies in China in seven core health care categories (structure, human resources, electronic health record system, finance, insurance, medications, and care quality). This article used data from China's national PHC survey, interviews with key stakeholders, frontline workers and policy makers, and an assessment of national and international examples. Figure 7.2 illustrates this.

Improvements in primary health care have been main action points for continued training for the PHC workforce with the use of, for example, online platforms, which cover large numbers of practitioners and can be easily adapted to the individual's skills. Another requirement is the use of PHC learning platforms for knowledge development and application, which are based on a digital database and cutting-edge technology. Understandable performance metrics with trustworthy data, in-depth analysis, and financial and non-financial incentive mechanisms should all be considered to promote accountability transparency (Li et al., 2020).

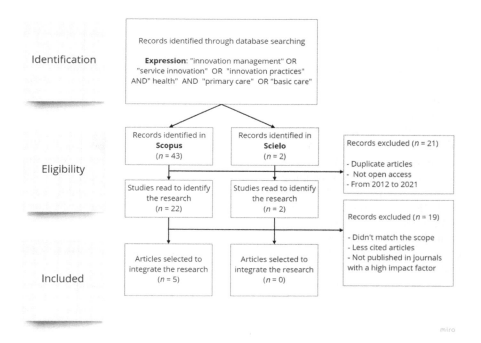

FIGURE 7.2 Domains of primary health care.
Source: The authors (2021).

Innovation Management Applied to Primary Care 121

According to Li et al. (2020), the current situation in China, the PHC system faces problems. One of them reported in the study is the low quality of diagnosis and treatment. Because it is a vast country, there is a variation in behavior in groups in large centers compared to villages or more remote areas and with low coverage of tests and recommendations for common diseases such as dysentery and angina.

Chronic disorders such as diabetes and hypertension have gaps in their treatment. These factors lead to low performance of PHC in the country, demonstrated by economic losses and compromised health, and high rates of hospital admission, which in the case of hypertension is the highest of all OECD countries. In the case of diabetes, the study found that only 37 percent of the patients were aware of their diagnosis, a result of low educational health coverage and lack of testing and treatment (Li et al., 2020).

As a structural challenge, the compensation and income system of PHC providers was identified. A payment system that creates financial incentives for diagnostic tests and prescriptions, without matching clinical need, and the failure to reward physicians for quality rather than quantity of care delivered can affect the PHC system. There has been an attempt to structure a new government drug policy, however, due to fragmented governance the implementation of these policies have failed (Li et al., 2020).

According to Li et al. (2020), health care coordination should be based on bidirectional mechanisms such as the formation of medical alliances or integrated systems. PC institutes and hospitals are still paid on a fee-for-service basis today, which means they compete for patients and have few incentives to work together. Hospital care is reimbursed more generously than PHC care, making it more difficult to function as gatekeepers. Another issue identified is that computerized patient records are not integrated and are rarely shared between PHCs and hospitals, indicating that there is a lack of workflow between health care programs.

Another work process factor that presents a challenge is continuity, which in the first place should be relational continuity of care, which allows contractual agreements to be formed with family physicians. The approach that would best address this challenge is the lifecycle approach to managing care.

Information systems must be improved to create a PC learning system. This is an electronically integrated system of each citizen's health history and is necessary to improve the quality and efficiency in PC throughout the health system. These systems should be available in PC institutions and their development and deployment needs to be centralized, with standardized data structure and definitions, as well as appropriate classification system, to ensure integration and interoperability. China's focus should be on building a data-driven, technologically enabled, high-performance real-time PHC learning system capable of capturing, organizing, and normalizing data from a variety of sources. Another requirement is to keep data safe by allowing limited access and having enough computer capacity to quickly evaluate data. For policymakers, researchers, physicians, and patients, the system must be able to produce understanding and discoveries (Li et al., 2020).

7.3.2 ANALYSIS OF ARTICLE 2

Public measures to promote primary care have been advocated in a number of countries (Avby et al., 2019). This qualitative study took place in the aftermath of Sweden's national health reform, in the county of Jönköping. The National Health Reform was fully implemented in Jönköping county in 2010. Swedish health care is funded by local taxes, and payment is sent to the organization rather than to individual doctors.

The three primary sources of revenue for primary health care centers (PHCCs) are fixed payments (82 percent), quality disbursements (6 percent), and special reimbursements (12 percent), such as those based on health indicators in predefined target groups. Each consultation is also subject to a nominal cost. In addition, the region has a quality improvement culture, which has been previously documented in other publications.

To find out what encourages innovation in primary care, the authors used a multiple case study method. Three public and two private PHCCs were purposefully chosen to ensure that a diversity of elements was included. To be considered a successful case, the PHCC must: (1) provide high-quality care based on national and regional comparisons (e.g., national patient surveys); (2) have low staff turnover; (3) have positive financial development; (4) have good leadership (in terms of reform goals); and (5) receive high scores in the regional quality incentive scheme.

The researchers performed 48 in-depth interviews with PHCC managers and professionals. One week before the interview, the participants were issued a questionnaire that included basic information such as age, education, job experience, occupation, number of years in the care unit, and function in the care unit.

The approach yielded a sample of 48 people ranging in age from 29 to 67 years old (33 women and 15 men). Sixty-one percent were admitted after the reform was implemented, and 39 percent had previously worked. Seventy-two percent of respondents had 11 or more years of healthcare experience. Four of the five managers (two women, three men) had nursing credentials, while one had a medical background.

From May to December 2015, all interviews were performed in person at the workplace. For this study, a semi-structured interview script was created and employed to boost reliability. The script covered five topics that have been linked to organizational innovation in prior studies: organizational culture, goals, change and improvement work, personal qualities, and leadership.

The content of the interviews was analyzed and in a first step changes in practices were identified. In the second step, these practices were categorized into three types of innovations: (1) service innovation, (2) process innovation, and (3) organizational innovation.

For the authors, a receptive environment for change is characterized by "an ability to absorb new knowledge, strong leadership, clear strategic vision, good relationship management, visionary staff in key positions, a climate conducive to experimentation and risk taking, and effective data capture systems." Moreover, the discussion of the results emphasizes a strong link between innovation and leadership in the public sector. The main resource in many public services, such as in primary health care, is undoubtedly the experience and problem-solving capacity of staff. Internal staff sources for motivation, such as autonomy, experience, and pro-social behavior to safeguard patient needs, tend to have a greater impact on behavior over time than external factors such as pay.

After the study and categorization, four primary characteristics of the practices emerged: (1) learning management, (2) performance monitoring, (3) adaptation to requirements, and (4) collaboration with others. This qualitative study identifies key qualities that, in practice, enable primary care innovation. Learning-oriented corporate culture and climate, entrepreneurship knowledge, leadership, transdisciplinary collaboration, performance measurements, and adaptability are all linked to innovation creation. The ability of the management to convert and integrate public policy with practitioner understanding and beliefs can help to encourage primary care innovation. The willingness, ability, and chance to innovate are, in the end, the most important aspects of innovation.

7.3.3 Analysis of Article 3

In low- and middle-income nations, strengthening health systems is critical. The abilities of health care executives enable the innovation that ensures the systems' durability, adaptability, and functionality.

The purpose of this study was to find out how managers felt about their engagement in and introduction of sophisticated facilities as part of the MIND project in the Western Cape region of South Africa. The MIND project is testing two techniques for integrating depression and hazardous alcohol use counseling in chronic illnesses including AIDS and diabetic patients. Three problem-solving motivational interviewing sessions are conducted by experienced lay counselors in primary health facilities as part of the intervention.

Individual interviews with these managers were undertaken as a study method in a convenience sample of 34 facilities with clinical services and sub-district level, totaling 28 people interviewed in a private workplace.

It was evident that managers play a vital role in incorporating innovation into routine practice and that the subject is under-researched. The constraints to the adoption of innovation can be defined in four main factors: Staff personalities, attitudes, and behavior that lead to change resistance; excessive workload due to resource limits and frequent policy changes that cause change resistance; and inadequate communication within healthcare organizations.

Technical abilities in participative management, communication, community participation, program monitoring, and evaluation skills were emphasized as measures used by managers to counteract these problems. Similarly, the non-technical skills listed were defined as the example and positive attitudes of the manager, understanding the personality of the team, establishing a relationship of trust, influencing the organizational climate, and the perceptions regarding innovation.

It has been determined that non-medical health professionals, as well as managers, require training and development. Managerial behavior and capacity appear to aid in coping with employee resistance to change. Poor communication around the introduction of new innovations to frontline personnel was noted as one of the health system's major flaws, as it appeared to need additional work to execute and maintain. A major roadblock to success appears to be resistance to change and a lack of common vision for improving health outcomes. Managers also reported significant mental health and stress difficulties in their roles, not simply in relation to implementing the innovation. Therefore mental health promotion and resilience building should be included in the intervention designed.

Limitations of the study were identified as selection bias due to participants being excluded who did not return the researchers' contacts and social desirability, which may have been present, and that managers' experiences are defined by resource constraints and this varies within a province itself.

7.3.4 ANALYSIS OF ARTICLE 4

The authors compared the attitudes of professionals working in primary care and those working in outpatient clinics in this study, which focused on the perceptions of health professionals working in a publicly funded health care system. Several strategies have been developed which emphasize the importance of patient-centered care and the new demands that are associated with this type of care. Much effort has been put, for example, into digital services that potentially improve information sharing and effective communication with patients.

An example of this type of service is PAEHRs, which have the potential to improve health care delivery and health outcomes with benefits such as better feedback and understanding of health information, greater adherence, and better communication between patients and healthcare professionals. However, there are differing reports about the effects that PAEHRs have on healthcare professionals and their daily work.

The study was conducted as an exploratory case study, with six participants in a workshop, six interviews, and 146 healthcare professionals responding to a survey. To begin with, the authors held a workshop to determine the anticipated benefits and downsides of PAEHR. The PAEHR project manager, an eHealth strategist, a communications director, two physicians, and the system owner were among those who attended the workshop. Participants were chosen based on their understanding of PAEHR and their capacity to represent various stakeholders within the company.

According to the findings, primary care physicians believe that enhanced patient information sharing can be advantageous. The majority of them see this as a means to improve patient compliance and clarify crucial information. This is also considered as a way for the patient to have more say over what is recorded in the primary health care records.

7.3.5 ANALYSIS OF ARTICLE 5

It is about HOBE Plus, a virtual professional community of practice (VCoP) designed to encourage and facilitate innovation in primary health care in the Basque area of Spain (Abos et al. 2013). It was designed for all primary care providers working for the Basque Public Health Service (Osakidetza) in Bizkaia and Araba regions. HOBE

PLUS is a VCoP that combines innovation management with idea nurturing in order to execute new ideas in primary health care.

The paper's goals were to assess the process of establishing and executing a VCoP open to all Osakidetza health professionals, as well as acceptance, participation, and use of the VCoP in the first 15 months after its October 2011 introduction. In addition, the VCoP's usefulness was assessed by a survey that solicited feedback from the experts engaged. The authors employed a case study approach, drawing on data from the VCoP technology platform as well as a poll of HOBE+ users. All primary care staff (all professional categories) in the provinces of Araba and Biscay in the Basque Country (Spain), who represent VCoP's target users, were the target population.

Even in nations with universal, tax-funded coverage and quasi-public monopoly providers (such as Spain's National Health Service), the urgent problems of chronic diseases and multimorbidities, as well as rapid technological and societal developments, jeopardize health care systems' long-term viability. In this industry, innovation is seen as crucial to long-term viability. The project demonstrated that it is possible to construct a virtual CoP for primary care innovation, in which experts from many categories contribute innovation ideas that may be adopted in the future.

HOBE + (Hobe, derived from the Basque word "Hobekuntza," which means "improvement") is an innovative online VCoP for primary care practitioners that was created to generate, recognize, and support innovation and improvement in the Basque Health System. Hobe+ is a platform for primary care workers in Osakidetza, Bizkaia, to identify, propose, define, and develop ideas that arise in their everyday work. It was launched in October 2011 by four main districts of Osakidetza and the Basque Institute for Health Innovation (O+berri). This project included the entire innovation management process, from idea generation through implementation. The Araba primary health care district joined the platform in June 2012.

On the other hand, from the outset of the project, the openness of HOBE + to other participants such as hospital personnel, patients, and other parties interested in the topic and belonging to the external community was discussed. The authors of this study consider it as a complicated topic, and they recognize that when applying CoP knowledge to other agents, this metric must be balanced.

The study, according to the authors, bears the limitations of a descriptive technique and lacks comparability in many ways. Because no analogous programs in primary care have been identified anywhere else in the globe, caution should be exercised in interpreting and extending the findings. Another significant flaw is the lack of an assessment of the economic benefit of implemented ideas in proportion to the cost of their development and implementation. The majority of the concepts that have been adopted have to do with business processes, process refinement, protocols, and time savings. Those relating to the implementation process include logistics, management changes, and inventories, all of which will take more time to fully implement and assess their impact.

7.4 INNOVATION MANAGEMENT AND PRIMARY HEALTH CARE

According to Canongia et al. (2004), one of the most significant features of competitive businesses is the ability to innovate. To this end, the systematic search for radical innovations, that is, those capable of creating new markets and generating rapid productive expansion and economic growth, as well as incremental innovations, identified with continuous improvement processes, or "doing better what was already done," is critical for business survival and government efficiency.

The current definition of competitiveness includes not only a company's or product's superior performance or technical efficiency, but also the ability to develop systematic processes for seeking new opportunities and overcoming technical and organizational obstacles through the production and application of knowledge. The goal of innovation management is to bring together the methods and instruments, as well as methodologies and organizational forms, that can secure an organization's ability to innovate.

Specifically in the target articles of this study, it is verified that there is an enormous interest in investing in innovation in public primary health care, which translates into finding new ways or improving protocols of patient care, improvements and advances in methodologies and management actions, in addition to adopting new technologies, aiming at advances in the quality of services provided and reduction of costs and expenses. It is noticed that such interest in innovation is better registered scientifically in international foreign cases and, even so, in a very small number, as it was verified when searching for articles in the scientific databases used.

The main premise with respect to innovation applied to PHC analyzed here refers to the promotion and intelligent application of ideas, processes and protocols for patient care, which are relevant to increasing the quality of services provided by the primary health care unit that adopts it, primarily for the benefit of the community served. This is one of the main challenges regarding innovation management in this category of organizations.

7.5 FINAL CONCLUSION

The challenges of primary health care (PHC) include several factors, including diagnosis and treatment, medical prescription, management of chronic diseases and multimorbidity, compensation and income system, coordination and integration of systems, and continuity of care.

Innovation management can be applied to primary care in public health with policies and guidelines for improving quality and care in PHC. This research has identified studies that point out innovative paths, with emphasis on innovations that enhance the exchange of tacit knowledge and the creation of new knowledge applicable to the reality of care in public health, creation of tools and processes to support clinical decision-making and quality monitoring and evaluation. Innovation management was also pointed out as necessary to improve the quality of training of PHC physicians professionals, to implement measurement and performance systems, and to integrate clinical care between PHC and hospital services, strengthening the coordination of patient care as they access different levels of care.

It was found that there is a lack of studies on innovation management in PHC. Thus, for future studies, we suggest themes on the management of health networks that seek to relate and apply the concepts, constructs, and practices of innovation management in the field of primary care as a strategy for its strengthening and consolidation in the context of public health policies.

REFERENCES

Abos Mendizabal, G., Nuño Solinís, R., & Zaballa González, I. (2013). HOBE+, a case study: A virtual community of practice to support innovation in primary care in Basque Public Health Service. *BMC Family Practice*, *14*(1), 168. doi:10.1186/1471-2296-14-168

Avby, G., Kjellström, S., & Andersson Bäck, M. (2019). Tending to innovate in Swedish primary health care: A qualitative study. *BMC Health Services Research*, *19*(1), 42. doi:10.1186/s12913-019-3874-y

Brooke-Sumner, C., Petersen-Williams, P., & Kruger, J. et al. (2019). "Doing more with less": A qualitative investigation of perceptions of South African health service managers on implementation of health innovations. *Health Policy and Planning*, *34*, i. 2, 132–140.

Canongia, C. (2004). *Gestão do Conhecimento e a Competitividade – Reflexão*. CGEE.

Cavalcante, P. (2017). *Inovação no setor público: teoria*, tendências e casos no Brasil.

Chesbrough, H.W. (2006). *New puzzles and new findings* (H.W. Chesbrough, W. Vanhaverbeke, & J. West, Eds.; pp. 15–34). Oxford University Press.

Kitchenham, B. & Charters, S. (2007). Guidelines for performing systematic literature reviews in software engineering, Technical Report EBSE 2007-001, Keele University and Durham University Joint Report.

Li, X., Krumholz, H., & Yip, W. et al. (2020). Quality of primary health care in China: Challenges and recommendations. *The Lancet*, 395, 1802–1812.

Machado, C.V. (2001). *Novos modelos de gerência nos hospitais públicos: as experiências recentes* (C.V. Machado, Ed.). Physis.

Macinko, J., Starfield, B., & Erinosho, T. (2009). The impact of primary healthcare on population health in low-and middle-income countries. *The Journal of Ambulatory Care Management*, *32*(2), 150–171.

Silva, S.A., Baitelo, T.C., & Fracolli, L.A. (2015). Avaliação da Atenção Primária à Saúde: a visão de usuários e profissionais sobre a Estratégia de Saúde da Família. *Revista Latino-Americana de Enfermagem*, 23, 979–987.

Silveira, J.P.M., & Zonta, R. (2020). Experiência de reorganização da APS para o enfrentamento da COVID-19 em Florianópolis. *APS Em Revista*, *2*(2), 91–96.

Starfield, B. (1998). *Primary care: Balancing health needs, services, and technology*. Oxford University Press.

Wass, S., & Vimarlund, V. (2019). Same, same but different: Perceptions of patients' online access to electronic health records among healthcare professionals. *Health Informatics Journal*, *25*(4), 1538–1548. doi:10.1177/1460458218779101

Whittemore, R., & K, K. (2005). The integrative review: Updated methodology. *Journal of Advanced Nursing*, Dec, *52*(5), 546–553.

8 The Spinner Innovation and Knowledge Flow for Future Health Scenarios Applications

Ronnie Figueiredo, Marcela Castro,
Pedro Mota Veiga and Raquel Soares

CONTENTS

8.1 Introduction 129
8.2 Methodology 130
8.3 Systematic Review and Meta-Analysis 131
    8.3.1 Systematic Review 131
    8.3.2 Meta-Analysis 136
8.4 How Can Spinner Innovation and Knowledge Flow Help Human Resources (HR) to Become Digital Human Resource Management? 138
8.5 Conclusion and Discussion 141
8.6 Implications 142
8.7 Future Research Directions 143
Acknowledgements 143
Appendices 144
References 145

8.1 INTRODUCTION

Digitalisation is now no longer a matter of choice but rather an imperative for organisations, presented with a set of opportunities and challenges that may and should impact their human resource management (Agrawal & Narain, 2018; Vardarlier, 2020). Given that people constitute the most important asset of any organisation, it becomes essential to invest in digital human resource systems (Pulyaeva, Kharitonova, Kharitonova, & Shchepinin, 2019), and adopting digital human resource management (digital HRM) rather than merely human resource management (HRM) (Vardarlier, 2020).

Through the spinner innovation flow or the spinner flow, companies such as knowledge-intensive business services (KIBS) firms may reach beyond the internal processes and equally act to accelerate the transformation process turning startups into SMEs. The spinner flow therefore represents a model designed to provide support to startups in this growth process, accelerating the passage to stages of maturity. This spans three dimensions configured in accordance with the spinner innovation model (Figueiredo & Ferreira, 2019): (i) knowledge creation; (ii) knowledge transfers and (iii) innovation through intensive knowledge usage. The present study seeks to grasp in what ways this model may impact on the development of new startups and their respective generation of innovation. Within the scope of fostering a new digital culture for business, the participation and engagement of human resources hold particular importance. Thus, this study answers the following research question: How can the spinner innovation flow help human resources to become digital human resource management?

The methodology proposed incorporates a systematic literature review based on the Scopus database and approaching studies on the main subject – "spinner innovation flow in digitalized HR" and a meta-analysis hierarchical classification. We here adopted the preferred reporting items for systematic reviews and meta-analyses/PRISMA methods.

The structure of this chapter is as follows: the second sectiondescribes the methodology applied; the third section sets out the systematic literature review and the meta-analysis undertaken; the fourth section provides a discussion of the results and conclusions before the fifth and sixth sections identify the implications and recommendations for future research, respectively.

8.2 METHODOLOGY

This scientific research approaches studies on the core subject, the "spinner innovation flow in digitalized HR", based on a systematic and hierarchical classification (Barroso et al., 2003; Cook, Mulrow, & Haynes, 1997; Dhammi & Haq, 2018; Echer, 2001; Gisbert & Bonfill, 2004; Robleda, 2019; Sardi, Idri, & Fernández-Alemán, 2017; Wright, Brand, Dunn, & Spindler, 2007), herein capturing new thinking on the theme reviewed over the course of time (Debajyoti & Lorusso, 2018; Evans & Pearson, 2001).

We took into consideration various studies that describe the steps necessary to undertaking an excellent and high-quality systematic literature review (Debajyoti & Lorusso, 2018; Dhammi & Haq, 2018; Ghezzi & Dramitinos, 2016; Khan, Kunz, Kleijnen, & Antes, 2003; Wright et al., 2007), with the study, for its initial planning structure, adopting the Preferred Reporting Items for Systematic reviews and Meta-Analyses/PRISMA methods (Page et al., 2021; Moher, Liberati, Tetzlaff, Altman, & Group, 2009).

The study developed in accordance with the QUOROM Statement, which is a protocol containing a flow diagram and a list of essential items (Liberati et al., 2009; Moher et al., 2009). While there are minor discrepancies as regards the stages and procedures, the majority of studies share many similarities and stem from the same context and structure (Dhammi & Haq, 2018; Wright et al., 2007).

This method was originally tailored to the health sector (Liberati et al., 2009) but easily and successfully adapts to other areas (Moyson, Raaphorst, Groeneveld, & Van de Walle, 2018), such as digital transformation.

A group of researchers developed this protocol and ensuring great methodological rigour and high quality in any systematic literature review (Debajyoti & Lorusso, 2018).

For the study in question, we adopted items 5–16 of the PRISMA method (Liberati, Altman, Tetzlaff, Mulrow, Gøtzsche, Ioannidis, & Moher, 2009), in accordance with the checklist detailed in Appendix 1.A.

The management platform for the references applied was that produced by Mendeley, in accordance with Dhammi and Haq (2018). The bibliographic study references included were also subject to analysis in order to identify any eventually relevant studies that may have been discarded in the earlier phases of exclusion (Dhammi & Haq, 2018).

In order to graphically map the studies, along with the division of their themes, we made recourse to the software programs: Microsoft Excel and Microsoft Word (both part of the Office 365 package, version 2019) and VOSViewer (Version 1.1.16). The definition of the propose scale for the evaluation of the spinner innovation flow stemmed from the reference (Commission Decision), technology readiness levels (TRL), (Mankins, 1995), Appendix 1.B, with a focus on efforts to apply the research to practice (RtP).

In accordance with the PRISMA checklist (5–16), we describe the items listed in Table 8.1 in accordance with all the rigour of the method presented.

In relation to the flow of identified and selected articles, Figure 8.1 illustrates this process.

The PRISMA method flow diagram started out by identifying articles according to the criteria and search methods of the Scopus database. The first search yielded a total of 214 articles. The selection phase approached these articles in terms of excluding duplicates and/or those inappropriate in content, thus defining a total of 200. Following this step, 150 articles emerged as eligible given they provided for full access. For inclusion, we rejected a further five for not adhering to the study and therefore leaving 145 articles for the systematic literature review.

Pay-to-access studies or those with incomplete texts were subject to automatic exclusion in the Cochrane data base, as were any duplicates (Dhammi & Haq, 2018; Sardi et al., 2017). Furthermore, the selection process also extended to analysis of the article title (Dhammi & Haq, 2018), thereby further defining the corpus.

To evaluate the quality of studies, this took into consideration the identification of the internal and external constructions, the validity and reliability of the conclusion as well as other characteristics of the respective study, such as the tools and measurements, procedures and techniques employed (Dhammi & Haq, 2018).

8.3 SYSTEMATIC REVIEW AND META-ANALYSIS

8.3.1 Systematic Review

The initial approach presented by Agrawal and Narain (2018) defines how the digitalisation process now impacts on practically every organisation worldwide and placing major pressure for change on organisations. This details how essential it is for managers to study the implications of digitalisation for both their organisation and its employees. The current scenario emphasises the importance of managing people and organisational issues to digital transformations.

TABLE 8.1
Items of Study Based on the PRISMA Checklist

Section/Topic	#	Checklist Item
Methods		
Protocol and registration	5	The model of analysis and the inclusion criteria arise from those referenced in the dimensions and variables of the Spinner innovation model, available at www.spinnercentre.com (Spinner Innovation Centre).
Eligibility criteria	6	This study considered indexed English language articles that are fully available.
Information sources	7	The identification of these studies took place through means of searching an electronic database and reading the list of article summaries of their main themes (Dhammi & Haq, 2018; Gisbert & Bonfill, 2004; Wright et al., 2007). This applied a restriction in terms of only accepting English articles with any non-English articles subject to translation. We applied this search to the Scopus database with the final search taking place on December 13, 2020.
Search	8	In order to define and certify that the studies feature all of the terms applied in the research (Gisbert & Bonfill, 2004; Sardi et al., 2017), we here deployed the Boolean search ("AND" or "OR"). We applied the following terms of research for searching the registration and databases of articles: (TITLE-ABS-KEY (digital AND innovation AND process) AND TITLE-ABS-KEY (human AND resources) OR TITLE-ABS-KEY (hr). Another parameter applied in this research was the inclusion of different types of alternative search terms (Sardi et al., 2017), such as searches of the text (TX), by author (AU), title (TI), terms of the subject to (SU), journal title/source (SO), summary (AB), ISSN (IS) or ISBN (IB).
Study selection	9	We evaluate eligibility on an independent basis but in a standardised, non-blind, approach by two reviewers. This resolved any disagreement between the revisors by consensus following the presentation of their considerations within the "spinner flow" framework. This defined the selection of criteria within the objective of minimising bias and excluding possibly irrelevant items that might impact on the study's quality (Wright et al., 2007) and with the purpose of identifying the most significant research outputs (Debajyoti & Lorusso, 2018; Dhammi & Haq, 2018; Gisbert & Bonfill, 2004; Sardi et al., 2017).
Data collection process	10	We produced a flow of extracted data and tested this on 10 randomly included studies based on the Spinner flow. One review author extracted the data from the studies included while the second author proceeded to verify the extracted data. The resolution of any disagreements followed discussions between the two review authors; whenever no agreement was obtainable, they turned to a third author for a casting decision. At the end of this process, all authors responded to the extracted data by demonstrating their agreement.

TABLE 8.1 (Continued)
Items of Study Based on the PRISMA Checklist

Section/Topic	#	Checklist Item
Data items	11	The information extracted from each article spanned: (1) digital innovation; (2) the spinner innovation model; (3) startups; (4) small and medium-sized companies (SMEs) and (5) knowledge-intensive business services (KIBS). To grasp the amplitude of this theme, we applied a temporal based selection, taking into account studies published through to 2020. Following the initial search, we filtered the results within the scope of identifying the most significant research (Debajyoti & Lorusso, 2018; Dhammi & Haq, 2018; Gisbert & Bonfill, 2004; Sardi et al., 2017).
Risk of bias in individual studies	12	In order to verify the validity of the data collected (articles), pairs of authors made independent reviews of the contents, considering diverse analytical flows. They determined the appropriateness of the analysis in accordance with the flow under presentation, the spinner flow.
Summary measures	13	The maturity phases in the technological/digital business model served as the main measure for evaluation in accordance with the spinner flow referenced by the TRL scale (1-9).
Synthesis of results	14	This study is a proposal for the organisation of human resources when striving to undertake the transformation from startups to SMEs over whatever period appears appropriate and viable. This seeks to stimulate human rights and provide incentives and support for professionals who lost their positions in startups during or after the COVID-19 period. Thus, as an initial and unpublished proposal, this requires testing and measuring in terms of the results produced.
Risk of bias across studies	15	For article subject to provisional selection, we made the comparison between the abstract and the items making up the spinner flow in order to verify their adherence and eventual contributions in addition to reducing the possible effects of generalisation.
Additional analyses	16	The review of the protocol through means of a blind review process took place, thereby boosting the effectiveness of the proposal and adherence to the origins of the flow, the spinner innovation flow as well as describing the flow applied to the systematic literature review (Figure 7.1).

Source: Authors.

The effects of digitalisation extend to various impacts on the economy as a whole hence generating both enormous opportunities and challenges for companies. In the current world of globalisation, digitalisation no longer represents a choice but is rather an imperative for all companies in every sector.

In the digital era, the concept of technology underwent powerful changes that led to the reconsideration of all the criteria by which companies may actually

FIGURE 8.1 Flow of information through the different systematic review phases.

be able to implement innovation. In relation to the last century, the tools available to companies have changed a great deal; however, there have above all been transformations in the management strategies susceptible for deployment in order to adopt paths leading to innovation and development as the study by Rangone (2020) highlights.

In recent discussions around the development of new information and communication technologies, we may observe a shift in paradigm reflecting the transition of work from a real, physical dimension into a virtual world as explained by Doloreux, Shearmur, and Rodriguez (2018).

In the last few years, the advances in infrastructures and technology implemented according to emerging paradigms, such as Cloud computing, Future Internet and SaaS (Software-as-a-Service), are driving the business system area to undergo progressive evolution resulting in significant transformations.

According to the approach by Novikov and Sazonov (2020), the development of digital transformation involves the effective and high-quality updating of product and business processes based on the widespread introduction of advanced innovations, a position reaffirmed Angelucci, Missikoff, & Taglino (2011) in keeping with the findings of their study.

Hence, we may understand the rising interest in research into big data, especially digital applications, over the last decade. The adoption of the paradigms behind cloud computing, artificial intelligence and the Internet of Things (IoT) in the business field potentially generates various opportunities for the human resource sector beyond the contribution made to continuous and systematic innovation in the business environment (Elhoseny et al., 2018).

This issue receives further emphasis by Laužikas and Miliūtė (2020) who maintain that, in addition to artificial intelligence and the modern technologies, talented communicators remain essential intermediaries in external communications and their talents require refining for a more technologically intuitive style of communication.

However, studies carried out by Ranky (2003) report that during the business digitalisation process, company leaders are discovering that, beyond minimising the costs of design and manufacturing and maximising quality, they are able to return competitive advantages through introducing new and innovative products that satisfy individual consumers on a global basis.

As a result, Pulyaeva, Kharitonova, Kharitonova, and Shchepinin (2019) highlight how within the context of digital economy training and the continuous introduction of innovations, coupled with a reduction in the permanent management costs of organisations through the automation of business processes, the automation of HR management takes on particular relevance given how people make up the core capital of any organisation.

According to Simonova, Lyachenkov and Kravchenko (2020), there is a great need to establish a strategic development model for companies deploying information systems that integrate the corporate knowledge base alongside intelligent systems for supporting decisions that shall eventually transform into individual human development.

Indeed, the findings of Jacobs and Webber-Youngman (2017) advocate the vital nature of organisations and individual operations having access to a platform with information relating to the technologies for consideration in future research and development.

In turn, Šuman, Poščić and Marković (2020) return results demonstrating the need to integrate the information and communication technologies (ICTs) into management, alongside other adaptations to the ongoing global changes while emphasising the importance of the leadership's motivation to adhere to planning and governance processes appropriate to high-quality projects.

Furthermore, the results attained by Edirisinghe (2019) define how the exponential rate of innovation in computation devices, communications technologies and application technologies is advancing swiftly and making increasing resource to intelligent technologies. This change in paradigm shifted from mobile computing to pervasive computing and then onto intelligent technologies.

The studies by Lepore, Metallo, Schiavone and Landriani (2018) stress how the adoption and effective utilisation of digital and computerised systems and registers in companies reflects a crucial means of boosting quality, security and the general results of human resource management.

The findings of Marcon et al. (2019) point to the scope for improving a product-service whether in terms of innovation processes, enabling coordination and cooperation as regards the results, which may themselves incorporate new contexts, digital technology functions and aggregate value through the provision of digital solutions that avoid potential market obstacles.

In the economic sphere, the service sector has recorded exponential growth throughout several decades and become the key source of wealth to some national economies. In turn, the ICT revolution brought about a transformation in production and fostered interconnected networks, and the globalisation of knowledge and business, thus expanding the access of clients and consumers of goods and services.

Similarly, the actors in these processes are no longer merely purchasers but have rather become sources and generators of ideas for innovation, potentially able to participate and contribute to the different stages of projects as highlighted in the study results produced by González & Nuchera (2019).

In the findings of Del Giudice, Garcia-Perez, Scuotto and Orlando (2019), managers may tend to concentrate on the installation of technology rather than planning a socio-technical system able to take into consideration the objectives of the organisation and nurturing authentic participation.

In terms of the environment, monitoring provides one of the leading tools applied to assist in understanding the natural environment and its respective conditions and thereby informing on the development of the most appropriate actions. Hence, Maffey, Irvine, Reed and van der Wal (2016) maintain that while the proposed digital system approaches the barriers perceived to better monitoring of the environment, this also reveals underlying concerns about the recourse and final purposes of monitoring. Such concerns reflect the scope for conflict between the scientific and management perspectives, which need taking into account and dealing with so that monitoring becomes more widely accepted as a tool for informing the management of human resources.

Furthermore, the results set out by Schmitt (2019) highlight how the capacities for their technical, information and social interpretations have to be able to accommodate diverse knowledge management (KM) models and cumulatively summarise a broad range of concepts and related perspectives.

In this process, the renowned model by Nonaka of socialise, externalise, combine, internalise and Ba is repurposed and extended to suggest a correspondence with distinct digital ecosystems, totally aligned with the diversity of the generative attributes therein introduced.

Within the scope of the Msiska and Nielsen (2018) study, the concepts of generative and relational technologies both focus on innovation. Furthermore, both relate to how innovation emerges as a result of exploring new ideas and opportunities for human actors.

In the global context, digital technology has the potential to reduce the distance between all of the stakeholders interested in the management as detailed in the study by Kataria and Ravindran (2018).

Hence, we may grasp how the findings by Avasilcəi and Rusu (2015) advocate fostering the development of innovative products and new technologies by companies now running open innovation systems, making recourse to online company platforms or social networks, generating and sharing solutions and creative ideas as the precise means of responding to the challenges companies face directly in terms of change.

Gürdür, El-khoury and Törngren (2019) describe various initiatives designed to actively drive change before identifying the need to make available the tools and human resources necessary. Furthermore, the level of cultural readiness, which concentrates on the acceptance of decisions taken based on data, ranks between high and very high. Simultaneously, the level of readiness of the information technology systems ranges between average and high with the exception of the telecommunications sector.

8.3.2 Meta-Analysis

Analysis of the co-authors according to the five leading countries demonstrates the United States ranks as the country with the largest total of connections, totalling 37 documents accounting for 273 citations (Table 8.2). The country with the second largest number is the Russian Federation, with 22 documents that received 26 citations followed by Germany on 15 documents that gained 77 citations. In turn, the United Kingdom emerges in fourth place, with 12 documents generating 77 citations

TABLE 8.2
Number of Clusters per Country: Documents and Citations

Cluster	Countries
1	Algeria, Colombia, Iran, Kenya, Mexico and United States
2	Australia, Brazil Canada, France, Netherlands and Thailand
3	Egypt, Saudi Arabia, Singapore, South Korea and Switzerland
4	Denmark, Ireland, North Macedonia and United Kingdom
5	Austria, Germany and Hungary
6	Italy and Romania
7	Israel

Source: Authors.

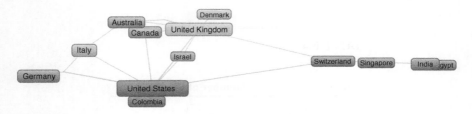

FIGURE 8.2 Representation of the clusters – Documents and citations by country.

TABLE 8.3
Co-occurrence – Keywords

Keywords	Frequency
Human Resource Management	42
Human	36
Innovation	32
Humans	28
Digital Storage	18

Source: Authors.

with Italy in fifth position on ten documents and 24 citations. Figure 8.2 depicts the respective clusters.

Analysis of the co-occurrence of keywords identifies "human resource management" as the single most common with a 42 repetition frequency (Table 8.3). The second term is "human" that gains a total of 36 repetitions followed by "innovation" on 32 repetitions. In turn, in fourth place comes "humans" accounting for a frequency of 28 repetitions with the fifth place taken by "digital storage" that amasses a total of 18 repetitions. Figure 8.3 sets out a portrayal of the clusters resulting.

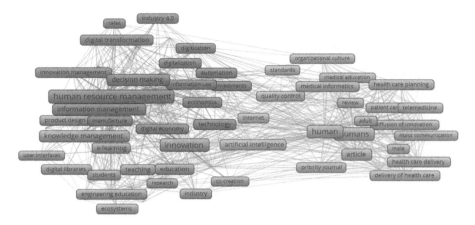

FIGURE 8.3 Representation of clusters – Co-occurrence – Keywords.

TABLE 8.4
Citations

Author	Number of Citations
Nobre, C. A.	171
Silva, J. S.	171
Sampaio, G.	171
Borma, I, S.	171
Cardoso, M.	171
Castilla-Rubio, J.C.	171

Source: Authors.

Analysis of these citations identifies how the leading six authors all return the same number of citations, with each individually receiving 171 citations (Table 8.4). Figure 8.4 depicts the representation of the other authors and their respective connections.

Analysis of the institutions conveys how the five leading universities all register the same number of published documents even while differing in terms of the citations returned by these documents (Table 8.5). Figure 8.5 depicts the other institutions and their respective connections.

8.4 HOW CAN SPINNER INNOVATION AND KNOWLEDGE FLOW HELP HUMAN RESOURCES (HR) TO BECOME DIGITAL HUMAN RESOURCE MANAGEMENT?

The original objective of the spinner innovation and knowledge flow (SIKF) was to support startups in their growth processes and accelerating their passage to stages

Spinner Innovation and Knowledge Flow for Future Health Scenarios

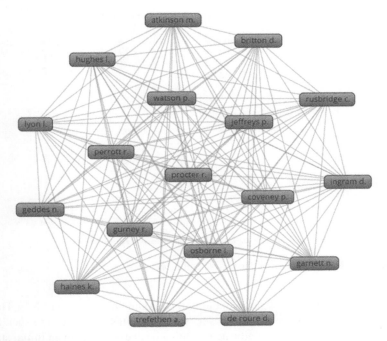

FIGURE 8.4 Representation of clusters – Connected authors.

TABLE 8.5
Institutions

Institutions	Number of Documents
South State University	02
Cheng Shiu University	02
National Changhua University of Education	02
National Penghu University of technology	02
National Yanlin University of Science and Technology	02

Source: Authors.

of maturity and the obtention of the scale of a small and medium-sized company (SME). The SIKF spans a set of three dimensions, configured in accordance with the spinner innovation model (Figueiredo & Ferreira, 2019; Figueiredo, Ferreira, Silveira, & Villarinho, 2019; Figueiredo & Bahli, 2021), correspondingly the creation of knowledge, the transfer of knowledge and innovation through the intensive usage of knowledge.

Each dimension presented incorporates an evolving set of measurements across three levels (1–9 levels), approaching (1–3) creation of knowledge, (4–6) transfer of knowledge and (6–9) innovation. Therefore, each measurement serves to evaluate the

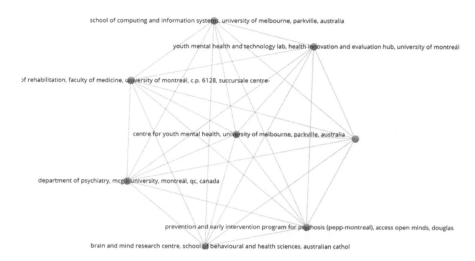

FIGURE 8.5 Representation of clusters – Institutions.

maturity of the startup through to attaining the maximum level of an SME. The creation of knowledge dimension thus covers stage 1 when startups require such knowledge creation as the means of advancing with a business model in an initial stage of market entry.

The transfer of knowledge spans stage 2 and simultaneously constitutes an intermediate process that contemplates the "trade-off" point in time when startups need to either progress to stage 3 or regress to stage 1 and therefore need to accelerate the new knowledge acquisition process through the intermediation of other sources, co-creating knowledge-intensive solutions, generally involving recourse to specific firms, whether technologically or professionally based, collectively referred to as "KIBS firms" or Knowledge-Intensive (Business) Services.

The innovation dimension bestows a level of maturity on the SMEs, developing new knowledge-intensive solutions to generate value and future business. At this stage, we may state the startups have grown and assumed the role of SMEs in the consumer market. Then, a leading question emerges over whether to remain in the host market or expand into international markets, deepening the capillarity of the business.

In the post-COVID SME context, the target remains to develop new startups that require launching into markets through means of producing innovative products and/or services. Hence, this requires the effective participation of human resources and installing a new digital culture into businesses.

According to Vardarlier (2020), human resource management (HRM) has now evolved into digital human resource management (digital HRM). Companies now run digital human resource systems within the scope of their core human resource functions. While companies launch many innovations in the digital field to consumers, their human resource management also applies similar levels of innovation to their employees and/or candidates.

Spinner Innovation and Knowledge Flow for Future Health Scenarios

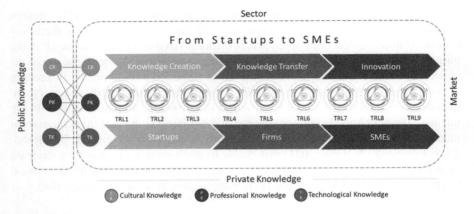

FIGURE 8.6 Spinner knowledge flow.

This reflects how, beyond their involvement in digital HRM applied to internal processes, knowledge-intensive business service providers, KIBS firms, may extend their support for startups to accelerating the SME transformation process through the intermediation of deploying the spinner innovation flow.

The scope of digital HRM has advanced into the acquisition of new knowledge, competences and/or soft or hard skills that support the creation of knowledge, the initial phase of any startup. As a second approach, this may monitor the same knowledge acquisition process while nevertheless focusing on stage 2, replacing the KIBS firms by knowledge specialists (mentors) that support the production and innovation processes. As another option, the spinner flow is susceptible for application by outplacement firms, collaborating following the launch of a startup in accelerating the maturity of the business model through to obtaining stage 9, as a PME. Figure 8.6 provides a complete overview of this approach.

8.5 CONCLUSION AND DISCUSSION

According to the results of the systematic review, innovation and digitalisation are fundamental to the generation of value as approached in a range of studies (e.g. Doloreux, Shearmur, & Rodriguez, 2018; Novikov & Sazonov, 2020; Ranky, 2003), furthermore highlighting the importance of integrating human resources into a new digital business culture (Agrawal & Narain, 2018; Elhoseny et al., 2018; Pulyaeva, Kharitonova, Kharitonova, & Shchepinin, 2019); Lepore, Metallo, Schiavone, & Landriani, 2018). The results of the meta-analysis developed demonstrates that as regards the number of documents and citations per national cluster, the United States takes first place with the largest connection index (37 documents and 273 citations). As regards the co-occurrence of the keywords, "human resource management" emerges with the greatest level of frequency, with 42 repetitions. As regards the citation analysis, this identifies six authors as generating the greatest frequency, with 171 citations apiece. Finally, as regards the institutions, the five leading universities all return the same number of published documents (2).

This study demonstrates the importance of the "spinner flow" to the human resource areas of organisations, acting as a model that supports startups in their process of evolution to becoming small and/or medium-sized companies, above all in the post-COVID reality. Therefore, digital innovation stands out as one of the decisions about which any organisation's human resource managers should ponder in keeping with the need for adaptation. Correspondingly, the spinner flow identifies a set of three dimensions – creation of knowledge, transfer of knowledge and innovation with the intensive usage of knowledge – configured in accordance with the spinner innovation model (Figueiredo & Ferreira, 2019; Figueiredo, Ferreira, Silveira, & Villarinho, 2019; Figueiredo & Bahli, 2021) that is able to assist organisations as they navigate such processes. This model contributes towards progress in terms of the theoretical integration and development of the field of digital human resource management.

8.6 IMPLICATIONS

In an era of digital disruption, there are various practical implications, particularly as regards the utilisation of different technologies within the framework of the spinner innovation flow in digital human resource management. The deployment of social media tools, such as Facebook, Twitter and LinkedIn, foster the creation of knowledge in terms of recruitment processes and the transfer of knowledge as regards internal market practices.

Publishing details about vacancies on company pages and encouraging employees to help in sharing them boosts the number of candidates attracted, verifying their social media profiles and the posts of potential members of staff may assist in determining whether such are suitable future employees. The recourse to social media platforms for communicating with employees, adopting internal marketing practices helps in improving the visibility of the employer and the satisfaction of the employees.

The implementation of technologies associated with data science, machine learning, artificial intelligence and big data accelerate processes of creating and transferring knowledge while stimulating outputs in the form of new products and services based on knowledge. The implementation of these technologies in HR management enables the automation of many routine operations and freeing recruitment processes, for example, while assisting in determining the optimal levels of labour required for particular projects.

The utilisation of automated curriculum vitae selection processes and interviews with automated robots may reduce a group of several hundred candidates to a list of the dozen or so with greatest potential and thus enabling greater speed and accuracy in recruiting the staff most appropriate to functions at the company.

The deployment of Cloud-based technologies also fosters the transfer of knowledge and accelerating new knowledge processes resulting from diverse sources and thereby enabling innovation. The migration of technologies to the Cloud centralises business data, especially HR-related data and ensuring the flexibility and ease of its updating, facilitating the ease of access to HR technologies and providing every member of the HR team with the scope for accessing relevant data at any time and in any place. The recourse to Cloud technologies also nurtures interdependence and collaboration while improving communications and transparency across the country.

The implementation of blockchain solutions for guaranteeing the security of HR information represents an important step. Through such technology, the data may be stored in different formats and values to ensure the management of secure access, blocking potential intruders and falsified information and again fostering transparency in the workplace. Blockchain shall enable the quality of HR organisation systems, deepening the collaboration ongoing among departments, stimulating talent, creation and the transfer of knowledge in ways that raise the propensity to develop innovative processes, products and services.

The application of virtual reality, augmented reality or chatbots, in association with gamification techniques, are able to perfect processes of creating and transferring knowledge, through networking and simulations of real scenarios. The gamification techniques may generate contributions towards the motivations and team feelings of staff as well as boosting their creativity.

8.7 FUTURE RESEARCH DIRECTIONS

Future research should undertake a survey of just which technologies are currently undergoing integration into company HR management practices and their respective impacts on the creation and transfer of knowledge and the level of innovation performance. In order to expand our discoveries, research should also extend to including those HR dimensions able to boost the acceptance of these technologies, for example in the cultural domain, across differences in age, education and personality.

The development, operationalisation and validation of a questionnaire for measuring the flows of knowledge creation and transfer and the propensity towards innovation within the scope of HR technological management practices also need to be subject to future research. These research studies require qualitative approaches to grasp the processing of these flows, the quantitative approaches enabling the theoretical validation of the model in conjunction with transversal and longitudinal approaches. There is equal importance to studying the ways in which HR technological management practices interlink with sources of knowledge (internal and external), the type of knowledge (tacit and explicit), the obtaining of sustainable competitive advantages and operational and financial performance results of organisations.

ACKNOWLEDGEMENTS

This study received financing from national funding through the the Foundation for Science and Technology (FCT) under the auspices of project UID/GES/04630/2020.

APPENDICES

APPENDIX TABLE 1A
Checklist PRISMA

Section/Topic	#	Checklist Item
Methods		
Protocol and registration	5	Indicate if a review protocol exists, if and where it can be accessed (e.g., web address), and, if available, provide registration information including registration number.
Eligibility criteria	6	Specify study characteristics (e.g., PICOS, length of follow-up) and report characteristics (e.g., years considered, language, publication status) used as criteria for eligibility, giving rationale.
Information sources	7	Describe all information sources (e.g., databases with dates of coverage, contact with study authors to identify additional studies) in the search and date last searched.
Search	8	Present full electronic search strategy for at least one database, including any limits used, such that it could be repeated.
Study selection	9	State the process for selecting studies (i.e., screening, eligibility, included in systematic review, and, if applicable, included in the meta-analysis).
Data collection process	10	Describe method of data extraction from reports (e.g., piloted forms, independently, in duplicate) and any processes for obtaining and confirming data from investigators.
Data items	11	List and define all variables for which data were sought (e.g., PICOS, funding sources) and any assumptions and simplifications made.
Risk of bias in individual studies	12	Describe methods used for assessing risk of bias of individual studies (including specification of whether this was done at the study or outcome level), and how this information is to be used in any data synthesis.
Summary measures	13	State the principal summary measures (e.g., risk ratio, difference in means).
Synthesis of results	14	Describe the methods of handling data and combining results of studies, if done, including measures of consistency (e.g., I2) for each meta-analysis.
Risk of bias across studies	15	Specify any assessment of risk of bias that may affect the cumulative evidence (e.g., publication bias, selective reporting within studies).
Additional analyses	16	Describe methods of additional analyses (e.g., sensitivity or subgroup analyses, meta-regression), if done, indicating which were pre-specified.

Source: Adapted from "The PRISMA statement for reporting systematic reviews and meta-analyses of studies that evaluate health care interventions: explanation and elaboration." De Liberati, A., Altman, D. G., Tetzlaff, J., Mulrow, C., Gøtzsche, P. C., Ioannidis, J. P. A., ... Moher, D., 2009, *Journal of Clinical Epidemiology*, 62(10), e1–e34.

APPENDIX TABLE 1B
Technology Readiness Levels (TRL) – Scale

Level	Description
TRL 1	basic principles observed
TRL 2	technology concept formulated
TRL 3	experimental proof of concept
TRL 4	technology validated in lab
TRL 5	technology validated in relevant environment (industrially relevant environment in the case of key enabling technologies)
TRL 6	technology demonstrated in relevant environment (industrially relevant environment in the case of key enabling technologies)
TRL 7	system prototype demonstration in operational environment
TRL 8	system complete and qualified
TRL 9	actual system proven in operational environment (competitive manufacturing in the case of key enabling technologies; or in space)

Source: Extract from Part 19 – Commission Decision C(2014)4995.

REFERENCES

Agrawal, P., & Narain, R. (2018). Digital supply chain management: An overview. In *IOP Conference Series: Materials Science and Engineering*, *455*(1), 012074. https://doi.org/10.1088/1757-899X/455/1/012074

Angelucci, D., Missikoff, M., & Taglino, F. (2011). Future internet enterprise systems: A flexible architectural approach for innovation. *Lecture Notes in Computer Science (Including Subseries Lecture Notes in Artificial Intelligence and Lecture Notes in Bioinformatics)*, *6656*, 407–418. https://doi.org/10.1007/978-3-642-20898-0_29

Avasilcəi, S., & Rusu, G. (2015). Innovation management based on proactive engagement of customers: A case study on LEGO Group. Part II: Challenge of engaging the digital customer. In *IOP Conference Series: Materials Science and Engineering*, *95*(1), 012144. https://doi.org/10.1088/1757-899X/95/1/012144

Barroso, J., Gollop, C. J., Sandelowski, M., Meynell, J., Pearce, P. F., & Collins, L. J. (2003). The challenges of searching for and retrieving qualitative studies. *Western Journal of Nursing Research*, *25*(2), 153–178. https://doi.org/10.1177/0193945902250034

Cook, D. J., Mulrow, C. D., & Haynes, R. B. (1997). Systematic reviews: Synthesis of best evidence for clinical decisions. *Annals of Internal Medicine*, *126*(5), 376–380. https://doi.org/10.7326/0003-4819-126-5-199703010-00006

Debajyoti, P., & Lorusso, L. N. (2018). How to write a systematic review of the literature. *Health Environments Research and Design Journal*, *11*(1), 15–30. https://doi.org/10.1177/1937586717747384

Del Giudice, M., Garcia-Perez, A., Scuotto, V., & Orlando, B. (2019). Are social enterprises technological innovative? A quantitative analysis on social entrepreneurs in emerging countries. *Technological Forecasting and Social Change*, *148*(August), 119704. https://doi.org/10.1016/j.techfore.2019.07.010

Dhammi, I., & Haq, R. (2018). How to write systematic review or meta-analysis. *Indian Journal of Orthopaedics*, *52*(6), 575–577. https://doi.org/10.4103/ortho.IJOrtho_557_18

Doloreux, D., Shearmur, R., & Rodriguez, M. (2018). Internal R&D and external information in knowledge-intensive business service innovation: Complements, substitutes or independent? *Technological and Economic Development of Economy, 24*(6), 2255–2276. https://doi.org/10.3846/tede.2018.5694

Echer, I. C. (2001). A revisão de literatura na construção Literature review in a scientific work. *Revista Gaúcha de Enfermagam, 22*(2), 5–20. Retrieved from www.lume.ufrgs.br/bitstream/handle/10183/23470/000326312.pdf?sequence=1

Edirisinghe, R. (2019). Digital skin of the construction site: Smart sensor technologies towards the future smart construction site. *Engineering, Construction and Architectural Management, 26*(2), 184–223. https://doi.org/10.1108/ECAM-04-2017-0066

Elhoseny, M., Abdelaziz, A., Salama, A. S., Riad, A. M., Muhammad, K., & Sangaiah, A. K. (2018). A hybrid model of Internet of Things and cloud computing to manage big data in health services applications. *Future Generation Computer Systems, 86*, 1383–1394. https://doi.org/10.1016/j.future.2018.03.005

Evans, D., & Pearson, A. (2001). Systematic reviews: Gatekeepers of nursing knowledge. *Journal of Clinical Nursing, 10*(5), 593–599. https://doi.org/10.1046/j.1365-2702.2001.00517.x

Figueiredo, R., & Bahli, B. (2021). *Service Business Growth: "A Spinner Innovation Model Approach."* Springer International Publishing. https://doi.org/10.1007/978-3-030-51995-7_9

Figueiredo, R., & de Matos Ferreira, J. J. (2019). Spinner model: Prediction of propensity to innovate based on knowledge-intensive business services. *Journal of the Knowledge Economy, 11*(4), 1316–1335. https://doi.org/10.1007/s13132-019-00607-2

Figueiredo, R., Ferreira, J. J. M., Silveira, R. G., & Villarinho, A. T. (2019). Innovation and co-creation in knowledge intensive business services: The Spinner model. *Business Process Management Journal, 26*(4), 909–923. https://doi.org/10.1108/BPMJ-10-2019-0424

Ghezzi, A., & Dramitinos, M. (2016). Towards a Future Internet infrastructure: Analyzing the multidimensional impacts of assured quality Internet interconnection. *Telematics and Informatics, 33*(2), 613–630. https://doi.org/10.1016/j.tele.2015.10.003

Gisbert, J. P., & Bonfill, X. (2004). ¿Cómo realizar, evaluar y utilizar revisiones sistemáticas y metaanálisis? *Gastroenterología y Hepatología, 27*(3), 129–149. https://doi.org/10.1016/s0210-5705(03)79110-9

Gürdür, D., El-khoury, J., & Törngren, M. (2019). Digitalizing Swedish industry: What is next?: Data analytics readiness assessment of Swedish industry, according to survey results. *Computers in Industry, 105*, 153–163. https://doi.org/10.1016/j.compind.2018.12.011

Herrera González, R. L., & Hidalgo Nuchera, A. (2019). Dynamics of service innovation management and co-creation in firms in the digital economy sector. Contaduría y administración, 64(SPE1), 0–0.

Jacobs, J., & Webber-Youngman, R. C. W. (2017). A technology map to facilitate the process of mine modernization throughout the mining cycle. *Journal of the Southern African Institute of Mining and Metallurgy, 117*(7), 637–648. https://doi.org/10.17159/2411-9717/2017/v117n7a5

Kataria, S., & Ravindran, V. (2018). Digital health: A new dimension in rheumatology patient care. *Rheumatology International, 38*(11), 1949–1957. https://doi.org/10.1007/s00296-018-4037-x

Khan, K. S., Kunz, R., Kleijnen, J., & Antes, G. (2003). Five steps to conducting a systematic review. *Journal of the Royal Society of Medicine, 96*(3), 118–121. https://doi.org/10.1258/jrsm.96.3.118

Laužikas, M., & Miliūtė, A. (2020). Impacts of modern technologies on sustainable communication of civil service organizations. *Entrepreneurship and Sustainability Issues*, *7*(3), 2494–2509. https://doi.org/10.9770/jesi.2020.7.3(69)

Lepore, L., Metallo, C., Schiavone, F., & Landriani, L. (2018). Cultural orientations and information systems success in public and private hospitals: Preliminary evidences from Italy. *BMC Health Services Research*, *18*(1), 1–13. https://doi.org/10.1186/s12913-018-3349-6

Liberati, A., Altman, D. G., Tetzlaff, J., Mulrow, C., Gøtzsche, P. C., Ioannidis, J. P. A.,Luis, R., González, H., Nuchera, A. H., Rica, U. D. C., & Rica, C. (2019). *Dinamica De La Innovación En Servicios Rafael Herrera*, *64*(1), 1–20.

Maffey, G., Irvine, R. J., Reed, M., & van der Wal, R. (2016). Can digital reinvention of ecological monitoring remove barriers to its adoption by practitioners? A case study of deer management in Scotland. *Journal of Environmental Management*, *184*, 186–195. https://doi.org/10.1016/j.jenvman.2016.09.074

Mankins, J. C. (1995). Technology readiness levels. White Paper, April 6, 1995.

Marcon, É., Marcon, A., Le Dain, M. A., Ayala, N. F., Frank, A. G., & Matthieu, J. (2019). Barriers for the digitalization of servitization. *Procedia CIRP*, *83*, 254–259. https://doi.org/10.1016/j.procir.2019.03.129

Moher, D. (2009). The PRISMA statement for reporting systematic reviews and metaanalyses of studies that evaluate health care interventions: explanation and elaboration. *Journal of Clinical Epidemiology*, *62*(10), e1–e34. https://doi.org/10.1016/j.jclinepi.2009.06.006

Moher, D., Liberati, A., Tetzlaff, J., Altman, D. G., & Group, T. P. (2009). Preferred Reporting Items for Systematic Reviews and Meta-Analyses: The PRISMA Statement. 6(7), E1000097. https://doi.org/10.1371/journal.pmed.1000097

Moyson, S., Raaphorst, N., Groeneveld, S., & Van de Walle, S. (2018). Organizational socialization in public administration research: A systematic review and directions for future research. *American Review of Public Administration*, *48*(6), 610–627. https://doi.org/10.1177/0275074017696160

Msiska, B., & Nielsen, P. (2018). Innovation in the fringes of software ecosystems: The role of socio-technical generativity. *Information Technology for Development*, *24*(2), 398–421. https://doi.org/10.1080/02681102.2017.1400939

Novikov, S. V., & Sazonov, A. A. (2020). Digital transformation of machine-building complex enterprises. *Journal of Physics: Conference Series*, *1515*(3), 032021. https://doi.org/10.1088/1742-6596/1515/3/032021

Page, M. J., McKenzie, J. E., Bossuyt, P. M., Boutron, I., Hoffmann, T. C., Mulrow, C. D., et al. (2021). The PRISMA 2020 statement: An updated guideline for reporting systematic reviews. *BMJ*, *372*(71). doi: 10.1136/bmj.n71. www.prismastatement.org/PRISMAstatement

Pulyaeva, V., Kharitonova, E., Kharitonova, N., & Shchepinin, V. (2019). Practical aspects of HR management in digital economy. *IOP Conference Series: Materials Science and Engineering*, *497*(1), 012085. https://doi.org/10.1088/1757-899X/497/1/012085

Rangone, A. (2020). Innovation and technology: The age of the digital enterprise. *Contributions to Management Science*, 49–68. https://doi.org/10.1007/978-3-030-31768-3_3

Ranky, P. G. (2003). A 3D web-enabled, case-based learning architecture and knowledge documentation method for engineering, information technology, management, and medical science/biomedical engineering. *International Journal of Computer Integrated Manufacturing*, *16*(4–5), 346–356. https://doi.org/10.1080/0951192031000089237

Robleda, G. (2019). How to analyze and write the results of a systematic review. *Enfermería Intensiva (English Ed.)*, *30*(4), 192–195. https://doi.org/10.1016/j.enfie.2019.09.001

Sardi, L., Idri, A., & Fernández-Alemán, J. L. (2017). A systematic review of gamification in e-Health. *Journal of Biomedical Informatics*, *71*, 31–48. https://doi.org/10.1016/j.jbi.2017.05.011

Schmitt, U. (2019). Designing decentralized knowledge management systems to effectuate individual and collective generative capacities. *Kybernetes*, *49*(1), 22–46. https://doi.org/10.1108/K-03-2019-0215

Simonova, M., Lyachenkov, Y., & Kravchenko, A. (2020). HR innovation risk assessment. *E3S Web of Conferences*, *157*, 1–7. https://doi.org/10.1051/e3sconf/202015704024

Šuman, S., Poščić, P., & Marković, M. G. (2020). Big data management challenges. International Journal of Advanced Trends in Computer Science and Engineering, *9*(1), 717–723. Available Online at www.warse.org/IJATCSE/static/pdf/file/ijatcse102912020.pdf Big Data Management Challenges.

Vardarlier, P. (2020). Digital transformation of human resource management: Digital applications and strategic tools in HRM. *Contributions to Management Science*, 239–264. https://doi.org/10.1007/978-3-030-29739-8_11

Wright, R. W., Brand, R. A., Dunn, W., & Spindler, K. P. (2007). How to write a systematic review. *Clinical Orthopaedics and Related Research*, *455*, 23–29. https://doi.org/10.1097/BLO.0b013e31802c9098

9 Reshaping Flows in Healthcare Systems?
Digital Technologies Are the Perfect Ally

*José Crespo de Carvalho and
Teresa Cardoso-Grilo*

CONTENTS

9.1 Introduction: Understanding Flows within Healthcare Systems 149
    9.1.1 Logistics, Operations Management and Supply Chain Management Concepts – How Do These Apply When Considering Both Goods and Patient Flows? 150
9.2 Tangible and Intangible Logistics in Healthcare Settings 153
    9.2.1 Tangible Logistics in Healthcare 153
    9.2.2 Intangible Logistics in Healthcare 155
9.3 Use and Impact of Digital Technologies in Healthcare Settings 157
    9.3.1 Artificial Intelligence (AI) 158
    9.3.2 Machine Learning as an Artificial Intelligence Technique 158
    9.3.3 AI Potential and Applications in Healthcare 158
        9.3.3.1 Robotics and Artificial Intelligence in Healthcare 159
        9.3.3.2 Internet of Things (IoT) 161
        9.3.3.3 3D Printing 162
        9.3.3.4 Big Data 163
        9.3.3.5 Cloud Services 164
        9.3.3.6 Blockchain 164
9.4 Conclusion 165
References 166

9.1 INTRODUCTION: UNDERSTANDING FLOWS WITHIN HEALTHCARE SYSTEMS

The reader of this book has entered a healthcare unit and is searching for a specific service, exam, treatment, medicine administration or visiting someone, among other possibilities. In most cases there is no preorganized track to follow from the moment this person enters the hospital to the moment he pays (if needed) and leaves. While walking through the healthcare unit, this person comes across a series of events (or activities) and at some point, healthcare collaborators are moving consumables or medicines, transferring patients, preparing drugs to be administered, replenishing shelves in the central or advanced warehouses, replacing instruments to be used in a surgery, fixing machinery, collecting money, calling patients, or disciplining queue lines. Several other activities may have been observed and the reader may have taken part in some of them.

A reflective approach to reader's cognitive system (type two system with reference to psychology manuals) will bring to the table at least two different situations:

i. The products, materials, medicines and consumables being warehoused, moved, reprovisioned, or administered are the result of a series of activities that follow a certain flow, from planning and collecting, in the early stages, to preparing and administering to patients, if this is the case, in the final stages.
ii. As a patient, visitor, or care companion, the reader will also follow a path (or is it a flow?) that is called a journey, essentially intended to make a match with the objective of being assisted in a service, taking a health exam, a treatment, collecting a medicine from the pharmacy or visiting someone ill. To obtain the final equation for the overall experience, a set of additional healthcare unit variables need to be considered, ranging from the waiting times to the availability and effectiveness of the treatment received.

Having this in mind, two types of flows can be considered straight away: product (e.g., medicines, consumables, devices) flows, which can be called tangible flows; and patient, visitor, or care companion flows, which can be called intangible flows.

For product flows (tangibles), it is usually applicable and an established language is used in the areas of logistics, operations, or supply chain management. Nevertheless, when considering the particular case of healthcare, in which both tangible and intangible flows need to be clearly distinguished, this language needs to be adapted.

9.1.1 Logistics, Operations Management and Supply Chain Management Concepts – How Do These Apply When Considering Both Goods and Patient Flows?

How can one define operations, logistics management and supply management?

Before proceeding, it is important to understand these are different concepts. For the purpose of this chapter, these concepts follow the definitions presented by the Council of Supply Chain Management Professionals (CSCMP), although somehow adapted to reflect its application to the healthcare setting.

According to the CSCMP (Vitasak 2013), supply chain management involves the planning and management of all the activities (i.e., operations inside the healthcare system), including sourcing and procurement, conversion and all the logistic management activities, but it also includes the coordination and collaboration with a network of interconnected entities and stakeholders, which can be suppliers, intermediaries, service providers and customers. When considering healthcare supply chains in particular, these comprise a multiplicity of suppliers, manufacturers (such as pharmaceutical companies), distributors, patient care units (which can include private or public hospitals, nursing homes or other healthcare units) and, finally, patients. health insurance companies, regulatory agencies, technology providers and the government also represent key stakeholders in healthcare supply chains.

There is clearly a high level of complexity when managing such a diverse network of partners, between which a multitude of products (such as devices, medicines and many other clinical and non-clinical consumables) and related information flow. But this complexity turns to be even higher when one adds to the analysis the numerous internal departments of hospitals, that is, when one moves beyond the external supply chain and considers the specificities of the internal chain – flows inside the internal supply chain will not be limited to products and information, also including patients flowing between different departments. This thus reflects the need to distinguish between tangible (or product) supply chains and intangible (or patient) supply chains within healthcare supply chains.

Logistics management, on the other hand, is an essential part of supply chain management, addressing the efficient and effective management of (forward and reverse) flows of goods, services and information, from the point of origin (i.e., suppliers) to the point of consumption (i.e., final consumers). And all these flows need to be managed while ensuring a safe, timely and reliable delivery of goods and services at a minimum cost and while conforming to customers' requirements.

Figure 9.1 illustrates all the before mentioned concepts considering as example a simple supply chain comprising the surgery department in a hospital composed

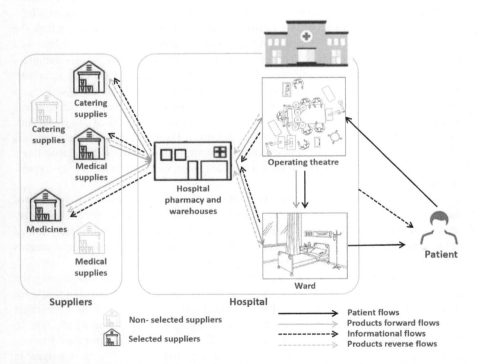

FIGURE 9.1 Patient (intangible) and product (tangible) flows in healthcare supply chains – the particular example of a surgery department in a hospital setting.

by three main tiers: suppliers (e.g., medical supplies, such as sutures and gloves; medicines; and catering supplies), the hospital and patients.

The following categories of flows can be distinguished in Figure 9.1:

i. Patient flows – solid black arrows, depicting flows between the operating theatre and the wards inside the hospital, as well as flows depicting the patient entrance and discharge;
ii. Product flows:
   a. Solid grey arrows for forward flows, depicting the delivery of supplies to the hospital pharmacy and warehouses (in which products are stored) and then to the operating theatre and wards (in which products are used and/ or administered);
   b. Dotted grey arrows for reverse flows, representing the return of surplus or defective supplies;
iii. Informational flows – dotted black arrows, corresponding to medical records or medical prescriptions (information sent by the hospital for the patient) or orders (flows between hospital services and between the hospital and the suppliers).

It is now easy to understand why it is growing the idea and language that brings the journey to the middle of the debate when the focus is on intangible flows. A journey is for all purposes a trip. Considering the patient as reference, a journey is an adaptable and mouldable trip according to the patient needs and circumstances, where the environment the customer faces, and the overall healthcare system available, will assume a very decisive role. This is easily understood using Figure 9.1 as example – consider that two different patients follow two distinct journeys within the same surgery department; these two patients may have received a cardiothoracic surgery in the same day, being afterwards moved to the ward, but while one of the patients was discharged in a few days, the other one may have suffered from surgical complications, requiring a new surgery after 24 hours in the ward. Furthermore, external elements like light, cleanness, design and decoration, layout, distances travelled, queue lines, how crowded was the healthcare unit, amongst others, can also influence the experience equation.

Summing up, while for tangibles it is quite simple to determine most of the steps of the process, usually including a plan-order-receive-stock-pick-distribute-administer-reverse (if needed) set of activities; for intangibles the steps are not completely closed or pre-determined and the journey can change according to the specific requirements of the patient and be adapted to momentary needs.

In any case, the idea present for tangibles should be something like having the proper products in the right place and time, in the right quantity at a minimal cost, whereas the idea present for intangibles, that is, patients, should also follow a parallel reasoning, such as being properly served when having suitable patients (adequate to the healthcare unit mission and objectives), following the most appropriate journey (i.e., being assisted on-time without unnecessary displacements), in manageable number at a minimal cost for the organization.

It is thus clear that attaining the minimum cost is a key objective for any healthcare unit when managing tangibles and intangibles flows. Nevertheless, when considering the case of intangibles, it entails a new level of complexity. In this case, the most important question to be answered is: should one serve patients or just serve the organization? And the appropriate answer should be both. Serving patients while being efficient and effective is in fact the desired equilibrium. But isn't it the same idea present with tangibles? While in tangibles we manage products to serve patients at a minimal cost for the organization, with intangibles we manage patients to serve them and, in parallel, at a minimal cost for the organization.

Having these first considerations in mind and the fact that there are at least two types of logistics – goods logistics and patient logistics. The major topics of this chapter will address deeper the processes and activities present in these two types of logistics (further detailed in the following section), with special focus on the use of technologies (final section). Although not a recent topic – technology is already recognized as the way forward for healthcare logistics for several years (Mckone-Sweet et al. 2005) – these technologies are now paving a new future for both types of logistics, by bringing elements from industry 4.0, big data, machine learning, artificial intelligence, cloud services, blockchain, self-guided vehicles or drones, among others. These technologies thus play a key role in the digital transition of what was and is a complex context and usually intensive in human labour.

9.2 TANGIBLE AND INTANGIBLE LOGISTICS IN HEALTHCARE SETTINGS

9.2.1 Tangible Logistics in Healthcare

As mentioned earlier in this chapter, the key activities characterizing the logistics system for tangibles in the hospital setting usually follow a plan-order-receive-stock-pick-distribute-administer-reverse sequence.

But how does all this process start? Similar to many other industries, it starts with planning, which should be carefully executed due to the widely acknowledged unpredictable nature of healthcare demand (Hof et al. 2015). Such unpredictability of demand is reinforced by the need to manage hundreds of different items – including perishable goods (such as medicines and blood), non-disposable goods (including a wide variety of instruments) and disposable goods (such as gloves and needles).

A key challenge in (tangible) healthcare logistics is in fact the management of this large variety of items with such an uncertain level of demand. And the starting point is to decide which supplier (or suppliers) should be contracted for the delivery of each item. Depending on the number of items and respective volumes, as well as on its costs (among other factors), several decisions need to be made – which suppliers should be selected for sourcing purposes? Should one resort to one single supplier, or to multiple suppliers, thus reducing the dependency and risks of shortages? Should one resort to consignment agreements in which the supplier remains as the owner of goods until these are used? The purchasing of goods can only start right after having all these decisions made.

As soon as the supplier base is defined and orders are placed, different approaches may be followed for managing the flow of goods sent by suppliers (Volland et al. 2017). A traditional delivery approach in which goods are received in a central warehouse, being afterwards redirected to the different point-of-use locations (such as a nursing unit, a long-term care unit or an operating theatre) is a common practice. As an example, consider, for instance, the delivery of bottled gas for use in a hospital setting. So as to reduce (or even eliminate) the risk of shortages on human health, these bottles are often over-ordered and kept as buffer inventory. Nevertheless, although not relevant from a medical point of view, these extra bottles will imply extra costs (i.e., waste) for the hospital.

An alternative approach that can be followed to reduce these wastes and inefficiencies has its foundations on the just in time (JIT) philosophy. Such an approach should rely on a close relationship between the hospital and the supplier, making it possible for the supplier to identify the needs at the different point-of-use locations, thus avoiding building buffer inventories. The success of such a (almost) stockless approach represents an additional challenge for healthcare organizations, since it implies a continuous flow of information between the hospital and the suppliers and enables the reduction of inventory costs.

Once delivered in the hospital facilities, new challenges arise related to the management of goods – how to organize the layout of the space and how to allocate the items in the shelfs, while ensuring the minimization of the time spent in movements inside the warehouse? How to ensure the handling and transportation of materials between services and hospital departments? How to ensure the tracking of goods from the moment in which these are received and stored, until being delivered and used/administered?

Another key challenge related to (tangible) logistics in healthcare is related to the management of the reverse flows of used items. How should one proceed to return medical devices that are damaged? Which should be the procedures when medicines are not administered or are out of date? How can one track reusable items (e.g., reusable gowns) from the moment these are first used until being returned, after the final usage? These reverse flows need to be carefully managed, and it is often the case that special conditions apply – for instance, devices with lithium-ion batteries require special conditions for its transportation, and these must be properly taken care of.

But are all these challenges really relevant? Why should we care about tangible logistics in healthcare? Past history shows that, when compared with other industries, tangible logistics was not a core concern for hospital managers – the key concern was always the effective delivery of care to patients, and all the logistics activities associated with the management of tangible goods were regarded as simple support activities (Volland et al. 2017). Nevertheless, the importance of achieving an efficient management of tangible flows within the healthcare sector is now well recognized (de Vries 2011), since it assures significant cost reductions while simultaneously ensuring the adequate delivery of care to patients.

And is it simple to achieve efficiency when managing tangible flows in healthcare? How can healthcare organizations deal with all the challenges that characterize the management of these flows? Finding solutions to these challenges may not be difficult – several recent advances in digital technologies are in fact recognized as key for the efficiency and effectiveness of healthcare logistics. Some examples are as follows: big data is gaining importance in the healthcare sector, for example, by helping organizations redesigning their flows and directing more resources where these are most needed (especially challenging when there is need to manage a high volume and variety of data, such as the ones found in healthcare settings); RFID technology is increasingly being used for tracking purposes; and a range of automated solutions is also available that can help healthcare organizations managing their processes in a more efficient way, while avoiding the reliance on highly manual processes, which often results in higher costs. These and many other examples will be further discussed in the final section of this chapter.

9.2.2 Intangible Logistics in Healthcare

It is not realistic to imagine intangible logistics in healthcare without delays, waiting lines, cancellations, changes and exchanges. These are in fact common practices in healthcare systems and the reason why patients and providers assume that, generally, waiting is always expected, although undesirable, in the care process. Dealing with these practices thus represents a key challenge when managing patient flows (Kane et al. 2019).

Healthcare systems usually respond to delays by adding either more resources – more buildings, rooms, beds, machinery or staff, among others, thus increasing installed capacity – or outsourcing and contracting parts of the system to external entities, both ways to deal with an increasingly underserved market.

But adding more resources costs money. Thus, this option increases almost automatically the inefficiencies of the healthcare system, because when adding more pieces without redrawing the system and processes the result will not benefit neither the healthcare system nor the patient.

Alternatively, externalizing part of the activity, core or non-core, may somehow help on refocusing the healthcare offer in some essential activities and build new approaches sustained on relations with some other entities. Although leveraging the network possibilities, it often implies the redrawing of its processes – otherwise, the targeted cost reduction may not be fully achieved.

While these types of approaches are commonplace in healthcare systems, the essence of problems will remain or even be more painful than before.

The topic is not only about adding or removing features (and consequently capacity) but rather managing flows and adjusting them to the existing capacity. Flow means the process, on the hospital side, or the journey, on the patient's side (both process and journey are convergent in interface zones, particularly when the service provided meets the service received). When the flow is not interesting enough from the point of view of time, cost or quality of service (contributing to service deterioration), mutually for the organization and the patient, the issue is not minor and the solution not as easy as one may think.

Smoother and timely flows of patients across hospital departments are not easy to obtain. To develop and implement methods to improve flows is also difficult. Additionally, the solution for the patient should also be reasonable and affordable for the healthcare organization. Specific areas where one can intervene include smoothing the flow of elective surgery, reducing hospital waits through emergency departments, rethinking and decentralizing waiting areas, redesigning schedules and lags to outpatients, achieving timely and efficient transfer of patients from the intensive care unit to medical/surgical units, and improving the flow from the hospital environment to long-term care facilities and external entities. Combinations of these proposed interventions and others can lead to more sustained results and better conditions both for the patient and the healthcare unit. All in all, this will be a capacity issue and managing capacity will be central to the healthcare unit.

Capacity (the output that a healthcare system can achieve within a period of time) management appears when an organization (such as healthcare organizations) is systematically improving constraints until they are no longer limiting factors of a process or a journey. Thus, the idea is to identify the most limiting factor, the constraint (also known as bottleneck), that stands in the way of achieving an objective. And all complex systems, such as healthcare systems, are likely to be worked, either in processes or in journeys, through approaches like this.

So, if processes have a constraint and the total process throughput can only be improved when the constraint is improved, one must tackle this constraint and spend some time improving it, avoiding extra time in refining alternative activities. And this should be pursued, at least until this one, the bottleneck, is not properly enhanced. An example of this can be presented in Figure 9.2.

Consider that a healthcare unit has four departments as shown in Figure 9.2, with the correspondent associated installed capacities, and also consider two different patient journeys: Patient Type A, that follows the path 1-2-3-4; and Patient Type B, that follows the path 1-3-2-4. One can see that capacity is limited to 300 patients per week at the X-ray constraint for both types of patients. This means that the process, from an organizational point of view, and the journey, considering the patient point of view, will have a bottleneck at the X-ray. Solutions can vary from increasing capacity at the X-ray to subcontracting X-ray activities. Alternatively, the process can be redrawn to only accept 300 patients per week of A and B types, creating certain days per week per each type of patient, or supposing that one of the types will be received during the day and the other during the night, in complementary hours. Other alternatives are also possible when redrawing the flows.

In summary, in healthcare units, patient logistics should be treated having in mind processes and journeys, according to the perspective, and be addressed as integrating flow perspectives in a capacity management approach.

Having this in mind, some projects are assuming a huge importance in this area when turning to technology as a way to improve flow. Examples can be found in many

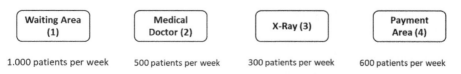

FIGURE 9.2 Example of departments in a healthcare system.

different settings – telemedicine systems used to cater chronic patients with difficult displacement problems; information of queue line length, including time and number of people, in real time; adjustments in patients' paths according to the departments capacity in use; suggested medical prescriptions according to symptom set in order to facilitate attendance; tagging patients in order to draw their natural journey and lead to improvements, among many others. More details about the use of digital technologies under these circumstances are explored in the following section.

9.3 USE AND IMPACT OF DIGITAL TECHNOLOGIES IN HEALTHCARE SETTINGS

The classical model of healthcare delivery, in which care is mainly provided at health facilities and in which information is basically delivered by medical staff, is nowadays undergoing serious changes. These changes are motivated by the use of digital technologies (strongly leveraged by the recent COVID-19 pandemic), together with the increasing demand for high-quality care by global populations.

The use of digital technologies in healthcare services (i.e., the application of Industry 4.0 principles in healthcare, also referred by some authors as Healthcare 4.0 (Tortorella et al. 2019)) is not only creating a shift from this traditional healthcare delivery model to a more accessible and flexible patient-centred model of care, but it is also key for achieving a more effective and efficient management of all the goods required to meet patients' needs.

In fact, digital technologies applied in healthcare settings can be used for a wide range of purposes – considering tangible flows, technologies such as radio-frequency identification (RFID) tags and readers (IoT devices) can be used to track the movement of medicines from the moment these are received in the central warehouse until being distributed for use inside the hospital, while also gathering and storing all the underlying information (Big Data and Cloud Storage) to better predict the future needs of these medicines; and when considering the patient journey, sensors can be used to monitor passages and patients' passage time in certain touch points (with information storing in the cloud), and digital consoles can be employed for patient admission to the hospital and data collection (again with Cloud storage).

It should also be highlighted that when one refers to the application of Industry 4.0 principles and technologies in the hospital context, it does not mean applying a single and individual technology but rather a set of several technologies. These technologies may include Artificial Intelligence (AI) and Advanced Robotics, Cloud Computing, Blockchain, Internet of Things (IoT), Big Data capture and analytics, digital customisable manufacturing (namely 3D printing), smartphones and other mobile devices and platforms that use algorithms to move motor vehicles.

It should also be highlighted that the usage of digital technologies in healthcare is nowadays recognized as a fundamental pillar whenever healthcare facilities aim at developing their processes and patient journeys to become a reference respectively in terms of flow and lived experience, thus resulting in benefits for both the facility and the patient. Most patients show greater satisfaction, as well as being more engaged, when it comes to evaluating the healthcare unit whenever it provides digital portals, mobile applications, online chats, or even simple answers by email. All in all, once the adoption curve is over and the experience curve is gained, technology becomes an ally of both customer satisfaction and engagement with healthcare units by providing better reliability and less consumption of resources (lower costs).

9.3.1 Artificial Intelligence (AI)

For this chapter, AI is defined as an area of computer science that deals with the development of systems, able to carry out cognitive functions, which one typically identifies with human minds (Duan et al. 2019). This involves fundamental abilities such as learning, understanding natural language, perception or reasoning, and these learning features are usually dependent on the existence of a large volume of data.

AI should not be regarded as a single technology but as an umbrella term for a variety of technological branches, which are often interrelated and built on top of each other. These include machine learning (ML), natural language processing (NLP), computer vision and robotics.

9.3.2 Machine Learning as an Artificial Intelligence Technique

The continuous development of digitization and integration of IT systems lead to constantly new achievements and establishing new trend topics. With the unstoppable progress in these research areas, more and more terms came into focus of the public which a few years ago were mainly discussed in blockbusters or novels – this is especially the case of machine learning (ML) under the guise of AI. Many immediately think of movies like *The Terminator* or *I, Robot* and fear the superior power of those technologies.

The reality, however, seems different: today self-learning algorithms are already integrated into everyone's products, like smartphones or cars. Private and business lives are increasingly determined by intelligent programs that learn from data and generalize what they have learned. Speech recognition and route planners are largely controlled by those algorithms as well as spam filters inside mail accounts. Clearly stated, humans are often in contact with learning systems, without knowing it.

This is also true when considering the particular case of healthcare. Algorithms built by computers with large amounts of data, and validated and adjusted by more and more data, can become powerful tools, for instance, to predict patient numbers and help designing the journeys of these patients. Collecting data with sensors and by means of various devices makes the formed algorithms richer and better able to mimic reality and create very reliable predictive scenarios. And this can be useful for many purposes: adjustments can be made to installed capacity, trend movements in stocks can be predicted and needs in materials and people to accompany patients can be adjusted.

9.3.3 AI Potential and Applications in Healthcare

Although, not limited to, AI shares with machine learning the ability to offer healthcare providers the opportunity to leverage data analytics to better manage workflows, patient flow, personnel flow and dynamically adapt capacity to demand in proactive ways (Secinaro et al. 2021).

As an example, AI systems can aid healthcare professionals in the clinical decision-making process, either by accelerating the process (it is easy to understand that humans have an analytical capacity far below the capacity of AI-based algorithms) and also by making better (and more informed) clinical decisions, with an obvious positive impact in the journey of the patients. And it is easy to foresee the usefulness of AI applications when the aim is to optimize the flows of goods (for both forward and reverse flows). In fact, AI can be useful to deal with a wide variety of data, including historical data related to the use of clinical and non-clinical supplies, and also socioeconomic, behavioural and/or clinical health data, so as to predict, for instance, which items are required in each service and for each group of patients. All this data will be extremely useful for aiding the decision-making process, thus informing planning decisions in the healthcare sector while simultaneously avoiding shortages or surplus of materials, both situations recognized as sources of extra costs in the sector. Accordingly, a more efficient and effective flow of goods will be assured, while also impacting the patient journey, and this because the required goods will be in the correct place, at the correct moment and in the correct quantities.

And it is not only the healthcare professionals and managers that can benefit from such AI-related capabilities – patients can also stay informed and engaged during the whole patient journey. Consider, as an illustrative example, a patient who can access information fully aware of a particular procedure that he will undergo and, if going to receive a prosthesis or valve, for example, the patient can also be properly informed about the pros and cons of 3D printing and automation in its robotic placement (assisted). In this way, the patient is aware both of what awaits him – in a prospective logic – and of what happened during and after surgery, so that all this can be part of his journey. All the follow-up of the process should be achieved through computerised access in the patient's room or via smartphone.

These AI capabilities are already being exploited in a wide variety of healthcare areas, including imaging – which is in fact a key area of AI application, allowing for a quick analysis and reading of images, and enabling accurate diagnosis – health monitoring, surgery, remote consultation, predictive modelling and decision support (Secinaro et al. 2021). Several examples of success can, in fact, be identified, impacting both the flow of goods and the flow of patients within healthcare settings. A particular successful example is related to the use of robotics.

9.3.3.1 Robotics and Artificial Intelligence in Healthcare

It is true that the vast majority of robots currently employed in industry are non-intelligent robots. Nevertheless, artificial intelligent robots – robots controlled by AI algorithms – are gaining a massive popularity across many different areas, such as happens in healthcare.

Self-guided vehicles (SGV) are a key example of AI-powered robots used in healthcare logistics, and are specifically designed to handle bulk material, medicines and laboratories samples, and also to transport food, linen or waste (biological or recyclable). But what is exactly an SGV? A SGV is any kind of vehicle that operates on its fullness without the direct input of a driver. It does not require any kind of pre-configured scripts to control the steering, acceleration and braking of the vehicle. Moreover, it operates under sensor devices and navigation algorithms combined with advanced AI technology that enables to interpret the outside environment and make the right decisions in the right moment, allowing to smoothly navigate from point A to point C, passing through point B or any other desired location. In parallel, the use of machine learning in this technology enables the vehicle to become more efficient and make the right decisions as it encounters obstacles or faces new situations.

A wide range of applications for SGVs can be found at a hospital facility (Pedan et al. 2017). Waste bins and trolleys can be carried out by these types of vehicles. The delivery of meals from the kitchen to wards and the return of empty trays to the kitchen can also be performed by SGVs, and the same applies for linen/laundry transportation (clean and soiled). Also, sterile supplies transportation can be also done by SGVs, and this also applies for drugs and other general supplies in the hospital (wards, pharmacies, laboratories, among others). These vehicles allow, among other things, to reduce the exposition of front-line medical and non-medical staff to infectious diseases (highly relevant during the recent COVID-19 pandemic), as well as to reduce the occurrence of human errors and allowing hospital staff to focus on patient care activities (Holland et al. 2021). The usage of these robots clearly has an impact on how goods flow inside the hospital (tangible flows), as well as on the patient experience (related to intangible flows).

Another successful example is still related to the use of AI-powered robots, but in this case for surgeries. These robots now can match or even exceed a human in dexterity, precision and speed. Using the same technologies that underpin self-driving cars, autonomous drones and warehouse robots, these robotisation approaches work to automate surgical counterparts – not to replace surgeons in the operating room, but to ease their load and increment success rates. They may even be applied to perform remote surgeries and can be integrated as a successful element of the patient journey, contributing to structure Industry 4.0 approaches to healthcare units.

Although not so often applied as in other industries, the role of autonomous flying robots like drones in healthcare is also undeniable. These unmanned vehicles (aeronautical, ground-based or naval) can be either remotely controlled by a human or act autonomously when relying on AI systems. And the potential of these flying robots when it comes to achieve a more efficient healthcare system is clear:

i. Small indoor drones can be used to deliver pharmaceutical products to the bedside of a patient from the pharmacy, thus eliminating some human steps and resulting in a more rapid and less error-prone administration of medications. Nurses and pharmacists can work more efficiently as supplies can be summoned to the bedside, having more time to dedicate to patient care, which surely affects the patient experience in a positive way. The patient journey may also benefit from small indoor drones that film and give instructions to patients in their journeys, providing that they can adequate their paths to adequate ones;

ii. Drones could also deliver medications and supplies to patients being cared for at home. The future will see more outpatient care, usually at home, replacing the need for inpatient care, and drone technologies may make it easier and

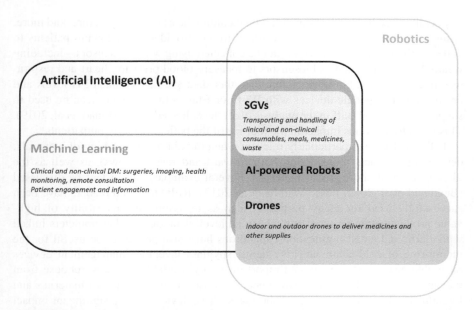

FIGURE 9.3 Examples of AI applications in healthcare. Legend: DM – decision making; SGV – self-guided vehicles.

safer to provide this home-based care. When a provider rounds on a home patient, blood can be drawn and immediately sent by drone to the lab to be tested. Medications, antibiotics and treatments ordered by the provider may also be delivered home by drone. In fact, similar outdoor drones have already been trialled for delivering food in areas affected by disasters, as well as to deliver blood products or to collect blood samples and distribute vaccines is more remote areas (Glauser 2018).

Figure 9.3 reviews the key concepts presented above, summarizing how AI is related with machine learning and robotics, and detailing key applications of these technologies in healthcare.

9.3.3.2 Internet of Things (IoT)

The Internet of Things (IoT) describes the network of physical objects that are embedded with sensors, software and other technological developments for the purpose of connecting and exchanging data with other devices and systems over the internet (Ahmadi et al. 2019). These devices range from current household objects to sophisticated industrial tools. In hospitals IoT "things", also commonly referred to as Internet of Healthcare Things (IoHT), are very diverse, and can vary from a patient bed to blood pressure machine.

Applications in hospitals are several, but remote patient monitoring is the most common application of IoT devices in healthcare (Senbekov et al. 2020). IoT devices can automatically collect a wide range of health metrics from patients who are not physically present in a healthcare facility, including heart rate, temperature, and more, eliminating the need for patients to travel to the providers, or allowing patients to collect it themselves. All this data can be collected using wireless sensors – including wearables sensors, such as biosensors to measure blood pressure, heart and respiratory rates; or ingestible sensors able to gather data such as stomach PH levels and enzymes – and specific devices with Wi-Fi or Bluetooth connection can be used to support the medical staff in the analysis of all the collected data (Ahmadi et al. 2019). This has a tremendous impact in the journey of the patient (usually outpatients).

IoT has also been particularly relevant when the aim is the tracking of materials, devices or products (considering both forward and reverse flows), as well as the identification of shortages of materials. Several examples can in fact be discussed (Bendavid et al. 2012, Fosso Wamba et al. 2013): RFID tags and readers have been successfully used for many purposes, such as to assure the traceability of high-value products often required at the hospital level; data logging thermometers linked with alarm and email notification capabilities have also proved to be useful for the monitoring of cold-chain sensitive items; spyglass inventory management devices have also been used to view and report inventory available in the warehouse from managers' own desk among many others. By ensuring the tracking of materials and by reducing the risk of associated shortages, these devices have a significant impact in the way goods are managed in healthcare settings, with obvious cost advantages for healthcare providers. Furthermore, the usage of these devices will also impact the patient journey inside the hospital – as an example, these devices can avoid delays or postponements of consultations or surgeries due to the lack of clinical or non-clinical consumables, or even medicines.

But what turns to be clear with all these examples is that IoT devices represent a key source of massive health-related data (Big Data), which makes it essential for the application of artificial intelligent solutions in health.

9.3.3.3 3D Printing
In the history of manufacturing, traditional manufacturing processes were subtractive. Although additive manufacturing processes are not completely new, since the first 3D printing attempts developed in the 1980s with the usage of a layer-by-layer approach, only the most recent developments in this technology allowed an important improvement – new design software, new printing methods, new materials – enabling the production of diverse products composed by multiple thin layers of material following instructions from a digital file.

Additive manufacturing offers medical industry great freedom of design, adaptability and functional integration, and this is especially true when combining 3D printing with the large volume of patient data gathered using IoT devices. For the manufacturers of dentures, medical and orthopaedic technology products, orthoses, valves and prostheses, this creates many far-reaching opportunities, with thousands of design customization (Aimar et al. 2019). There is complete control over the shapes, materials and designs based on patient specific requirements – specific data which makes it possible to offer more individual treatments, simplifies biomechanical reconstruction and enables innovative therapy methods to be implemented rapidly.

Such a technology also offers new opportunities for the preparation of personalised medicines (Awad et al. 2021) – 3D printers can be used in pharmacies and hospitals, enabling the production of individual units based on specific patients' characteristics, such as age, gender and weight. All in all, all these alternatives can have a tremendous impact in the overall experience coming from the health journeys of patients, as well as in the management of flows and storage of medicines and other products.

9.3.3.4 Big Data

As it can be easily understood from the previous examples, large volumes of data can be gathered using IoT devices, and this impressive amount of data is a step forward for the implementation of advanced AI applications. AI applications can in fact unlock information that may be relevant but hidden in the currently existing massive amount of clinical and non-clinical data (Jiang et al. 2017).

But what is big data and how can we obtain and deal with it in general?

Companies are gathering voluminous information about their entire value chain, including, but not limited to, their customers, suppliers and operations. This data collection is supported by sensors collecting data through physical devices such as mobile phones, wearables, vehicles, hospital engines home appliances and industrial machines. Simultaneously, data is collected through non-physical processes in the web. Meaning, consumer behaviour can be tracked and identified through purchasing, chatting, sharing and searching patterns. Thus, the physical and non-physical data collection is continuously expanding into new areas, providing data with greater variety, increasing volumes, arriving at ever-higher velocity (known as the 3Vs). As a result, data sets are too large or complex for traditional data-processing application software to adequately deal with – for instance, considering only clinical data gathered during the lifetime of a single person (including data related to diagnosis, treatments, medicines, lab results, among others), it has potential to generate 0.4 terabytes of information (Senbekov et al. 2020). These data sets are referred to as big data.

Let us now consider the specific example of healthcare. All the data related to the patient journey, including information generated from entries and exits of a health unit, distances travelled, paths taken, the clinical acts one underwent and the order and times in which they occurred, along with health data, can be considered, as an example, part of this universe of big data.

But how can one collect and take advantage of all this data? Specific examples that can directly affect the experience of patients include the use of wearables (IoT) devices that can continuously collect patients' health data and send this data to the cloud, generating real-time alerts from patients out of the hospital (important to redesign the journey including out of hospital components); or telemedicine practices, which are among the best available applications of big data to improve healthcare processes and patient journeys.

On the other hand, big data can also be used to foster a more efficient and effective management of tangible flows in healthcare. Following the same rational previously presented for AI technologies, patient predictions can be used as input for planning purposes, such as for improved staffing decisions or for managing capacity adjustments (beds, instruments, devices, among others) in different moments. Data dashboards are also useful to manage and report on supply levels of goods, including medicines and a wide range of clinical and non-clinical consumables. Such dashboards allow to track and monitor the levels of inventories, notify if these levels fall below a critical level, and it can also be useful to build demand forecasts for healthcare supplies.

9.3.3.5 Cloud Services

An additional issue that also deserves a special attention when dealing with Big Data is related to its storage and processing. How can one store, clean and process such a massive amount of data (either clinical or non-clinical, depending on whether we aim at collecting medical data from patients or non-clinical data related to the flow of goods inside the hospital)? Cloud services appear as a potential solution, thus leveraging the use of IoT and AI in healthcare settings (Ma et al. 2020).

It is considered a Cloud service any service made available to users on demand via the Internet from a cloud computing provider's server, in opposition to being provided from a company's on-premises, also known as company's data centres. The services provided through Cloud computing are very diversified, from merely sending an email or editing documents online, to more complex tasks such as hosting a website or creating a new app, for example.

The word "Cloud" is commonly used as a metaphor for the Internet, which translates Cloud computing or Cloud service into Internet-based computing or Internet-based service, respectively. It means that different services, including servers, storage and applications are delivered to an organization's computers and devices through the Internet.

Remote accessibility of cloud data is possibly one of the biggest advantages that cloud storage of data offers (Elmisery et al. 2019). The combination of Cloud computing with the massive healthcare data currently available has the potential to improve several healthcare-related functions such as telemedicine, post-hospitalization care plans and virtual medication adherence. It also improves access to healthcare services through telehealth.

Telemedicine apps add the element of convenience to healthcare delivery while upgrading the patient experience. Cloud-based telehealth systems and applications allow easy sharing of healthcare data, improve accessibility and provide healthcare coverage to the patients during preventative, treatment and recovery phases.

9.3.3.6 Blockchain

Blockchain is also a technological option with an increasingly relevance in healthcare.

Blockchain technology, also known as a distributed ledger technology, is, as the name implies, a chain of blocks. Blocks record a transaction, or group of transactions, and are available to all the participants in a network, making blockchain a peer-to-peer digital ledger. Participants, otherwise called nodes, have access to an identical real-time updated copy of the ledger and validate the blocks inserted into the chain. Therefore, blockchain is a distributed database as all nodes have access to the same information (Ray et al. 2021).

Distributed ledger technology and smart contracts can deliver ground-breaking solutions for problems in the healthcare industry (Ray et al. 2021). Combined with a digital identity, blockchain technology can boost efficiency and effectiveness in healthcare for providers, payers and patients, and provides data security as a basic need for trust – that is, blockchain-encrypted digital identity management contributes to place humans in the centre of medical data management. The critical issues are the availability of medical data and trust of patients and stakeholders in data security and quality before, during or after any journey in a health system (and the issues of data security and quality are in fact a key issue of concern when large-scale IoT networks store data using cloud services), which can be addressed, precisely, by using blockchain technology.

Several examples can be discussed concerning the use of blockchain in healthcare (Musamih et al. 2020, Ray et al. 2021). Using blockchain to make the process of drug management more transparent is one of those examples – this transparency is a key issue when managing medicines, especially considering the large number of people that is affected by the consumption of counterfeit drugs. Ensuring a permanent, immutable and decentralized record of medicines is expected to reduce the frauds, impacting both patient experience and also the entire underlying logistic system.

9.4 CONCLUSION

At this point, what can be said about the role of digital technologies in healthcare? It is undeniable that digital technologies have already had, and will continue to have, a huge impact in the way healthcare systems are organized all over the world.

In fact, different technologies have been positively impacting the patient experience, enabling more flexible and patient-centred models of care and irreversibly changing the way patients and healthcare providers interact. These technologies bring new and powerful opportunities, such an increasing number and variety of remote care solutions that can aid healthcare delivery for different patient groups, such as for the elderly and chronically ill patients.

Nevertheless, the impact of technologies goes well beyond the patient journey, since an increasing number of technological applications are already in place for ensuring a more efficient and effective management of all the devices, medicines and clinical and non-clinical consumables (tangibles). For instance, when the aim is to achieve more intelligent operations within healthcare settings, AI-enabled robots (either for indoor or outdoor use) appear as a promising technology.

But all these achievements can only be maintained, or even surpassed, in the future if healthcare organizations continue to move towards a full integration of different technologies (see Figure 9.4). A wide implementation of IoT devices is the key requirement for gathering the wide range of clinical and non-clinical data that flow within healthcare systems (including information related to patient flows or tangible flows), and all this massive amount of information (Big Data, which can be stored either via cloud services or blockchain) can afterwards be used for many purposes: to inform clinical and non-clinical decision-making; to foster a higher patient engagement; to be used as input for AI solutions that can improve both patient and tangible logistics; and also to promote a higher level of customization when using 3D printing.

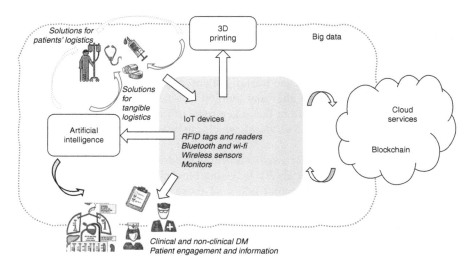

FIGURE 9.4 Global overview of technologies' use in healthcare settings. Legend: IoT – Internet of Things; DM – decision making.

But the long-term successful implementation of this wide variety of promising technologies is highly dependent on the ability shown by healthcare systems (together with governments, developers of digital health technologies and other entities) to deal with some critical challenges. For instance, it is essential to invest in the digital literacy of healthcare professionals so as to ensure they have the required competencies to correctly use these smart devices and digital tools. Also, these technologies are often associated with expensive equipment, which represents one of the key barriers for significant advances in the sector. The accuracy and reliability of medical devices is an additional concern, and not less important, there is also data privacy and confidentiality issues that must be taken care of. And, of course, there is always a certain resistance to change that may hinder the successful application of technologies in real practice.

There is for sure a lot of improvements and advances to achieve when the aim is to move towards a more technological-oriented healthcare delivery, but there is also a vast amount of knowledge already in place that should be used to inform future advances in the area. But these future developments should not only be guided by previous experiences and knowledge – it should also depart from an in-depth analysis related to the specific features of the healthcare environment under the digitization process, and this because there is certainly not a one-size-fits-all strategy for the digital transition of healthcare systems on a global scale.

REFERENCES

Ahmadi, H., Arji, G., Shahmoradi, L., Safdari, R., Nilashi, M. and M. Alizadeh. 2019. The application of internet of things in healthcare: a systematic literature review and classification. *Universal Access in the Information Society*. 18: 837–869.

Aimar A., Palermo, A. and B. Innocenti. 2019. The role of 3D printing in medical applications: a state of the art. *Journal of Healthcare Engineering.* 5340616: 1–10. https://doi.org/10.1155/2019/5340616

Awad, A., Fina, F., Goyanes, A., Gaisford, S. and A.W. Basit. 2021. Advances in powder bed fusion 3D printing in drug delivery and healthcare. *Advanced Drug Delivery Reviews.* 174: 406–424.

Bendavid, Y., Boeck, H. and R. Philippe. 2012. RFID-enabled traceability system for consignment and high value products: a case study in the health care sector. *Journal of Medical Systems.* 36: 3473–3489.

de Vries, J. 2011. The shaping of inventory systems in health services: a stakeholder analysis. *International Journal of Production Economics.* 133: 60–69.

Duan Y., Edwards, J.S. and Y.K. Dwivedi. 2019. Artificial intelligence for decision making in the era of Big Data – evolution, challenges and research agenda. *International Journal of Information Management.* 48: 63–71.

Elmisery, A.M., Rho, S. and M. Aborizka. 2019. A new computing environment for collective privacy protection from constrained healthcare devices to IoT cloud services. *Cluster Computing.* 22: 1611–1638.

Fosso Wamba, S., Anand, A. and L. Carter. 2013. A literature review of RFID-enabled healthcare applications and issues. *International Journal of Information Management.* 33: 875–891.

Glauser, W. 2018. Blood-delivering drones saving lives in Africa and maybe soon in Canada. *Canadian Medical Association Journal.* 190: 3.

Hof, S., Fügener, A., Schoenfelder, J. and J.O. Brunner. 2015. Case mix planning in hospitals: a review and future agenda. *Health Care Management Science.* 20: 207–220.

Holland, J., Kingston, L., McCarthy, C., Armstrong, E., O'Dwyer, P., Merz, F. and M. McConnell. 2021. Service robots in the healthcare sector. *Robotics.* 10: 1–47.

Jiang, F., Jiang, Y., Zhi, H., Dong, Y., Li, H., Ma, S., Wang, Y., Dong, Q., Shen, H. and Y. Wang. 2017. Artificial intelligence in healthcare: past, present and future. *Stroke and Vascular Neurology.* 2: 230–243.

Kane, E.M., Scheulen, J.J., Püttgen, A., Martinez, D., Levin, S., Bush, B.A., Huffman, L., Jacobs, M.M., Rupani, H. and T.D. Efron. 2019. Use of systems engineering to design a Hospital Command Center. *The Joint Commission Journal on Quality and Patient Safety.* 45: 370–379.

Ma, Y., Ping, K., Wu, C., Chen, L., Shi, H. and D. Chong. 2020. Artificial Intelligence powered Internet of Things and smart public service. *Library Hi Tech.* 38: 165–179.

McKone-Sweet, K.E., Hamilton, P. and S.B. Willis. 2005. The ailing healthcare supply chain: a prescription for change. *Journal of Supply Chain Management.* 41: 4–17.

Musamih, A., Salah, K., Jayaraman, R., Arshad, J., Debe, M., Al-Hammadi, Y. and S. Ellahham. 2021. A blockchain-based approach for drug traceability in healthcare supply chain. *IEEE Access.* 9: 9728–9743.

Pedan, M., Gregor, M. and D. Plinta. 2017. Implementation of automated guided vehicle system in healthcare facility. *Procedia Engineering.* 192: 665–670.

Ray, P.P., Dash, D., Salah, K. and N. Kumar. 2021. Blockchain for IoT-based healthcare: background, consensus, platforms, and use cases. *IEEE Systems Journal.* 15: 85–94.

Secinaro, S., Calandra, D., Secinaro, A., Muthurangu, V. and P. Biancone. 2021. The role of artificial intelligence in healthcare: a structured literature review. *BMC Medical Informatics and Decision Making.* 21: 125.

Senbekov, M., Saliev, T., Bukeyeva, Z., Almabayeva, A., Zhanaliyeva, M., Aitenova, N., Toishibekov, Y. and I. Fakhradiyev. 2020. The recent progress and applications of digital technologies in healthcare: a review. *International Journal of Telemedicine and Applications.* 2020: 1–18.

Tortorella, G.L., Fogliatto, F.S., Vergara, A.M.C., Vassolo, R. and R. Sawhney. 2019. Healthcare 4.0: trends, challenges and research directions. *Production Planning & Control.* 31: 1245–1260.

Vitasak, K. 2005. Supply chain and logistics terms and glossary. Council of Supply Chain Management Professionals.

Volland, J., Fugener, A., Schoenfelder, J. and J.O. Brunner. 2017. Material logistics in hospitals: a literature review. *Omega.* 69: 82–101.

10 Work with Me, Don't Just Talk at Me

When "Explainable" Is Not Enough

Brian Pickering

CONTENTS

10.1	Introduction	169
10.2	The Disruptive Introduction of Technology	171
10.3	The Ethics of Medical Care	176
10.4	"Big Data" in Data-Driven Advanced Technologies	177
10.5	Explainable AI	179
10.6	A Note on Clinicians and Work	181
10.7	The Likelihood of Advanced Technology Acceptance	182
10.8	Recommendations	185
10.9	Conclusions	186
Acknowledgements		186
References		187

10.1 INTRODUCTION

With advances in medical care enabling ever more complex interventions and care provision across phases – primary and secondary, and on occasion tertiary sectors – unforeseen interactions between what is recommended and in particular prescribed at different points may ultimately have an adverse effect on the patient. Looking at medication, it has been recognised for some time that adverse side effects and interactions need careful monitoring. At the simplest levels, pharmaceutical companies provide supplementary information about potential side-effects and what to look out for which the patient is advised to read. Whether they are capable or willing to read all of this information may not be the case especially if unwell and looking solely for a way to relieve their symptoms. However, as information becomes available at all stages of drug design, testing and manufacturing, perhaps a better way would be to look for patterns of interactions and adverse effects. Indeed, researchers such as Tatonetti, Ye, Daneshjou, and Altman (2012) used data analysis techniques to do just that. Using over 1.8 million adverse event reports, they were able to identify potentially confounding effects significantly improving patient outcomes.

Applying such techniques would perhaps pay dividends in other areas of medicine. There are *quasi* limitless sources of data, from individual care plans and interactions, to data from clinical trials. Analysing all or even just a portion of such data would not be feasible if it had to be done manually. Even in the Tatonetti et al. (2012) study, 1.8 million reports would take a whole team a significant time just to read. As they had done, techniques for data processing at scale were needed. Large corporations like IBM, Oracle and SAP have been providing insights, training, products and services which support the data scientists to approach the processing of the vast amounts of data available (see Khan et al., 2014). So, these big data technologies, including artificial intelligence applications (AI), promise much exploiting the vast range of data available to support and improve healthcare (Belle et al., 2015) though there are some caveats (Mehta & Pandit, 2018; Metcalf & Crawford, 2016). Apart from technology and the complexities of data analytics, all such activities are situated against a background of tight regulatory oversight and control.

Health data are regarded in Europe as so-called special category personal data (European Commission, 2016, Art. 9), requiring additional governance measures regarding data protection.[1] Further, data subjects (those that the data identify) have the right to object to automated processing (European Commission, 2016, Art. 18, Art. 22). Unless their data are truly anonymous, therefore, developing predictive models for healthcare may not be possible. There are additionally concerns around the ethical treatment of private individuals in the context of data analytics and advanced technologies. The Toronto Declaration (Amnesty International and AccessNow, 2018), for instance, situates fundamental human rights within the context of machine learning. In a similar vein, both the European Union[2] and the UK Government Digital Services (UK Government, 2020) are developing guidance on the ethical treatment of personal data, including obligations for all those involved in developing and exploiting big data solutions. Like the Toronto Declaration, the focus is the continued preservation of individual rights to privacy and equitable treatment. Further, the European Union published voluntary guidance on Trustworthy AI in general (European Commission, 2019) and the UK Government Health and Social Care Department has developed a code of conduct for data usage in healthcare and similar public authorities (UK Government, 2021). Although the focus is initially on the secure handling of data, which is already covered by data protection legislation, there is also a concern for the equitable introduction and use of advanced, AI-enhanced technology in the domain.

For healthcare professionals, especially clinicians, the main concern is on the well-being of their patients. Patients rely on their clinician's competence to help manage and hopefully improve their health status. Up until now, the clinician uses the knowledge they gain from training along with the experience they have gathered in practice. Advanced technologies should support this state of affairs, allowing clinicians access to information from a much larger cohort of patients and conditions than they would typically come across in their day-to-day duties. Such technologies may even help individual patients manage and understand their own condition. This is not simply through unsupervised access to the Internet and potentially unmoderated information, but also personal devices collecting and analysing personal and personalised data.

This complex background to the introduction of advanced technologies into healthcare raises multiple questions. How does the introduction of technology into healthcare affect the existing relationships between clinician and patient, for instance? Further, how can the clinician continue to satisfy the regulatory and ethical requirements of their profession when advanced technology is introduced? Are they able, for example, to understand and work with these technologies in the same way that they may have responded to training or consultation with their peers or their professional bodies? This chapter begins by considering technology as introduced within healthcare before moving on to specific issues around the ethics of healthcare, interacting with advanced technology, and some empirical evidence of user reaction to data-driven technologies. The chapter concludes with a set of recommendations for the future development and deployment of advanced technologies.[3]

10.2 THE DISRUPTIVE INTRODUCTION OF TECHNOLOGY

In their Lofoten Islands study, Dyb and Halford (2009) describe the introduction of supportive technologies for care during pregnancy. Given the remoteness of the islands and the challenges of maintaining local specialist support, the motivation was to allow Lofoten midwives to consult directly with specialist clinicians on the mainland. This would effectively provide the same level of care and treatment to the locals as if they were under the care of the nearest mainland hospital, it was hoped. The trial was not a success despite the sophistication of the technology and efficiency of the underlying infrastructure. For Dyb and Halford, the reasons lay principally in place: islanders wanted on-site support and reassurance from clinicians they knew and who were part of the community. The midwives themselves reported an unwelcome disconnect. It was difficult, they claimed, to report via video link what you felt as you examined a woman and her unborn child. However, there is an alternative explanation and one which Dyb and Halford (2009) touch on. Introducing technology into an existing socio-technical system, the actor network they refer to, has a disruptive effect. Relationships change not least trust between patient and clinician, and the latter's reliance on the expertise of shared knowledge on clinical practice.

An actor network (Latour, 2005; Law, 1992) seeks to encapsulate the various constructs involved within a socio-technical system. Constructs may be human (individual agents and stakeholders), institutions or technologies. Although the role, essentially the agency, of technology may not be uniformly agreed, an important characteristic of the original actor network concept assumes that both human and non-human components have agency and influence the workings and thereby the overall effectiveness of the network.

Figure 10.1 summarises a simple actor network for patient–clinician interactions. A *Patient* trusts the *Clinician* to provide them with appropriate and effective treatment on an individual basis, during a consultation for primary or secondary care, for example, or over an extended period including specialist treatment for a specific condition or set of circumstances like pregnancy and childbirth for the Lofoten islands. The *Patient* is not alone, however. Patient advocates and both government and non-government agencies (referred to as *Community* in the figure) fulfil a representative role, taking patient interests forward to government to lobby *Regulators*

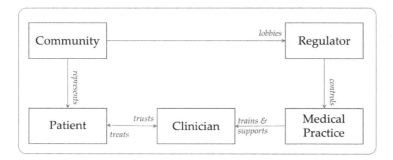

FIGURE 10.1 A schematic actor network for clinical care based on patient–clinician interactions.

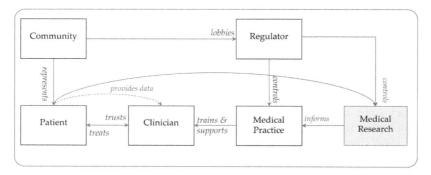

FIGURE 10.2 The influence of medical research.

for appropriate safeguards and control. In turn, *Regulators* oversee *Medical Practice*, introducing standards and governance to what can and cannot be done. Finally, the *Clinician* relies initially on the training, and later on guidance from *Medical Practice* to remain current and satisfy their responsibilities to treat their *Patients*. This type of representation of clinical care has persisted for many years. The *Clinician* must demonstrate competence against prevailing *Medical Practice* and thereby develop trustworthiness vis-à-vis their *Patient*. The *Clinician* has a high level of autonomy effectively and is expected to make on-the-spot judgements for the wellbeing of their *Patients*.

Medical Practice is, of course, informed by significant *Medical Research*. This may take the form of clinical trials, including drug trials, or individual case studies. In either case, the success of *Medical Research* depends on the data received from *Patients*. This may come directly from the *Patients* themselves, or secondary use via the *Clinician(s)* treating them. This is summarised in Figure 10.2.[4] Given oversight from one or more *Regulators*, it is assumed that the quality of *Medical Practice* and *Medical Research* is such that *Clinicians* may rely on them completely.

However, with the advent of new technologies and a growing ease amongst the general population with technologies as part of everyday life, it is evitable that the starting point in Figures 10.1 and 10.2 needs to be reconsidered along with any particular consequences. Figure 10.3 provides an example where the basic healthcare network is

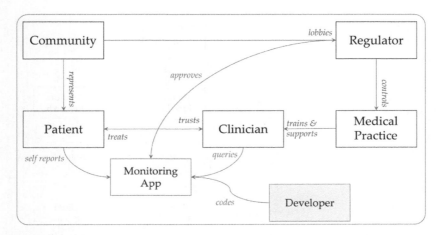

FIGURE 10.3 The effect of introducing a monitoring app into healthcare.

extended to include self-reporting monitoring apps. These may include general apps to count the number of steps taken daily or other metrics such as heart rate and blood pressure. These may or may not be subject to regulation. Increasingly, there are also specialised apps used as part of the overall healthcare regime for the patient to monitor a particular activity such as blood sugar levels under the specific guidance of a clinician. Data collected in this way may form part of a face-to-face consultation. Finally, with increasing familiarity with Internet use and search engines, patients may seek initial information about symptoms or a diagnosis to take to the clinician.

In such scenarios, the simple *Patient–Clinician* interaction is disrupted to the extent that the *Patient* may now come to a consultation with additional information. There is no guarantee that this is helpful or constructive (though Van Riel, Auwerx, Debbaut, Van Hees, & Schoenmakers, 2017, suggest it tends to be beneficial). More importantly, the *Clinician* may have no direct input or influence on the technology which is now producing data that they are expected to explain. Further, where such apps are regulated, there is still no guarantee that the *Developer* is medically trained or subject to the same oversight from *Medical Practice*. For the *Patient–Clinician* interaction to remain intact based on *Clinician* trustworthiness and successful *Patient* outcomes, that is satisfactory treatment, the *Clinician* must also be able to interpret the output from the *Monitoring App* at least to the satisfaction of the *Patient*.

The situation is exacerbated with the advent of advanced technologies, such as Machine Learning (ML) and Artificial Intelligence (AI) more generally.[5,6] The actor network here might be seen as a decision-support system, whereby given a set of symptoms or the collection of drugs the *Patient* currently takes the *Clinician* may search for likely diagnoses or conflicts between medications that should be avoided. To a large extent, of course, the *Clinician* retains ultimate responsibility for the outcome with the *Patient*. Indeed, the *Patient* may be unaware of any semi-automated decision-support system and assume that the *Clinician* either knew already how to treat them or had consulted other professionals as necessary. The trust relationship, therefore, remains between *Patient* and *Clinician*. It is up to the *Clinician* to satisfy themselves that such technology has been appropriately developed and tested.

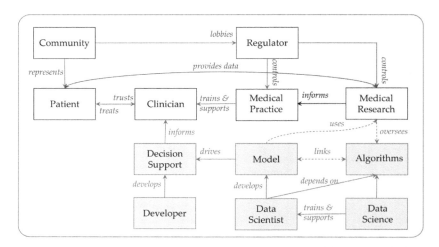

FIGURE 10.4 A suggested actor network for advanced technology introduction into healthcare.

On that note, there are multiple entities within the overall socio-technical system who must provide input to generate such a decision-support system. There must be a *Developer* in the first instance who codes the system itself, including the appropriate user interface for the *Clinician* to query. Such a system is assumed to be robust and reliant (on reliance versus trust, see Lankton, McKnight, & Tripp, 2015). The *Clinician* may quickly adapt to and understand the limitations of the system, exploiting it to support them in much the same way that previously they may have consulted literature or their peers. Understanding the system enough will allow them to identify and compensate for potential failings (Lee & Moray, 1992; Lee & See, 2004).

The simplest decision-support systems may encode heuristic experience and so simply provide an efficient way of accessing and perhaps reasoning over that knowledge. However, if heuristics are replaced with more advanced techniques such as ML or other AI approaches, then the decision-support system the *Developer* codes becomes a less sophisticated component. Rather than executing the heuristics as before based on parsing the query presented to it, such systems would instead provide a relatively simple query interface to a much more sophisticated *Model*. Such *Models* must be created by specialists, *Data Scientists*, who use the *Algorithms* of *Data Science* to process the data available to them, typically from different types of *Medical Research*.[7] The *Models* typically identify patterns in large datasets ("big data") to provide predictions (Chmiel et al., 2020; Wang, Kung, & Byrd, 2018).

There is no reason, of course, why multiple technologies may not be deployed in tandem. Figure 10.5 illustrates the case where a *Monitoring App* for patients to manage their own health and well-being to some degree (see also Figure 10.3) is used at the same time a clinician uses a decision-support system (as seen in Figure 10.4). In this scenario, data may also be available from the app itself to inform the model in the decision-support system. For instance, a patient may be monitoring their heart rate on a semi-permanent basis. This information can be given directly to the clinician, of

Work with Me, Don't Just Talk at Me

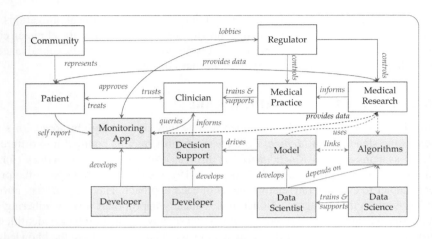

FIGURE 10.5 Complete advanced technology enabled healthcare socio-technical system.

course, for them to use in the next consultation. But the data may also be aggregated with information from others to develop, for example, a model for the early prediction of cardiac arrest or the risk of stroke. Such additional information would then be available to the clinician for the next consultation. Recent examples of each type – decision support and self-monitoring apps – include the modelling activities of advisory groups like Scientific Advisory Group from Emergencies (SAGE) in the UK[8] and the Centers for Disease Control and Prevention (CDC) in the United States,[9] and individual track and trace specific to the management of coronavirus transmission (Parker, Fraser, Abeler-Dörner, & Bonsall, 2020; Rowe, Ngwenyama, & Richet, 2020; Velicia-Martin, Cabrera-Sanchez, Gil-Cordero, & Palos-Sanchez, 2021; Walrave, Waeterloos, & Ponnet, 2020) respectively. SAGE and the CDC were both dependent on data-driven advanced technologies to be able to evaluate evidence and formulate guidance for policy makers in managing the pandemic. Track-n-trace or contact tracing apps were seen as one significant source of data in developing such models as well as providing direct broadcast of advice to private citizens (Hinch et al., 2020).

In this section, the socio-technical systems supporting patients and clinicians have been presented to show the potential effects of increased technology deployment. Other examples are possible, of course; though only decision-support systems for the clinician and self-monitoring apps for the patient were considered here. What becomes clear as the number of technology components increase is that the simple trust relationship between clinician and patient is no longer a straightforward reliance by the patient on the competence of the clinician. Instead, the clinician is expected to maintain their trustworthiness while relying on complex technology developed by other specialists in other disciplines and which may not be completely transparent to them. Further, a patient using a self-monitoring app may well welcome the additional information being made available to the clinician almost passively. At the same time, however, there is a danger that they lose control and oversight of how their data are used and integrated into a broader technological healthcare solution. This raises questions of how the clinician can continue to meet their ethical obligations in treating their patients, dealt with in the following section, and of how the clinician can be expected to understand and interpret the results coming from data-driven advanced technologies, which will be dealt with in the subsequent sections.

10.3 THE ETHICS OF MEDICAL CARE

Clinicians are regulated and constrained in how they operate both by *Regulators* as well as the standards of *Medical Practice*. Standards may be codified into different ethical systems such as *Rawls' theory of justice, utilitarian or virtue ethics* (for a summary of these approaches, see Parsons, 2019). It's beyond the scope of the present chapter to evaluate the relative merits of each approach. Having said that, *virtue ethics* assumes that an agent will act in an ethically appropriate way by adhering to standards that agents *should* comply with in that society. Such compliance identifies a virtuous agent, therefore. For the clinician, this would mean they have the knowledge and experience of what is considered appropriate behaviour.

The four principles of medical ethics as outlined by Beauchamp and Childress (2019) extend the original Hippocratic concept of *non-maleficence* ("do no harm" to the *Patient*) to include a commitment to do what's best for them (*beneficence*), to seek and respect their opinion and choice (the *right to autonomy*) and for any advantages of intervention or treatment to be available to all (*justice*). These or similar principles continue to be taught during medical training (Eckles, Meslin, Gaffney, & Helft, 2005), albeit with varying degrees of effectiveness (Carrese et al., 2015; Maxwell et al., 2016). There are those who continue to support this simple framework as a touchstone for clinical practice (Dawson & Garrard, 2006; Gillon, 1994, 2014). For healthcare which includes data-driven advanced technologies, the clinician must still be able to adhere to these principles.

There are those, however, who take issue with the four principles. They are too abstract and lacking a basis in common morality (Christen, Ineichen, & Tanner, 2014; Karlsen & Solbakk, 2011; Muirhead, 2011; Rhodes, 2015), though as suggested they align with a *virtue ethics* approach. Further, the four principles may not provide practical support for split-second and potentially life-changing clinical decisions (Cowley, 2005; Page, 2012). That being said, they are accepted at least in part by some of the patient population (Chung, Lawrence, Curlin, Arorar, & Metzler, 2012). Further in a study on the privacy of health data, Liyanage et al. (2016) found that experts across multiple European countries agreed with *autonomy* as well as a fundamental need for trust across the socio-technical ecosystem. So there seems to be some value to the principles especially in connection with community-wide, multi-stakeholder discussion (Boyd, 2013). Indeed, similar principles exist in related fields (APA, 2010; BPS, 2009, 2014), not least in research (Department of Health, 2014). The real question here though is whether clinicians can internalise these principles and continue to apply them when supported by advanced data-driven technologies.

One specific concern about the four principles relates to primacy. The assertion is that a *right to autonomy* should be viewed as more important than the other three (Campbell, 2003; Gillon, 1994, 2014). But this is not without issues. First, religious conviction (Campbell, 2003; Page, 2012), especially in relation to minors or those temporarily or permanently unable to make such decisions (Muirhead, 2011; Rubin, 2014), highlights the obvious conflict between patient autonomy and the clinician's motivation for *non-maleficence* and *beneficence*. This in turn raises questions of cultural applicability, and thereby the relative importance of individual and society (McLaughlin & Braun, 1998; Nijhawan et al., 2013). This is not simply about collectivist versus individualist cultures (Hofstede, Hofstede, & Minkov, 2010). As the SARS-CoV-2 pandemic has shown, there are times when community or general societal imperatives may be deemed to outweigh individual rights. In the UK, for instance, the COPI Regulations specifically require some data sharing (UK Government, 2002). Clinicians would be expected at least to consider the cultural and temporal consequences of their decisions even though a priori in alignment with the four principles.

Irrespective of the particular ethical framework that applies or practical concerns about any set of principles, the clinician is faced with a significant problem regarding the introduction of advanced technologies into their clinical practice. How such technologies influence their ability to commit to such principles is affected by their dependence on the *Developer* but more especially on the *Data Scientist* responsible for those technologies (as shown in the figures in the previous section) and *Data Science*. To approach an answer to this question, it is important to step back and consider the characteristics of data-driven technology.

10.4 "BIG DATA" IN DATA-DRIVEN ADVANCED TECHNOLOGIES

Advanced technologies which depend on the processing of large amounts of data, or *big data*, share similar properties and pose common problems in terms of data storage and curation (Hashem et al., 2015; Khan et al., 2014). For some time, big data has been described in terms of specific characteristics: commonly summarised as the *four Vs*,[10] namely *Volume*, *Variety*, *Velocity* and *Value*; the current discussion will also include *Veracity*. Each affects the quality of the outcomes of data-driven advanced technologies and therefore will influence the soundness of the conclusions the clinician may draw.

Volume refers to the amount of data available and used for the predictions made by advanced technologies. It is an issue for *Data Science* to determine how much data are required particularly how such data are shared between training and validation. The clinician, however, must be aware of how the *volume* of data affects the accuracy and generalisability of any result from the decision-support system. In consequence, it may be difficult to guarantee *justice* (ensuring that all patients are treated equitably), *non-maleficence* in consequence, and *autonomy* or *beneficence* (in that individual rights and concerns cannot be singled out from an extensive amount of data, nor in return specific benefit to individuals). Further, the clinician has no oversight of any data automatically shared from individuals via a self-monitoring app. Their sole reliance would be regarding an approved self-monitoring app. The trust a patient develops in response to the trustworthiness shown by the clinician is therefore dependent on the trust or reliance a clinician has in *Data Science*, the *Developer/Data Scientist* and/or the appropriateness of approval from the *Regulator*.

Variety relates to the number of different sources of data. In practical terms, this may include data from different cohorts across different locations and geographies, and perhaps with different conditions and comorbidities. In addition, it would also relate in the figures above to data from the self-monitoring apps and from direct patient–healthcare interactions, not all of which the clinician may know about. The challenges for the clinician are therefore the same. They cannot guarantee alignment with any of the ethical principles outlined above without a detailed understanding of the data-driven advanced technology. What is more, there needs to be some agreement about how and when model updates are available, what they include and why, and how to introduce or at least represent the specific sub-cohort of patients that they are directly responsible for. This again introduces challenges to the four principles. For instance, depending on the nature of different types of data to be aggregated, it may not be possible to ensure *justice* – or equitable treatment for all patients. If environmental factors were included that are relevant only to a subset of patients in the dataset, then this may produce results which are not relevant to the rest.

Velocity refers to the rate at which data become available from whatever source. Self-monitoring apps, for instance, may be expected to receive data regularly. In the case of track and trace apps, this would be *quasi* constant as new infections are registered and those potentially in contact have been alerted. For decision-support systems, *Velocity* will depend on the rate at which data from different sources (see *Variety*) are received. This also means that data may need synchronising on an ongoing basis. Again, the clinician is totally dependent on *Data Science*. Further, although they may make decisions about what data are relevant for their patient, the *Data Scientist* may at the simplest level rely on what appears to be greater robustness and richness of the model they develop. The *Data Scientist* is now dependent on the insights of the clinician to reason over the results the model generates. Here, it is not only a challenge for the clinician to adhere to the four principles for instance, but also for them to acknowledge and be able to satisfy any dependence from *Data Science* on their insights and experience. The main ethical issues here are much the same. If relevant data are not available at the right time, then this may prejudice predictions for some patients. *Beneficence* and even *non-maleficence* may be affected, for instance.

Value means the benefit derived from the data-driven models, especially unexpected results. Ultimately, this will be seen on a case-by-case basis: although a clinician's patient list only represents a fraction of the data required and used to build a given model if the clinician and those patients do not see identifiable benefit back to them. To some degree, this is a common issue in healthcare, namely the balance between idiopathic presentation versus common, nomothetic manifestations of a condition. For the clinician to make this judgement, once again they must understand the technology and its outputs to make a critical evaluation of the *Value* for the individual cases of their patients. Fulfilling ethical obligations is again a challenge. Demonstrating *Value* in specific (i.e., per patient) and general (i.e., across all practice) requires co-operation and understanding between *Clinician* and *Data Scientist*. This will vary between primary care, focused on individual patients, and secondary care, where clinicians will have a perhaps broader experience of specific pathological presentation. This also requires, of course, a willingness on the part of the clinician to reassess how they deal with outcomes from the data-driven advanced technology especially when updated model versions are released. This has implications, of course, for all of the ethical principles for medical care. There has to be value to all patients from what is found from data modelling otherwise *justice* at the very least will be compromised.

Veracity refers to the accuracy and general quality of the data. Good data are essential to ensure a robust and reliable model (notwithstanding the requirement for enough data, appropriate *Volume*). Quality issues may include missing data, mistakes in data input, possible corruption in transit and so forth. Now, data from *Medical Science* should be accurate: appropriate and effective treatment is crucially dependent on the accuracy of data such as readings and laboratory testing. However, including other sources (*Variety*) such as patient self-reporting, or sources such as environmental data, may not be so reliable. Further, in general data cleaning is seen as a statistical problem such as imputation[11] and outlier handling rather than a specifically medical one. In consequence, truly idiopathic or unusual symptom presentation may be removed from a dataset meaning it may never be represented in the output model. This is one of the main concerns regarding big data characteristics. If the data are not accurate, because insufficient data are available, outliers removed or inappropriate imputation applied, then the clinician cannot meet their obligations towards their patients.

A summary review of the main characteristics of data-driven advanced technology (the four or five *Vs*), or at least the *big data* component, highlights specific issues with a clinician's responsibility to respect the ethical standards of their profession. More importantly, though, they highlight the different dependencies of the clinician on the outputs of any model they may rely on. Given the assumed benefit of advanced technologies for healthcare and for deriving benefit from the vast and ever-increasing amounts of data available (e.g., Belle et al., 2015), there is a clear need for clinicians to alter their way of working. They can no longer rely on clinical competence alone but must engage with advanced technologies to ensure compliance with ethical standards and thereby fulfil their obligations to their patients. Notwithstanding issues associated with *Volume*, *Variety*, *Velocity*, *Value* and *Veracity*, there has been much work done on understanding and trying to deal with how understandable the underlying technologies might be, especially for the non-specialist stakeholder.

10.5 EXPLAINABLE AI

Taking the discussion around artificial intelligence as an important example of data-driven advanced technology, the question arises how users of such technology – in this case clinicians – react to the technology and what their expectations might be. The concept of Explainable AI dates back to the beginning of the millennium at least (Adadi & Berrada, 2018) and has been formalised in research programmes in the United States (Gunning & Aha, 2019) and Europe (European Commission, 2021). DARPA define the concept thus:

> Systems that can explain their rationale to a human user, characterize their strengths and weaknesses, and convey an understanding of how they will behave in the future.
>
> (Gunning & Aha, 2019: 44)

This definition is concerning. Anthropomorphising technology – seeing humanness in the data-driven advanced technology – will influence user response to that technology (Weitz, Schiller, Schlagowski, Huber, & André, 2019). A clinician may make assumptions about the technology, no longer feel the need to compensate for algorithmic oddities (Lee & Moray, 1992; Lee & See, 2004), and shift their perceptions away from reliance to trust (Lankton et al., 2015). DARPA's call for "explanation-informed acceptance" (Gunning & Aha, 2019: 51) of advanced technologies within the socio-cognitive context of its use (in their case gaming) would seem a valid goal. But is it premature?

There is some inconsistency about how to characterise explainable AI. Common attributes include explainability, transparency, fairness and accountability (Adadi & Berrada, 2018; Arrieta et al., 2020; Taylor et al., 2018). There may, of course, be different reasons to look for explainability: for control, to justify outcomes, to enable improvement and finally to provide insights into human behaviours ("explain to discover") (Adadi & Berrada, 2018). For clinicians, justifying outcomes would provide them with a narrative to reassure their patient, and improvement would lead to better care (responding at least to a responsibility for *beneficence*). Interestingly, though, providing new insights would effectively form part of the clinician's ongoing training, and ad hoc continued professional development. Explainable data-driven advanced technologies would therefore become not only a mechanism to support day-to-day clinical care for patients but could potentially offer learning opportunities for the clinicians themselves. At the very least, explainable AI in this context would in essence provide the clinician with an automated colleague to consult: they could potentially use the advanced technology as a source of confirmation or counter-explanation associated with how the patient presents, whereas in the past they may have discussed with a colleague at some later date. The danger, of course, is that the clinician mistakes the advanced technology for a colleague, changing reliance on the robustness of technical performance into a *quasi* inter-personal trust relationship. This may be what the DARPA program aims at (see the definition above).

This still assumes the output from the advanced technology to be correct, however that is defined. While some researchers seek to identify formalised, objective methods to classify and validate performance (Samek, Wiegand, & Müller, 2017), others come back to a perceived need to quantify subjective responses to explainability and interpretability (Došilović, Brčić, & Hlupić, 2018). Indeed, there is some evidence that acceptance of advanced technologies includes an emotive element (Khrais, 2020). If clinicians therefore develop a collegial attitude to the technology – exploiting "explain to improve" and "explain to discover" (Adadi & Berrada, 2018) as they would a colleague whose competence they trust – then the technology itself must provide assurances (Israelsen & Ahmed, 2019) in much the same way that trustworthiness needs to be developed and maintained by the clinician to encourage patient trust. So, as the patient–clinician relationship (see Figure 10.1) is predicated on the co-construction of a trusting relationship, so the clinician's perceptions of the advanced technologies they use (the *Decision Support* system in Figure 10.4) needs to derive from similarly socially constructed trust development (Rohlfing et al., 2020). The question is whether the co-construction of trust depends on human actors (i.e., the clinician and the data scientist) as researchers like Weber and Carter (2003) would maintain for personal relationships, or can be directly between clinician and an anthropomorphised advanced technology as a logical next step from DARPA (Gunning & Aha, 2019) including deliberate attempts to humanise the technology (Weitz et al., 2019). If that is the case, namely that clinician-to-advanced-technology interaction effectively derives from a socially constructed trust relationship, there is indeed a need for a new kind of ethics to ensure responsible AI (Taylor et al., 2018). Whereas official initiatives target the developers (European Commission, 2019; UK Government, 2020, 2021) as does much of the Explainable AI literature (Arrieta et al., 2020), there still needs to be equal attention to how consumers of these technologies react to them.

10.6 A NOTE ON CLINICIANS AND WORK

In the previous sections, the changing network of relationships in clinician–patient interactions with the introduction of data-driven advanced technologies has been outlined, a generic set of ethical principles for medical practice introduced, and some of the features of advanced technologies and attempts to render such technologies acceptable and understandable to clinicians in their work highlighted. It is worth considering briefly how this is likely to affect the clinicians themselves. There have been many surveys of clinician well-being in different cultures and clinical environments (see Shanafelt, Sloan, & Habermann, 2003; Whitley et al., 1994). Spickard Jr, Gabbe, and Christensen (2002) highlight contributory factors such as control and workload (p.1448) as well as finding work meaningful (p.1449), and external factors like regulation and working practices (p.1448). Introducing data-driven advanced technologies may well contribute to this.

General psychological models of work and work motivations provide some insights here. The Job Characteristics Model (Fried & Ferris, 1987; Hackman & Oldham, 1976) as originally conceived, for instance, positions high job satisfaction and performance as a result of job characteristics such as skill variety and autonomy in combination with psychological responses such as the perceived responsibility for outcomes, meaningfulness of the work and understanding of results (see Hackman & Oldham, 1976: 256). Subsequently, Pierce, Jussila, and Cummings (2009) added psychological ownership of the task. According to one of the conceptions of the Job Demand-Control Model (Karasek Jr, 1979), increasing demands (e.g., workload) and decreasing control over the work both increase stress. Later, Fila, Purl, and Griffeth (2017) assert that for occupations like healthcare feelings of control decrease as the demands of the job increase.

Combining the two, the conclusion would be that reduction in the responsibility for outcomes (from Job Characteristics) and perceptions of decreasing control with increasing demands (from Job Demand-Control) would negatively affect job satisfaction and performance. Introducing data-driven advanced technology into healthcare may well result in such adverse effects. As described in the preceding sections, regulation, ethical standards and understanding the technology itself need to be considered. In moving forward, it is essential to evaluate these factors. Introducing advanced technology may not only be disruptive to the socio-technical system as illustrated in the figures above but may also have a direct influence on how clinicians both perceive (is it meaningful to them?) and manage (is it too demanding?) their work.

10.7 THE LIKELIHOOD OF ADVANCED TECHNOLOGY ACCEPTANCE

Four recent studies[12] are relevant to the introduction of different types of technology within healthcare.[13] The first one asked students about taking part in a novel diagnostic test and illustrates that technology – or in this case a process for collecting samples, biological analysis, and reporting back to the student – is seen not simply in terms of efficacy or ease of use, but also within a broader socio-technical context. The next two explore app users' or potential users' responses to self-monitoring apps, once more showing a more nuanced though favourable response to technology use. The final one asked different primary actors how they felt about data-driven advanced technology, highlighting among other things that there needs to be open communication and understanding between the different actors.

During the SARS-CoV-2 pandemic, a cohort of over 900 students were asked to take part in a new diagnostic test for COVID. Although the test was easy and non-invasive (perceived ease-of-use), the accuracy of outcomes came into question (perceived usefulness). Traditional models of technology acceptance (Davis, 1989; Venkatesh, Morris, Davis, & Davis, 2003) would therefore predict general adoption of the intervention, that is the diagnostic test, assuming accuracy could be improved. For advanced technologies, therefore, a decision-support system or a self-monitoring app should be easy to use and produce predictable or understandable results. Is that enough? The students also reported two major concerns. First, if results were not accurate, then people could be socialising with others unaware that they are also potentially infecting them. This is not just a caveat about test accuracy (perceived usefulness) but also contextualises outcomes against a broader social backdrop, namely community obligations. Second, they reported that an inhibitor against engagement with the test was a fear that if they tested positive this would have a detrimental effect on those closest to them, not so much health wise but in terms of restricting their lives. So, their willingness to engage with the testing is mediated by how they perceive the consequences for those around them. For data-driven advanced technologies, it is not simply their efficacy and transparency which would be important, therefore. It is also the effect they would have on the existing socio-technical context.

Another potential effect might simply be to undermine the quality of existing care structures. Looking at trust in technology, the TRIFoRM project interviewed five self-selecting rheumatoid arthritis sufferers and one clinician about the potential use of a self-monitoring and reporting app (Hooper, Pickering, Prichard, & Ashleigh, 2015). In response to the statement:

> Technology, services, and systems can be relied upon to disclose risks of using them and disclose errors when they occur.

all participants disagreed. They are aware that there is some way to go on transparency and explainability with technology. Against this, they recognised the utility of technology so long as it enhanced and did not replace their existing care. They

TABLE 10.1
Responses to Some Statements about Using Healthcare Apps

	Agree	Disagree
I don't trust healthcare apps will get it right for me	186	214
I don't feel the health service always has enough time for me	250	150
Using a healthcare app on my own means I'm being fobbed off	84	316
Using healthcare apps means I'm not so alone	203	197

expressed little concern, however, about sharing their data provided such sharing would benefit fellow sufferers. They also reported an appreciation that assistive technology would help record important information when they were unable to because of the severity of their pain, and that this would help improve and support actual consultation with the clinician. Finally, they identified that having access to a lot of information might help the clinician understand their particular condition. Once again, therefore, introducing technology is set in the context of a community of sufferers, not just as perceived usefulness, so long as this did not undermine existing relationships with their clinician.

In a separate survey, 400 UK private citizens[14] were asked about their perceptions of using healthcare apps. Some 143 reported not using a healthcare app; 179 reported they did; the others did not comment about app usage. Table 10.1 summarises responses to some of the statements.

Regarding app accuracy, a substantial minority (186/400 or some 46.5 percent) are suspicious of technology in this setting. Remembering that this was an anonymous online survey, this suggests that private citizens tend to be wary of healthcare-related apps. However, there is also a feeling that access to healthcare is not always straight forward: 250/400 (62.5 percent) of respondents felt that the health service did not have enough time for them as individuals. Whether this reflects perceptions specifically during the SARS-CoV-2 pandemic (the survey was run in July 2021) is not clear, since the overwhelming majority (316/400 or 79 percent) did not think introducing a self-monitoring app was an indication that the health service wished to avoid caring for them. Finally, when asked whether using an app would help mitigate against feelings of isolation, respondents were split almost 50/50 (203/400 or 50.8 percent agree to the statement). This is only an extract of responses; there were 28 assertions in all, most derived from a common model of healthcare adoption, which the study validated. What we see, however, is that attitudes towards healthcare apps are not simple or uniform across respondents. There may be different motivators and different responses to app accuracy and how an app might fit into healthcare. Across these three studies, it is apparent that not one explanatory factor – technology accuracy, socio-technical context, perceptions of healthcare delivery – obtains. Patient responses to technology vary as much perhaps as they do themselves. This complexity has led other researchers (Greenhalgh et al., 2017; May et al., 2009 as discussed below) to develop multi-layered adoption frameworks to guide developers and innovators around healthcare.

The fourth and final study focused more directly on clinicians and data scientists, as well as other stakeholders, and their perceptions and attitudes to advanced technologies such as decision-support systems. Forty-six collaborators on a European H2020 project[15] responded voluntarily to an anonymous survey. The questionnaire contained 30 assertions across four domains (requirements, design & responsibility, ethics & governance, and transparency) derived from a Delphi study with domain experts (Taylor et al., 2018). Participants identified themselves as clinicians (4), data scientists (18), social scientists (5), vendors (5), other (14). Domain and role both turned out to have a significant main effect on responses, with domain accounting for some 50 percent of the variance in responses (partial η^2 was 0.501) and role for 8.5 percent (partial η^2 was 0.085). Ethics and governance provoked the least agreement with the experts in the Taylor et al. (2018) study, while other domains tended to show agreement; clinicians and vendors showed most agreement with the experts, and social scientists the least. Despite the limited scope of this study, there are clear indications that different actors within the healthcare network (see above) hold different views, and that different aspects of data-driven advanced technology as represented by these five domains may require more or less attention. Overall, participants agreed that these technologies can be disruptive and that there is a clear need for cross-disciplinary discussion and agreement. A set of 18 recommendations were derived. These may be summarised as follows:

> Technology in healthcare is one component within a broader, complex network. Focus needs to be given to how technology can affect even disruption [*sic.*] existing relationships... where advanced technologies are to be deployed, all main actors (those directly involved with the technology) and all other stakeholders (those affected by the technology) should be consulted.

and in a free form comment left by one respondent:

> We need to start viewing the world as a socio-technical system where humans and technologies are networked together and inseparable [from] each other.

For different actors in healthcare networks, these four studies indicate that consumers of healthcare (patients) have multiple and nuanced responses to technologies in general. Further, different actors in the healthcare ecosystem all agree that there needs to be open, cross-domain discussion to understand issues and ensure the successful development and deployment of advanced technologies.

To complete this section, it is important to consider how this relates to the acceptance and possible adoption of advanced technologies. Through my own work, it is apparent that traditional models of technology acceptance are insufficient. Instead, potential users of technology will rely on their own intuitions and create a personal narrative rather than responding blindly to technology-centric factors like ease-of-use and perceived usefulness (Pickering, Bartholomew, Nouri Janian, Lopez Moreno, & Surridge, 2020; Pickering, Boletsis, Halvorsrud, Phillips, & Surridge, 2021; Pickering et al., 2019). Given the challenges of predicting intervention and technology uptake within healthcare as stated above though, other researchers have suggested empirically based adoption frameworks. Greenhalgh and her colleagues focus on the complexity of an intervention as well as continuous and ongoing discussion among stakeholders (Greenhalgh & Abimbola, 2019; Greenhalgh et al., 2017). The NASSS (non-adoption, abandonment, scale up, spread, sustainability) would suggest that data-driven advanced technologies would not be adopted readily into the healthcare context. If stakeholders – especially patients, their advocates and clinicians – are suspicious of how algorithms work either because they lack transparency or because of increasing nervousness about what they do (Cheney-Lippold, 2017; O'Neil, 2016), NASSS predicts that data-driven advanced technology adoption is unlikely. Normalisation process theory (May et al., 2009), by contrast, focuses on "cognitive participation" as a major construct. This would mean that all major stakeholders, especially patients and clinicians, would need to embrace and understand at least something of the technology and be prepared to develop a narrative around it (see, for instance, Pickering et al., 2019). As the empirical studies in this section have shown, there may already be a willingness from those stakeholders to engage, to try to understand the benefits they may derive from the advanced technology, and thereby to appreciate how it will work for them.

10.8 RECOMMENDATIONS

Given the previous sections, it is important to record what might be done in future to help the major actors understand and integrate data-driven advanced technology especially patients and clinicians but also data scientists.

- *Deployment of disruptive technology*: All major stakeholders must discuss and agree how technology is deployed, where in the healthcare regime it should sit and how it should be used. This will require ongoing communication between the main players and the development of a common language between them as well as agreed checks.
- *Ethics*: Given the potential of data-driven advanced technologies to improve outcomes for patients as well as support clinicians and effectively provide them with data and information they would not otherwise have access to, there needs to be a new perspective on governance and ethics. The four principles of medical ethics would be a good starting point to encourage data scientists and those using their technology to validate that they continue to respect the rights and expectations of patients. In addition, it is important to consider the implications of trust between patient and clinician, and whether it is appropriate for encouraging a trusting relationship rather than traditional reliance on advanced technology.
- *Technology development*: Validation and approval of technology should be overseen by relevant stakeholders. This may include the extension of regulatory oversight and accreditation by relevant medical authorities. Most importantly, responsibility cannot rest solely with the technologists; patients and especially clinicians must be involved in testing and approval, including accepting responsibility for the acceptability of the technology, given their understanding during communication as outlined above.

- *Explainable AI*: There needs to be an effort not only to improve the transparency and explainability of advanced technologies, but also a mechanism for users – clinicians and also the patients they treat – to be able to provide input and encourage "co-construction" of technology transparency.
- *Clinician activities*: With greater understanding of the technology through their involvement and through regular consideration of how the technologies might best be used, clinicians need to be supported and encouraged to identify the value of technology and what they need in future as data science progresses. In this way, they might develop an understanding of how they can make meaningful use of technology to act autonomously and benefit their overriding responsibility to their patients.

10.9 CONCLUSIONS

The emphasis in this chapter has been on collaboration and communication. Introducing advanced technologies into clinical practice involves many different challenges. To meet them all requires concerted effort by all parties to understand one another and continuously co-create acceptable innovation. To begin with, it is essential to understand where technology fits and the effects that its introduction has on existing healthcare processes and especially healthcare relationships. This extends to the technology and how it is developed. Clinicians in particular need to be able to understand but also engage with technologists to ensure that what is produced is appropriate and does not compromise the clinician's ethical duties towards their patients. Patients and clinicians should engage with data scientists and other technologists to monitor and understand how advanced technologies work to be able to make informed decisions about their use. Cross-disciplinary communication is never easy, but in this case, there is no alternative if all parties are to feel they can engage effectively with the technology. There is already empirical evidence indicating a willingness to engage and to embrace new technologies. It is important moving forward to encourage and develop that willingness to share responsibility in how data-driven advanced technologies are developed and deployed in future.

ACKNOWLEDGEMENTS

This work was funded in part by the European Union's Horizon 2020 research and innovation programme under grant agreement No 780495 (project BigMedilytics). Disclaimer: Any dissemination of results here presented reflects only the author's view. The Commission is not responsible for any use that may be made of the information it contains.

NOTES

1. In the United States, the US Department of Health and Human Services and the Health Insurance Portability and Accountability Act of 1996 (HIPAA) define individual rights specifically for healthcare www.hhs.gov/hipaa/index.html.
2. https://digital-strategy.ec.europa.eu/en/policies/strategy-data.

3 Throughout this chapter, I do not distinguish Machine Learning, Artificial Intelligence, Big Data, and Data Analytics. Since they tend to provide predictions dependent on significant amounts of data, I use the term *data-driven advanced technologies* as an umbrella term to include any and all of those technologies.
4 In this and subsequent actor network diagrams, new constructs are shaded. They are associated with new connections back to the existed boxes.
5 The addition of *Medical Research* to inform *Medical Practice* and controlled by the *Regulator* could have been included in the previous networks but were left out for simplicity.
6 For now, the *Monitoring App* has been removed to focus on the effects of introducing advanced technologies like ML and AI.
7 Occasionally aggregated with data from other sources, such as environmental indicators such as pollution and weather.
8 www.gov.uk/government/organisations/scientific-advisory-group-for-emergencies.
9 www.cdc.gov/coronavirus/2019-ncov/science/science-and-research.html.
10 It is beyond the scope of this chapter to provide a detailed review of the history of *big data* and terminology.
11 Simply put, imputation involves interpolation of missing values.
12 All of the work cited here was approved by a Faculty Research Ethics Committee at the University of Southampton, UK: ERGO/FSS/61445.A1, ERGO/FPSE/14892, ERGO/FEPS/65003 and ERGO/FEPS/65194.A1 respectively.
13 The full results of these studies are reported elsewhere. What is presented here focuses only on specifically relevant work.
14 Participants were balanced across relevant demographic categories for the UK.
15 BigMedilytics (Grant Agreement 780495).

REFERENCES

Adadi, A., & Berrada, M. (2018). Peeking inside the black-box: A survey on explainable artificial intelligence (XAI). *IEEE Access*, *6*, 52138–52160. doi:10.1109/ACCESS.2018.2870052

Amnesty International and AccessNow. (2018). *The Toronto Declaration: Protecting the right to equality and non-discrimination in machine learning systems*. Retrieved from www.torontodeclaration.org/wp-content/uploads/2019/12/Toronto_Declaration_English.pdf

APA. (2010). *Ethical Principles of Psychologists and Code of Conduct*. Retrieved from www.apa.org/ethics/code/principles.pdf: American Psychological Association.

Arrieta, A. B., Díaz-Rodríguez, N., Del Ser, J., Bennetot, A., Tabik, S., Barbado, A., ... Herrera, F. (2020). Explainable Artificial Intelligence (XAI): Concepts, taxonomies, opportunities and challenges toward responsible AI. *Information Fusion*, *58*, 82–115. doi: https://doi.org/10.1016/j.inffus.2019.12.012

Beauchamp, T. L., & Childress, J. F. (2019). *Principles of Biomedical Ethics* (8th ed.). Oxford: Oxford University Press.

Belle, A., Thiagarajan, R., Soroushmehr, S. M. R., Navidi, F., Beard, D. A., & Najarian, K. (2015). Big data analytics in healthcare. *BioMed Research International*, 1–16. doi:10.1155/2015/370194

Boyd, K. M. (2013). Autonomy, patients' preferences, and leaving decisions to doctors. *Journal of the Royal College of Physicians, Edinburgh*, *43*, 136. doi:10.4997/JRCPE.2013.210

BPS. (2009). *Code of Ethics and Conduct*. Leicester: The British Psychological Society.

BPS. (2014). *Code of Human Research Ethics*. Leicester: The British Psychological Society.

Campbell, A. V. (2003). The virtues (and vices) of the four principles. *Journal of Medical Ethics*, *29*(5), 292–296. doi:10.1136/jme.29.5.292

Carrese, J. A., Malek, J., Watson, K., Lehmann, L. S., Green, M. J., McCullough, L. B., ... Doukas, D. J. (2015). The essential role of medical ethics education in achieving professionalism: The Romanell Report. *Academic Medicine*, *90*(6), 744–752. doi:10.1097/ACM.0000000000000715

Centers for Disease Control and Prevention (n.d.) *COVID-19* www.cdc.gov/coronavirus/2019-ncov/science/science-and-research.html

Cheney-Lippold, J. (2017). *We Are Data: Algorithms and the Making of Our Digital Selves*. New York: New York University Press.

Chmiel, F. P., Burns, D. K., Pickering, B., Blythin, A., Wilkinson, T. M. A., & Boniface, M. (2020). Retrospective development and evaluation of prognostic models for exacerbation event prediction in patients with chronic obstructive pulmonary disease using data self-reported to a digital health application. Retrieved from www.medrxiv.org/content/10.1101/2020.11.30.20237727v1

Christen, M., Ineichen, C., & Tanner, C. (2014). How "moral" are the principles of biomedical ethics? – a cross-domain evaluation of the common morality hypothesis. *BMC Medical Ethics*, *15*(47). https://doi.org/10.1186/1472-6939-15-47

Chung, G. C., Lawrence, R. E., Curlin, F. A., Arorar, V., & Metzler, D. O. (2012). Predictors of hospitalised patients' preferences for physician-directed medical decision-making. *Journal of Medical Ethics*, *38*(2), 77–82. doi:10.1136/jme.2010.04

Cowley, C. (2005). The dangers of medical ethics. *Journal of Medical Ethics*, *31*(12), 739–742. doi:10.1136/jme.2005.011908

Davis, F. D. (1989). Perceived usefulness, perceived ease of use, and user acceptance of information technology. *MIS Quarterly*, 319–340. doi:10.2307/249008

Dawson, A., & Garrard, E. (2006). In defence of moral imperialism: Four equal and universal prima facie principles. *Journal of Medical Ethics*, *32*(4), 200–204. doi:10.1136/jme.2005.012591

Department of Health, Education, and Welfare. (2014). The Belmont Report. Ethical principles and guidelines for the protection of human subjects of research. *The Journal of the American College of Dentists*, *81*(3), 4.

Došilović, F. K., Brčić, M., & Hlupić, N. (2018). *Explainable Artificial Intelligence: A Survey*. Paper presented at the 2018 41st International Convention on Information and Communication Technology, Electronics and Microelectronics (MIPRO), Opatija, Croatia.

Dyb, K., & Halford, S. (2009). Placing globalizing technologies: Telemedicine and the making of difference. *Sociology*, *43*(2), 232–249. doi:10.1177/0038038508101163

Eckles, R. E., Meslin, E. M., Gaffney, M., & Helft, P. R. (2005). Medical ethics education: Where are we? Where should we be going? A review. *Academic Medicine*, *80*(12), 1143–1152.

European Commission. (2016). Regulation (EU) 2016/679 of the European Parliament and of the Council of April 27, 2016.

European Commission. (2019). Ethics guidelines for trustworthy AI. Retrieved from https://ec.europa.eu/digital-single-market/en/news/ethics-guidelines-trustworthy-ai

European Commission. (2021). *New rules for Artificial Intelligence – Questions and Answers*. Retrieved from https://ec.europa.eu/commission/presscorner/detail/en/QANDA_21_1683

European Commission (2022) *A European Strategy for data*. Shaping Europe's digital future https://digital-strategy.ec.europa.eu/en/policies/strategy-data

European Commission (n.d.) *Horizon 2020*. Research and Innovation https://research-and-innovation.ec.europa.eu/funding/funding-opportunities/funding-programmes-and-open-calls/horizon-2020_en

Fila, M. J., Purl, J., & Griffeth, R. W. (2017). Job demands, control and support: Meta-analyzing moderator effects of gender, nationality, and occupation. *Human Resource Management Review, 27*(1), 39–60. doi:10.1016/j.hrmr.2016.09.004

Fried, Y., & Ferris, G. R. (1987). The validity of the job characteristics model: A review and meta-analysis. *Personnel Psychology, 40*(2), 287–322. doi:10.1111/j.1744-6570.1987.tb00605.x

Gillon, R. (1994). Medical ethics: Four principles plus attention to scope. *British Medical Journal, 309*(6948), 184–188. Retrieved from www.ncbi.nlm.nih.gov/pmc/articles/PMC2540719/pdf/bmj00449-0050.pdf

Gillon, R. (2014). Defending the four principles approach as a good basis for good medical practice and therefore for good medical ethics. *Journal of Medical Ethics, 41*(1), 111–116. doi:10.1136/medethics-2014-102282

Greenhalgh, T., & Abimbola, S. (2019). The NASSS framework: A synthesis of multiple theories of technology implementation: A knowledge base for practitioners. In P. Scott, N. De Keizer and A. Georgiou (Eds.), *Applied Interdisciplinary Theory in Health Informatics* (Vol. 263, pp. 193–204). Amsterdam, the Netherlands: IOS Press.

Greenhalgh, T., Wherton, J., Papoutsi, C., Lynch, J., Hughes, G., Hinder, S., ... Shaw, S. (2017). Beyond adoption: A new framework for theorizing and evaluating nonadoption, abandonment, and challenges to the scale-up, spread, and sustainability of health and care technologies. *Journal of Medical Internet Research, 19*(11), e367. doi:10.2196/jmir.8775

Gunning, D., & Aha, D. W. (2019). DAPRA's explainable artificial intelligence program. *AI Magazine, 40*(2), 44–58.

Hackman, J. R., & Oldham, G. R. (1976). Motivation through the design of work: Test of a theory. *Organizational Behavior and Human Performance, 16*, 250–279.

Hashem, I. A. T., Yaqoob, I., Anuar, N. B., Mokhtar, S., Gani, A., & Ullah Khan, S. (2015). The rise of "big data" on cloud computing: Review and open research issues. *Information Systems, 47*, 98–115. doi: https://doi.org/10.1016/j.is.2014.07.006

Hinch, R., Probert, W., Nurtay, A., Kendall, M., Wymant, C., Hall, M., ... Stewart, A. (2020). *Effective configurations of a digital contact tracing app: A report to NHSX*. Retrieved from The Conversation.com

Hofstede, G., Hofstede, J. G., & Minkov, M. (2010). *Cultures and Organizations: Software of the Mind* (3rd ed.). New York: McGraw-Hill.

Hooper, C., Pickering, B., Prichard, J., & Ashleigh, M. (2015). *TRIFoRM Final Report: TRust in IT: Factors, metRics, Models*. Retrieved from www.itutility.ac.uk/

Israelsen, B. W., & Ahmed, N. R. (2019). "Dave ... I can assure you ... that it's going to be all right ..." A definition, case for, and survey of algorithmic assurances in human-autonomy trust relationships. *ACM Computer Survey, 51*(6), Article 113. doi:10.1145/3267338

Karasek Jr, R. A. (1979). Job demands, job decision latitude, and mental strain: Implications for job redesign. *Administrative Science Quarterly, 24*(2), 285–308.

Karlsen, J. R., & Solbakk, J. H. (2011). A waste of time: the problem of common morality in Principles of biomedical ethics. *Journal of Medical Ethics, 37*, 588–591. doi:10.1136/medethics-2011-100106

Khan, N., Yaqoob, I., Hashem, I. A. T., Inayat, Z., Mahmoud Ali, W. K., Alam, M., ... Gani, A. (2014). Big data: Survey, technologies, opportunities, and challenges. *The Scientific World Journal, 2014*, 712826. doi:10.1155/2014/712826

Khrais, L. T. (2020). Role of artificial intelligence in shaping consumer demand in e-commerce. *Future Internet, 12*. doi:10.3390/fi12120226

Lankton, N. K., McKnight, D. H., & Tripp, J. (2015). Technology, humanness, and trust: Rethinking trust in technology. *Journal of the Association for Information Systems, 16*(10), 880–918. doi:10.17705/1jais.00411

Latour, B. (2005). *Reassembling the Social: An Introduction to Actor-Network-Theory*. Oxford: Oxford University Press.

Law, J. (1992). Notes on the theory of the actor-network: Ordering, strategy, and heterogeneity. *Systems Practice, 5*(4), 379–393. doi:10.1007/BF01059830

Lee, J. D., & Moray, N. (1992). Trust, control strategies and allocation of function in human–machine systems. *Ergonomics, 35*(10), 1243–1270. doi:10.1080/00140139208967392

Lee, J. D., & See, K. A. (2004). Trust in automation: Designing for appropriate reliance. *Human Factors: The Journal of the Human Factors and Ergonomics Society, 46*(1), 50–80. doi:10.1518/hfes.46.1.50_30392

Liyanage, H., Liaw, S.-T., Di Iorio, C., Kuziemsky, C., Schreiber, R., Terry, A., & de Lusignan, S. (2016). Building a privacy, ethics, and data access framework for real world computerised medical record system data: A Delphi study. *Yearbook of Medical Informatics, 25*(01), 138–145. doi:10.15265/IY-2016-035

Maxwell, B., Tremblay-Laprise, A.-A., Filion, M., Boon, H., Daly, C., van den Hoven, M., ... Walters, S. (2016). A five-country survey on ethics education in preservice teaching programs. *Journal of Teacher Education, 67*(2), 135–151. doi:10.1177/0022487115624490

May, C. R., Mair, F., Finch, T., MacFarlane, A., Dowrick, C., Treweek, S., ... Montori, V. M. (2009). Development of a theory of implementation and integration: Normalization process theory. *Implementation Science, 4*(1), 29. doi:10.1186/1748-5908-4-29

McLaughlin, L. A., & Braun, K. L. (1998). Asian and Pacific Islander cultural values: Considerations for health care decision making. *Health & Social Work, 23*(2), 116–126.

Mehta, N., & Pandit, A. (2018). Concurrence of big data analytics and healthcare: A systematic review. *International Journal of Medical Informatics, 114*, 57–65. doi:10.1016/j.ijmedinf.2018.03.013

Metcalf, J., & Crawford, K. (2016). Where are human subjects in big data research? The emerging ethics divide. *Big Data & Society, 3*(1), 1–14. doi:10.1177/2053951716650211

Muirhead, W. (2011). When four principles are too many: Bloodgate, integrity and an action-guiding model of ethical decision making in clinical practice. *Clinical Ethics, 38*, 195–196. doi:10.1136/medethics-2011-100136

Nijhawan, L. P., Janodia, M. D., Muddukrishna, B. S., Bhat, K. M., Bairy, K. L., Udupa, N., & Musmade, P. B. (2013). Informed consent: Issues and challenges. *Journal of Advanced Pharmaceutical Technology & Research, 4*(3), 134–140. doi: https://doi.org/10.4103/2231-4040.116779

O'Neil, C. (2016). *Weapons of Math Destruction: How Big Data Increases Inequality and Threatens Democracy*. New York: Crown.

Page, K. (2012). The four principles: Can they be measured and do they predict ethical decision making? *BMC Medical Ethics, 13*(10). doi:10.1186/1472-6939-13-10

Parker, M. J., Fraser, C., Abeler-Dörner, L., & Bonsall, D. (2020). Ethics of instantaneous contact tracing using mobile phone apps in the control of the COVID-19 pandemic. *Journal of Medical Ethics, 46*(7), 427–431. doi: https://doi.org/10.1136/medethics-2020-106314

Parsons, T. D. (2019). *Ethical Challenges in Digital Psychology and Cyberpsychology*. Cambridge: Cambridge University Press.

Pickering, B., Bartholomew, R., Nouri Janian, M., Lopez Moreno, B., & Surridge, M. (2020). *Ask me no questions: Increasing empirical evidence for a qualitative approach to technology acceptance*. Paper presented at the Human-Computer Interaction. Design and User Experience. HCII 2020. Lecture Notes in Computer Science.

Pickering, B., Boletsis, C., Halvorsrud, R., Phillips, S., & Surridge, M. (2021). It's not my problem: How healthcare models relate to SME cybersecurity awareness. Paper presented at the HCI International 2021: 23rd International Conference on Human-Computer Interaction, Washington DC, USA.

Pickering, B., Nouri Janian, M., Lopez Moreno, B., Micheletti, A., Sanno, A., & Surridge, M. (2019). Seeing Potential is more important than usability: Revisiting technology acceptance. Paper presented at the Design, User Experience, and Usability. Practice and Case Studies. HCII 2019, Orlando, FL.

Pierce, J. L., Jussila, I., & Cummings, A. (2009). Psychological ownership within the job characteristics model. *Journal of Organizational Behavior, 30*(4), 477–496. doi:10.1002/job.550

Rhodes, R. (2015). Good and not so good medical ethics. *Journal of Medical Ethics, 41*, 71–74. doi:10.1136/medethics-2014-102312

Rohlfing, K. J., Cimiano, P., Scharlau, I., Matzner, T., Buhl, H. M., Buschmeier, H., ... Wrede, B. (2020). Explanation as a social practice: Toward a conceptual framework for the social design of AI systems. *IEEE Transactions on Cognitive and Developmental Systems*, 1–1. doi:10.1109/TCDS.2020.3044366

Rowe, F., Ngwenyama, O., & Richet, J.-L. (2020). Contact-tracing apps and alienation in the age of COVID-19. *European Journal of Information Systems, 29*(5), 545–562. doi: https://doi.org/10.1080/0960085X.2020.1803155

Rubin, M. A. (2014). The collaborative autonomy model of medical decision-making. *Neurocritical Care, 20*(2), 311–318. doi:10.1007/s12028-013-9922-2

Samek, W., Wiegand, T., & Müller, K.-R. (2017). Explainable artificial intelligence: Understanding, visualizing and interpreting deep learning models. *ITU Journal: ICT Discoveries*, Special Issue No. 1, 39–48..

Shanafelt, T. D., Sloan, J. A., & Habermann, T. M. (2003). The well-being of physicians. *American Journal of Medicine, 114*(6), 513–519. doi:10.1016/s0002-9343(03)00117-7

Spickard Jr, A., Gabbe, S. G., & Christensen, J. F. (2002). Mid-career burnout in generalist and specialist physicians. *JAMA, 288*(12), 1447–1450.

Tatonetti, N. P., Ye, P. P., Daneshjou, R., & Altman, R. B. (2012). Data-driven prediction of drug effects and interactions. *Science Translational Medicine, 4*(125), 125ra131–125ra131. doi:10.1126/scitranslmed.3003377

Taylor, S., Pickering, B., Boniface, M., Anderson, M., Danks, D., Følstad, A., ... Wollard, F. (2018). *Responsible AI – Key themes, concerns & recommendations for European research and innovation*. Retrieved from https://zenodo.org/record/1303253

The Health Service (Control of Patient Information) Regulations 2002. (2002).

UK Government. (2020). *Data Ethics Framework*. Retrieved from www.gov.uk/government/publications/data-ethics-framework/data-ethics-framework-2020

UK Government. (2021). *A guide to good practice for digital and data-driven health technologies*. Retrieved from www.gov.uk/government/publications/code-of-conduct-for-data-driven-health-and-care-technology/

UK Government (n.d.) *Scientific Advisory Group for Emergencies* www.gov.uk/government/organisations/scientific-advisory-group-for-emergencies

Van Riel, N., Auwerx, K., Debbaut, P., Van Hees, S., & Schoenmakers, B. (2017). The effect of Dr Google on doctor–patient encounters in primary care: A quantitative, observational, crosssectional study. *BJGP Open, 1*(2). doi:10.3399/ bjgpopen17X100833

Velicia-Martin, F., Cabrera-Sanchez, J.-P., Gil-Cordero, E., & Palos-Sanchez, P. R. (2021). Researching COVID-19 tracing app acceptance: Incorporating theory from the technological acceptance model. *PeerJ Computer Science*, *7*, e316. doi: https://doi.org/10.7717/peerj-cs.316

Venkatesh, V., Morris, M. G., Davis, G. B., & Davis, F. D. (2003). User acceptance of information technology: Toward a unified view. *MIS Quarterly*, 425–478. Retrieved from www.jstor.org/stable/30036540

Walrave, M., Waeterloos, C., & Ponnet, K. (2020). Ready or not for contact tracing? Investigating the adoption intention of COVID-19 contact-tracing technology using an extended unified theory of acceptance and use of technology model. *Cyberpsychology, Behavior, and Social Networking*. doi: https://doi.org/10.1089/cyber.2020.0483

Wang, Y., Kung, L., & Byrd, T. A. (2018). Big data analytics: Understanding its capabilities and potential benefits for healthcare organizations. *Technological Forecasting and Social Change*, *126*, 3–13. doi:10.1016/j.techfore.2015.12.019

Weber, L. R., & Carter, A. I. (2003). *The Social Construction of Trust*. New York: Springer Science+Business Media.

Weitz, K., Schiller, D., Schlagowski, R., Huber, T., & André, E. (2019). *"Do you trust me?": Increasing User-Trust by Integrating Virtual Agents in Explainable AI Interaction Design*. Paper presented at the IVA '19: Proceedings of the 19th ACM International Conference on Intelligent Virtual Agents, Paris, France.

Whitley, T. W., Allison, E. J., Jr., Gallery, M. E., Cockington, R. A., Gaudry, P., Heyworth, J., & Revicki, D. A. (1994). Work-related stress and depression among practicing emergency physicians: An international study. *Annals of Emergency Medicine*, *23*(5), 1068–1071. doi:10.1016/s0196-0644(94)70105-9

11 Application of Mobile Technologies in Healthcare During Coronavirus Pandemic Lockdown

Edeh Michael Onyema, Nwafor Chika Eucheria, Ugboaja Samuel Gregory, Nneka Ernestina Richard-Nnabu, Akindutire Opeyemi Roselyn, Emmanuel Chukwuemeka Edeh and Ifeoma Ugwueke

CONTENTS

11.1	Introduction	194
11.2	Conceptual Framework	195
	11.2.1 Mobile Technology	195
	11.2.2 Mobile Technology in Healthcare – Mhealth	196
11.3	Material and Method	198
11.4	Results	198
	11.4.1 Profile of Participants	198
	11.4.1.1 Do You Own a Mobile Technology Device?	198
	11.4.2 Are You Permitted to Use Your Mobile Devices at Workplace During Clinical Practice?	199
	11.4.3 To What Extent Do Participants Use Mobile Technologies for Health Purposes During the COVID-19 Lockdown?	200
	11.4.4 Barriers to the Use of Mobile Technologies in Health Setting	201
11.5	Discussion of Findings	202
11.6	Conclusion	203
11.7	Future Work	203
Competing Interest		204
References		204

DOI: 10.1201/9781003227892-11

11.1 INTRODUCTION

The COVID-19 disease has caused many harms to humanity. The outbreak of COVID-19 overwhelmed the world health system, leading to mass deaths and destruction. The speed of transmission of COVID-19 was fast, and the mortality rate was devastating to the entire world. The world knew nothing much about COVID-19 in late December 2019, when the outbreak of the disease happened (CNBC, 2020). The virus spread rapidly, the number of cases soared on daily basis, and world health systems were seriously overstretched (Onyema et al., 2020a). As of May 23, 2020, the number of COVID-19 deaths had risen to more than 340,000 and over five million cases worldwide (Worldometers, 2020). The easy of spread of COVID-19 are upsetting, causing many countries across the world to adopt severe measures to contain the disease (Onyema et al., 2020b). Many cities were on lockdown and social distancing became the new normal. The COVID-19 lockdown altered the lifestyles of people, and many had to depend largely on mobile technology solutions to access health services and to ensure a balance of life.

Mobile tools and platforms played a critical role in facilitating public medical access and disseminations of coronavirus-related information and health guidelines during the coronavirus pandemic lockdown in Nigeria. Not only did mobile devices enhanced public access to the internet and online resources, but it also served as a means for individuals and families to communicate with their loved ones during the COVID-19 lockdown, particularly those who were on the frontlines and those who were distanced from their families due to the lockdown. Mobile phones and other handheld devices facilitated public sensitization on dangers of COVID-19. For instance, Nigeria Centre for Disease Control (NCDC) leveraged on the wide penetration of mobile devices among the Nigerian population to send daily customized COVID-19 text messages and updates to the public through the mobile telecommunication networks, as seen in Figure 11.1. Smartphone app called Aarogya Setu was launched in India for pandemic information dissemination (DW News, 2020). Iranian authorities developed a mobile app for COVID-19 self-diagnosis checks (Gilbert, 2020). South Korea also created an app with the ability to track individual level of exposure to COVID-19 (Lyons, 2020). Similarly, BBC News reported that "millions of people in Britain will soon be asked to download a smartphone app that will monitor who they come in contact with as part of the contact tracing programme to combat COVID-19" (BBC News, 2020). Mobile devices such as TytoHome Remote Exam Kits, Mobile EKG, ScanWatch, Wireless Smart Glucometer, Brain Sensing Headband and smart temporal thermometer, were useful for basic medical examinations at home during the COVID-19 lockdown (travelaway, 2020).

Mobile applications and computer models aided the prediction and understanding of COVID-19 dynamics, thereby enhancing decision making relating to the mitigation of the disease. Also, authorities, organizations and individuals leveraged on the features of mobile apps and platforms to express their support to frontline health workers during the COVID-19 pandemic. Considering the potential of mobile technology in healthcare settings, this chapter examines its applications to enhance medical activities.

Application of Mobile Technologies in Healthcare 195

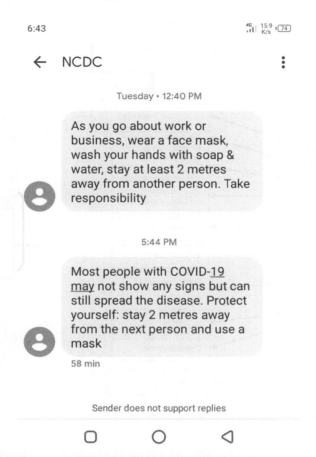

FIGURE 11.1 Screenshot of a customized COVID-19 awareness message sent by NCDC to all mobile phone users in Nigeria during the COVID-19 pandemic (*source*: authors).

11.2 CONCEPTUAL FRAMEWORK

11.2.1 MOBILE TECHNOLOGY

Mobile technologies encompass all forms of handheld, wearable, portable and mobile devices, applications, platforms and networks (Figure 11.2). It includes mobile phones, tablets, notebooks, e-readers, MP3, laptops and other personal digital assistants (PDAs). Stephen (2018) and Onyema (2019) jointly defined mobile devices as handheld computer or any other electronic device that is designed with portability in mind. There is growing evidence on the potentials of mobile technologies in different sectors, including health and education (Onyema, Deborah, Alsayed, Noorulhasan, & Sanober, 2019). The penetration of mobile technologies across the world is overwhelming and it has become essential part of human daily activities (Onyema, Chime, Faluyi, & Chinecherem, 2019). Mobile devices are used in the

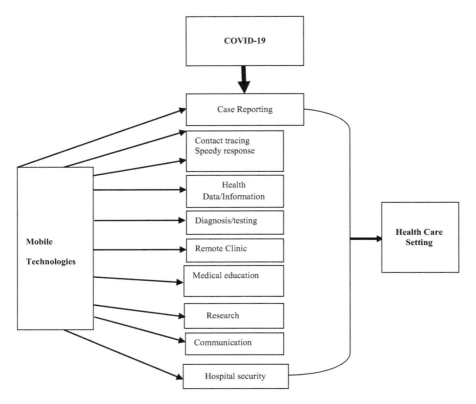

FIGURE 11.2 Conceptual model (*source*: authors).

society today more than ever before and it is hard to imagine a world without mobile (Stephen, 2018). Mobile devices/platforms/networks facilitated consultations and communications between physicians and patients at home during the COVID-19 lockdowns. With the aid of the internet and mobile technologies, the first ever Virtual World Health Assembly was held in May 2020.

Mobile technology evolution has promoted global internet access. According to International Telecommunication Union (ITU), there are more people now online; about 53.6 percent of the the world representing 4.1 billion persons are now in the cyberspace as shown in Figure 11.3 (ITU, 2020). With the aid of mobile technology devices, access to the internet has increased and made easier for both medical workers and their patients/clients. From the aforementioned, it seems the world has gone mobile and online, hence, the future of healthcare could be potentially influenced by mobile technology.

11.2.2 Mobile Technology in Healthcare – Mhealth

Technology is playing a critical role in the health sector. Several mobile health information systems, software and applications/platforms are being used across the world to provide health services (Figure 11.4).

Application of Mobile Technologies in Healthcare 197

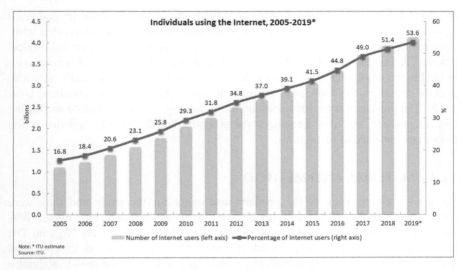

FIGURE 11.3 Individuals using the Internet, 2005–2019 (ITU, 2020).

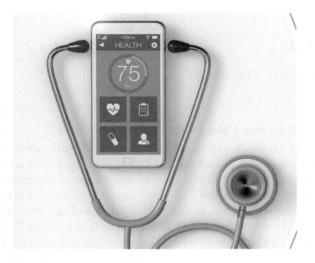

FIGURE 11.4 Mhealth (lmcins.com, 2020).

Mobile technology evolution catalysed the concept of remote clinic or medicine, particularly eHealth, mHealth and telemedicine. mHealth is the application of mobile technologies to healthcare services. mHealth depends largely on mobile technology to facilitate communication (real-time or delayed) between doctors and patients that are not in the same physical location for the purpose of medical evaluation, diagnosis and treatment (lmcins.com, 2020). Mobile technologies were used to promote and monitor compliance to Coronavirus mitigation measures during the lockdown (Onyema et al., 2020b). Contact tracing apps that run through mobile devices were also useful in combat against the disease (DW News, 2020). Despite the perceived positive effects of mobile devices in healthcare, some medical institutions often introduce restrictive policies that regulate staff usage of mobile devices during clinical practice. According to Preetinder, Ashwini, & Tejkara (2012), the use of smartphones in healthcare work is promising. Mobile devices facilitate medical staff's access to pool of resources for knowledge acquisition and training (healthcarebusiness.com, 2016). The present chapter advocates mobile technology use in workplace and during clinical practice, but also recognizes the potential health, security and distraction risks in the absence of restrictions.

11.3 MATERIAL AND METHOD

We used questionnaires as the primary instrument for the study and they were administered to 100 participants online consisting of physician/doctors, nurses, hospital support/administrative staff and others selected from Southwest Nigeria. The instrument was designed using 2-Linkert scale – "Yes" and "No", and it contained questions relating to the application and stumbling blocks to using mobile technologies for medical reasons during the Coronavirus pandemic lockdown. Also, secondary data were generated from review of relevant articles, media publications, newspapers and reports.

11.4 RESULTS

11.4.1 PROFILE OF PARTICIPANTS

See Tables 11.1 and 11.2.

11.4.1.1 Do You Own a Mobile Technology Device?

As seen in Table 11.3, when the participants were asked whether they have mobile devices, 98 percent of them answered "Yes" while only 2 percent said "No". We conclude here that ownership of mobile devices was prevalent among the participants. This is consistent with an earlier study by Ramesh et al. (2008), which found that mobile devices are widespread among medical staff and and that of Nicole, Olga, & Christine (2013), which found that 91 percent of healthcare professionals owned mobile devices of which most of them (87 percent) used it to support their clinical practices.

TABLE 11.1
Distribution of Respondents by Gender

Gender	Freq.	Percent
Masculine	55	55
Feminine	45	45
Total	100	100

TABLE 11.2
Participants' Designations

Designations	Frequency	Percentage
Physicians/doctors	35	35
Nurses	24	24
Hospital support/administrative staff	15	15
Others	26	26
Total	100	100

Note: This table indicates that 35 percent of the participants were physicians/doctors, 24 percent were nurses, and 15 percent were Supportive/Administrative staff, while others were 26 percent. This implies that majority of the participants were Physicians/Doctors.

TABLE 11.3
Ownership/Spread of Mobile Devices among the Participants

Options	Frequency	Percentage (%)	Cumulative percent
Valid Yes	98	98	98
No	2	2	100
Total	100	100	

TABLE 11.4
Use of Mobile Devices at Workplace during Clinical Practice

Options	Frequency	Percentage (%)	Cumulative percent
Valid Yes	94	94	94
No	6	6	100
Total	50	100	

11.4.2 ARE YOU PERMITTED TO USE YOUR MOBILE DEVICES AT WORKPLACE DURING CLINICAL PRACTICE?

As seen in Table 11.4, when the participants were asked if they were allowed to use their mobile devices at workplace during clinical practice, 94 percent of them answered "Yes" while only 6 percent said "No". This implies that majority of the participants were permitted to use their mobile devices by their employers at workplace and during clinical practice. The result supports the assertion by Preetinder, Ashwini and Tejkaran (2012) that the use of mobile devices should be allowed in medical workplace and complete ban should not be an option. It is also consistent with an earlier findings by Ramesh et al. (2008) and Ajami and Torabian (2013), which also proved the growing applicability of mobile tools in health service delivery.

11.4.3 TO WHAT EXTENT DO PARTICIPANTS USE MOBILE TECHNOLOGIES FOR HEALTH PURPOSES DURING THE COVID-19 LOCKDOWN?

Table 11.5 shows the extent of usage of mobile technologies for health purposes by the participants during the COVID-19 lockdown. As seen in Table 11.5, 80 percent of the participants concurred that mobile devices/apps aided their activities and work during the COVID-19 lockdown. Seventy-five percent of the participants

TABLE 11.5
Extent of Usage of Mobile Technologies by Participants during the COVID-19 Lockdown

Items	Yes (%)	No (%)
Mobile devices/apps aids my activities and work during the COVID-19 lockdown	80	20
I use my mobile devices to make calls, send and receive messages and e-mails from my patients or Physicians during the COVID-19 lockdown	75	25
I frequently use my mobile devices to access the internet during the COVID-19 lockdown	95	5
Mobile devices/apps/platforms enhanced healthcare delivery and services during the COVID-19 lockdown	80	20
I use my mobile phone to communicate with patients/physicians during the COVID-19 lockdown	77	23
I use my mobile devices for remote education during the COVID-19 lockdown	60	40
Mobile technologies enhanced data mining and management during the COVID-19 lockdown	84	16
Mobile technologies support patients' monitoring and counseling during the Coronavirus lockdown	74	26
Mobile devices/platforms were used as tools to disseminate health information and enlightenments to the public during the COVID-19 lockdown	97	3
The use of mobile phones, laptops and other PDAs facilitated timely reporting and responses to COVID-19 cases and other urgent medical issues during the COVID-19 lockdown	89	11
Mobile devices/apps facilitated remote clinics, medical tests and diagnosis during the COVID-19 lockdown	86	14
The use of mobile technologies facilitated remote/virtual medical meetings and programmes during the COVID-19 lockdown	70	30
The use of mobile devices and apps facilitated contact tracing during the COVID-19 lockdown	60	40
I used my mobile devices/apps for research activities during the COVID-19 lockdown	65	35
Mobile devices and apps enhanced hospital security during the COVID-19 lockdown	67	33

used their mobile devices for phone calls, sending and receiving messages and e-mails from patients or physicians. Ninety-five percent of the participants used their mobile devices for internet access. Eighty percent of the participants believed mobile devices/apps/platforms enhanced healthcare delivery and services during the COVID-19 lockdown. Seventy-seven percent used their mobile devices and platforms to establish contact with patients/physicians during the COVID-19 lockdown. Sixty percent of the participants used their mobile devices/apps/platform for remote education. Eighty-four percent of the participants believe the use of mobile enhanced medical data collection. Seventy-four percent agreed that mobile devices/apps supported patients' monitoring and counseling. Ninety-seven percent of the participants used mobile technologies to disseminate health information and public interest messages during the COVID-19 lockdown. Eighty-nine percent agreed that mobiles facilitated timely reporting and responses to urgent medical issues. Eighty-six percent of the participants agreed that mobile technologies facilitated remote clinics, medical tests and diagnosis. Seventy percent of participants agreed that mobile technologies facilitated remote/virtual medical meetings and programmes during the COVID-19 lockdown. Sixty percent of the participants agreed that mobile devices were used for contact tracing during the COVID-19 lockdown. Sixty-five percent of the participants used their mobile devices and apps for research activities. Finally, 67 percent of the participants responded that enhanced clinic security during the COVID-19 lockdown.

11.4.4 Barriers to the Use of Mobile Technologies in Health Setting

Security (privacy and confidentiality) concerns, as shown in Table 11.6, restrict the adoption of mobile devices in healthcare settings, according to 67 percent of the participants. The use of mobile technologies in healthcare settings is hampered by network and affordability issues, according to 88 percent of the participants. Institutional restrictions, according to 52 percent of participants, inhibit the usage of mobile devices in healthcare settings. Inequalities – unequal access and availability difficulties – limit the usage of mobile devices in healthcare settings, according to 75 percent of the participants. The lack of supportive facilities such as electricity, according to 90 percent of the participants, hinders the adoption of mobile technologies in healthcare settings. Furthermore, 60 percent of the participants agreed that the use of mobile technologies in healthcare is limited due to a lack of digital skills and training. Fear of mobile devices interfering with medical equipment is cited by 55 percent of participants as a barrier to using mobile technologies in healthcare settings. The use of mobile technologies in healthcare settings is hampered by health concerns such as cross infection, radiation exposure and so on, according to 70 percent of the participants. The perceived distractions connected with the use of mobile technology may hinder their adoption in healthcare settings, according to 80 percent of the participants. Finally, 85 percent of the participants felt that the adoption of mobile technologies in healthcare is hampered by time restrictions. This means that 100 percent of the participants believed that there are barriers to the effective use of mobile technology in healthcare settings.

TABLE 11.6
Barriers to the Use of Mobile Technologies in Health Setting

Items	Responses Yes (%)	No (%)
Security (privacy and confidentiality) concerns obstruct the use of mobile tools in healthcare	67	33
Network and affordability issues hinder mobile use in Health settings	88	12
Institutional policies limit the use of mobile for medical activities	52	48
Inequalities – Unequal access and availability issues impede the use of mobile technologies in health settings	75	25
Lack of supportive infrastructures e.g electricity, limits the negatively affect the application of mobile devices	90	10
Poor digital skills and trainings limits the use of mobile technologies in health setting	60	40
Fear of interference with medical equipment hampers the use of mobile technologies in health setting	55	45
Health concerns (e.g cross infection, exposure to radiations etc) hinders the use of mobile technologies in health setting	70	30
Perceived distractions associated with mobile devices discourage the use of mobile technologies in health setting	80	20
Time constraints hinder the use of mobile technologies in health setting	85	15

11.5 DISCUSSION OF FINDINGS

Tables 11.3 and 11.4 clearly show that majority of the participants (98 percent) owned mobile devices, and most of them (94 percent) were allowed by their employers to use it at workplace and during clinical practice. Similarly, the responses of the participants, shown in Table 11.5, indicates that mobile apps and devices enhanced efficiencies and general healthcare delivery during the COVID-19 crisis. Most of the participants as seen in Table 11.5, agreed that mobile technologies were largely used in the healthcare setting during the COVID-19 lockdown for internet access, dissemination of health information, communication, medical education, medical data mining/access and management, patients' monitoring, COVID-19 case reporting and contact tracing, speedy medical responses, remote clinics/telemedicine, testing/examination and diagnosis, virtual health meetings and programmes, research and hospital security.

The result validates that of Onyema, Udeze & Chinecherem (2019), which affirmed that the use of mobile devices facilitates access to the internet and remote education, and a study by Thakre and Thakre (2015), which found that majority of medical students surveyed (56.41 percent) use mobile devices for academic purposes. This chapter shows that the use of mobile technologies was essential to healthcare delivery during the COVID-19 lockdown. The use of different mobile applications and platforms, such as Zoom, GoToMeeting, Google Meet, CISCO WebEx,

Bluejeans and Slack aided the activities of medical health workers, patients, educators/ students and health institutions during the Coronavirus crisis. Thus, the potential of mobile technologies, if properly harnessed, can help to improve healthcare delivery particularly during the pandemic.

Also, Table 11.6 shows that the mobile adoption and application can be hindered by security (privacy, confidentiality), network and affordability issues, unequal access and availability issues. Also, participants agreed that poor digital skills, lack of supportive infrastructures (e.g. electricity), health concerns (e.g cross infection, exposure to radiations), time constraints, fear of interference with medical equipment and perceived distractions hinders the use of mobile technologies in healthcare settings.

The chapter proves that mobile devices, applications/networks and platforms significantly enhanced the activities of frontline health workers and accessibility to healthcare during the COVID-19 lockdown. Case reporting and contact tracing were made possible by mobile technology use, and physicians and patients were able to communicate and engage remotely during the COVID-19 lockdown through the use of various mobile technologies. From our findings, we infer that mobile technologies are fast becoming part of medical equipment, and essential component of the health sector. Thus, if this potential is properly harnessed, it could effectively enhance patient care and the efficiencies of healthcare workers. The use of mobile technologies should be allowed in workplaces, particularly in healthcare settings to reduce the stresses of health workers, enhance their digital competencies and improve general healthcare response and delivery.

11.6 CONCLUSION

The chapter highlights the medical importance and potential of mobile technologies, particularly during the COVID-19 pandemic. It establishes that mobile technology solutions were very useful and largely applied to mitigate the gap in the healthcare setting during the Coronavirus pandemic lockdown. It is important to note that a mobile world has become real, and healthcare organizations have to develop ways to optimize the potential benefits of emerging mobile health technologies. Thus, there is need for medical organizations to train their staff on the effective use of mobile devices to support their activities, and on the need to maintain professionalism, and observe mobile hygienic practices and safety/security precautions when using mobile devices. This would go a long way to reduce mobile device abuse and possible mobile-related microbial infections. Also, health authorities have to put measures in place to address the emanating security concerns that accompanies the growing use of mobile technologies in the health sector.

11.7 FUTURE WORK

Our next project is assessing the patients' attitudes towards the use of mobile technologies by healthcare professionals during clinical practice, and also the effects of mhealth on medical quality and confidentiality.

COMPETING INTEREST

We declare that there are no competing interests for the chapter.

REFERENCES

Ajami, S. & Torabian, F. (2013). Mobile technology in healthcare. *Journal of Information Technology & Software Engineering*, S7: e006. doi:10.4172/2165-7866.S7-e006

BBC News (19 May 2020). Coronavirus: How does contact tracing work and is my data safe? Retrieved online from www.bbc.com/news/explainers-52442754. Accessed May 19, 2020.

Braun, R., Catalani, C., Wimbush, J. & Israelski, D. (2013). Community health workers and mobile technology: A systematic review of the literature. *PLoS One*, 8(6). doi: http://dx.doi.org/10.1371/journal.pone.0065772.

CDC. (2019). Novel coronavirus, Wuhan, China. www.cdc.gov/coronavirus/2019-nCoV/summary.html. Accessed May 10, 2020.

Chib, A., Lwin, M.O., Ang, J., Lin, H. & Santoso, F. (2008). Midwives and mobiles: Using ICTs to improve healthcare in Aceh Besar, Indonesia 1. *Asian Journal of Communication*, 18(4): 348–364.

CNBC. (2020). Worldwide coronavirus cases reach 1million doubling in a week. Retrieved from Online via: www.cnbc.com/2020/04/02/worldwide-coronavirus-cases-reach-1-million-doubling-in-a-week.html

Darrell, M.W. (2013). Improving Healthcare through Mobile Medical Devices and Sensors. Center for Technology Innovation at Brookings, 1–13. Available at: www.brookings.edu/research/improving-health-care-through-mobile-medical-devices-and-sensors/ Accessed May 21, 2020.

DW News. (April 27, 2020). Coronavirus tracking apps: How are countries monitoring infections? www.dw.com/en/coronavirus-tracking-apps-how-are-countries-monitoring-infections/a-53254234. Accessed May 19, 2020.

Ekong, I., Chukwu, E. & Chukwu, M. (2020). COVID-19 mobile positioning data contact tracing and patient privacy regulations: Exploratory search of global response strategies and the use of digital tools in Nigeria. *JMIR Mhealth Uhealth*, 8(4): e19139, 1–7. doi: 10.2196/19139.

File:SARS-CoV-2 without background. (2020). Retrieved April 21, 2020, from Wikipedia website: https://en.wikipedia.org/wiki/File:SARS-CoV-2_without_background.png

GSMA Intelligence. (2017). Number of Mobile Subscribers Worldwide Hits 5 Billion. www.gsma.com/newsroom/press-release/number-mobile-subscribers-worldwide-hits-5-billion/. Accessed May 19, 2020.

Gilbert, D.V. (2020). Iran Launched an App That Claimed to Diagnose Coronavirus. Instead, It Collected Location Data on Millions of People. www.vice.com/en/article/epgkmz/iran-launched-an-app-that-claimed-to-diagnose-coronavirus-instead-it-collected-location-data-on-millions-of-people https://tinyurl.com/yddt2p38. Accessed May 19, 2020. www.mobius.md/blog/2019/02/benefits-mobile-devices-in-healthcare/. Accessed May 20, 2020.

Güler, K. & Cigdem, T. (2015). Mobile technology applications in the healthcare industry for disease management and wellness. *Procedia – Social and Behavioural Science*, 195: 1–6.

Healthcarebusiness.com. (October 3, 2016). Latest on Mobile Use in Hospitals. www.healthcarebusinesstech.com/latest-mobile-use/

lmcins.com (2020) www.lmcins.com/blog/what-are-the-benefits-of-telemedicine. Accessed May 20, 2020.

ITU. (2018). Half of the World Populations Now Using the Internet. Available online via: https://news.itu.int/itustatistics-leaving-no-one-offline Accessed May 20, 2020.

ITU. (2020). Statistics. Available online via: www.itu.int/en/ITUD/Statistics/Pages/stat/default.aspx. Accessed May 20, 2020.

Linda, V., David, H., Justine, D. & Wanjiku, K. (2015). Application of mHealth to improve service delivery and health outcomes: Opportunities and challenges, *African Population Studies Special Edition*. pp. 1683–1698.

Lyons, K. (2020). Governments around the world are increasingly using location data to manage the Coronavirus. *The Verge*. Available online via: https://tinyurl.com/y75xuwyz. Accessed May 19, 2020.

Nicole, K., Olga, V. & Christine, M. (2013). Healthcare professionals' use of mobile phones and the internet in clinical practice. *Journal of Mobile Techchnology in Medicine*, 2(1): 1–12. DOI:10.7309/jmtm.76

Onyema, E.M. (2019a). Integration of emerging technologies in teaching and learning process in Nigeria: The challenges. *Central Asian Journal of Mathematical Theory Computer Science*, 1(August): 35–39.

Onyema, E.M. (2019b). Opportunities and challenges of use of mobile phone technology in teaching and learning in Nigeria – A review. *International Journal of Research Engineering Innovation*, 3(6): 352–358. http://doi.org/10.36037/IJREI.2019.3601.

Onyema, E.M., Chime, I.P., Faluyi S.G. & Chinecherem, D.E. (2019), Impact of mobile phone technology on job performance of human resource managers in Nigeria. *International Journal of Research Engineering Innovation*, 3(6): 387–391. https://doi.org/10.36037/IJREI.2019.3606

Onyema, E.M., & Deborah, E.C. (2019) Potentials of mobile technologies in enhancing the effectiveness of inquiry-based learning. *International Journal of Education*, 2(1): 1–25. https://doi.org/10.5121/IJE.2019.1421

Onyema, E.M., Deborah, E.C., Alsayed, A.O., Noorulhasan, Q. & Sanober, S. (2019). Online discussion forum as a tool for interactive learning and communication. *International Journal Recent Technology and Engineering*, 8(4): 4852–4859. https://doi.org/10.35940/ijrte.d8062.118419

Onyema, E.M., Eucharia, A.U., Gbenga, F.S., Roselyn, A.O., Daniel, O. & Kingsley, N.U. (2020). Pedagogical use of mobile technologies during coronavirus school closure. *Journal of Computer Science Applications*, 27(2): 97–110. https://dx.doi.org/10.4314/jcsia.v27i2.9

Onyema, E.M., Eucheria, N.C., Obafemi, F.A., Sen, S. Atonye, F.G., Sharma, A. and Alhuseen, O.A. (2020). Impact of coronavirus pandemic on education. *Journal of Education Practice*, 11(13): 108–121. https://doi.org/10.7176/JEP/11-13-12

Panigutti, C., Tizzoni, M., Bajardi, P., Smoreda, Z. & Colizza, V. (2017). Assessing the use of Mobile phone data to describe recurrent mobility patterns in spatial epidemic models. *Royal Society Open Science*, 4(5): 160950. doi:10.1098/rsos.160950

Preetinder, S.G., Ashwini, K. & Tejkaran, S.G. (2012). Distraction: An assessment of smartphone usage in healthcare work settings. *Dove Press J. Risk Mgt. Healthcare Policy*, 5: 105–114.

Ramesh, J., Carter, A.O., Campbell, M.H., Gibbons, N. Powlett, C., Moseley Sr. H., Lewis, D. & Carter, T. (2008). Use of mobile phones by medical staff at Queen Elizabeth Hospital, Barbados: evidence for both benefit and harm. *Journal of Hospital Infection*, 70: 160–165. doi:10.1016/j.jhin.2008.06.007

Rosenfield, D., Hébert, P.C., Stanbrook, M.B., MacDonald, N.E. & Flegel, K. (2011). Being smarter with smartphones. *CMAJ*, *183*(18): E1276–E1276.

Stephen, P. (2018). Mobile devices: Examples, impact and trends. Retrieved from Online via: https://study.com/academy/lesson/mobil-devices-examples-impact-trends.html. Accessed May 17, 2020.

Thakre, S.S. & Thakre, S.B. (2015). Perception of medical students for utility of mobile technology use in medical education. *International Journal of Medicine and Public Health*, *5*: 305–311.

travelaway. (2020). 18 portable health gadgets that can change your life. Available online via: www.travelaway.me/portable-health-gadgets. Accessed May 20, 2020.

Vsee.com. (2020). What Is Telemedicine? Available online via: https://vsee.com/what-is-telemedicine/. Accessed May 22, 2020.

WHO. (2020a). COVID-19 questions and Answers. Retrieved from www.emro.who.int/health-topics/corona-virus/questions-and-answers.html. Accessed May 10, 2020.

WHO. (2020b). Coronavirus Disease (COVID-19) – Events As They Happen. Available online via: www.who.int/emergencies/diseases/novel-coronavirus-2019/events-as-they-happen. Accessed May 19, 2020.

WHO. (2011). mHealth: New horizons for health through mobile technologies. Available online via: www.who.int/goe/publications/goe_mhealth_web.pdf. Accessed May 20, 2020.

Worldometers. (2020). Coronavirus Updates Live. Retrieved from online via: www.worldometers.info/coronavirus/. Accessed May 23, 2020.

Zhang, O. (2020). Inside China's smartphone 'health code' system ruling post-coronavirus life. Available online via: https://time.com/5814724/china-health-code-smartphones-coronavirus/. Accessed May 19, 2020.

12 Application of Pattern Recognition in Taste Perception for Healthcare
An Exploratory Study

Dipannita Basu, Anusruti Mitra and Ahona Ghosh

CONTENTS

12.1 Introduction 208
12.2 Motivation 209
12.3 Contribution 209
12.4 Related Works 210
12.5 Pattern Recognition Models Used in Taste Perception Analysis 213
    12.5.1 Supervised Algorithm 213
        12.5.1.1 Logistic Regression 213
        12.5.1.2 Linear Discriminant Analysis 215
        12.5.1.3 Soft Independent Modelling of Class Analogy 217
    12.5.2 Unsupervised Algorithm 218
        12.5.2.1 Hierarchical Clustering 218
        12.5.2.2 Principal Component Analysis 220
12.6 Application Areas 220
    12.6.1 Application in Healthcare 221
    12.6.2 Application in Culture 221
    12.6.3 Application in Agriculture 221
    12.6.4 Application in Food Science 222
12.7 Sensors Used to Acquire Taste Perception 222
    12.7.1 Introduction to Electroencephalogram 222
    12.7.2 Introduction to Electronic Tongue 223
    12.7.3 Introduction to Glassy Carbon Electrode (GCE) 224
    12.7.4 Multichannel Sensor 224
    12.7.5 Smart Chemical Sensor 224
    12.7.6 Introduction to Electrogustometry 225
    12.7.7 Introduction to Functional Magnetic Resonance
        Imaging (fMRI) 225

DOI: 10.1201/9781003227892-12

| 12.8 | Conclusion and Future Scope | 225 |
| References | | 226 |

12.1 INTRODUCTION

The automatic identification of patterns and regularities in data is basically known as pattern recognition. It has applications in machine learning, computer graphics, data compression, bioinformatics, information retrieval, image analysis, signal processing and statistical data analysis. Mindful eating and a healthy diet are the keys to a good lifestyle. Individual taste quality responds to nutritional status and health. Taste is an important factor for evaluating food quality. Our gustatory and observation system identifies what food it is, but the tongue as the sensing organ gives the last decision that the food is consumable or not. There are taste receptors which are located on human taste buds, the upper facet of the trachea. When a stuff in the muzzle acts chemically with taste bud cells, a commotion gives rise to is known as taste. There are five sensory qualities (sweet, sour, bitter, salty, umami), which can differentiate the taste of food. Nutrients of sweet, salty and umami are helpful when we consume it at a low rate but avoidable in high-rate consumption. It is said that bitter and sour tastes are harmful more often. Sour taste denotes acid detection and also harmful to take excess, which will maintain pH of our body. Bitter taste can protect people from consuming poisons and toxins. Studies on family and twins help us to find their choice of proteins, fat and carbohydrates. The fundamental tastes are:

Sweet: Sugar and its derivatives like fructose or lactose are the inception for sweetness. Rather than sugar other types of individuals can also stimulate our taste bud to ally sugariness.

Sour: Sour agitation is generated by hydrogen ions, the chemical symbol is H+, decoupled by a pungent disintegrated in a water-based fluid. Acidic solution like lemon juice or organic acids gives an effect of sourish.

Salty: The chemical composition of sodium and chloride gives a chemical form of salt crystal. Dish containing table salt is what we perceive as briny. Commotion of granularity caused also by mineral salts like salt of potassium and magnesium.

Bitter: Many different mediums also bring bitter taste. Our taste bud reacts to pungent matter of about 35 different proteins. Different bitter genres of plants may be poisonous as explained previously. Using pattern recognition algorithm in taste perception, prediction is done whether the species are good for human consumption.

Umami: Glutamic acid or aspartic acid are the source of umami or savoury taste, which is akin to meat broth taste. These two amino acids are nutritious for human beings, which is also available in plants.

COVID-19 is a contagious illness caused by a novel coronavirus that spread all over the world (Callejon et al., 2021). There are many symptoms like fever, continuous cough, disappearance of tang and scent, and breathing difficulties. The study has described how disappearance of tang and scent are effective to predict COVID-19. The survey focuses on a thorough model using pattern recognition

Application of Pattern Recognition in Taste Perception for Healthcare

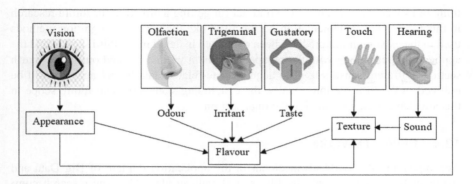

FIGURE 12.1 Generalized block diagram of food related sensory perception analysis for healthcare.

algorithms to gauge the prophesied worth of Anosmia and Ageusia. Anosmia, found in the early stages of COVID-19, may be explained by damage to the olfactory nerve caused by SARS-CoV-2 invasion and multiplication. As a result, anosmia or ageusia may be more common in the COVID-19 sick persons than in sufferers with other inhaling viral germs. Ageusia may be a side effect of olfactory dysfunction (Figure 12.1).

12.2 MOTIVATION

What we eat is 80 percent of our body. Mindful eating is the key to a healthy life. Many researchers have worked on food computing and taste perception. Many only focussed on the dichotomy of sweet–bitter taste. While surveying, we found prediction of tastes of a molecule using pattern recognition algorithms. Our work will aim based on people's taste. If any restaurant or food brand can tailor their items, that will grow soon. It is crucial to touch on what the present experiments had not obtain to regulate channel capability for the orbit of taste stimuli.

12.3 CONTRIBUTION

The organisation of the chapter can be summarized as follows:

  i. Analysis with respect to existing pattern recognition algorithms applied in taste perception is itself a novelty since no one has carried out research in this domain before.
  ii. A possible road map for the future researchers in the concerned domain.
  iii. Detailed analysis of different contributing sensors in the domain of taste perception for healthcare.

The next section describes and analyses the state of the art in the concerned domain. They have been compared by their performance evaluation and the drawbacks or limitations have been attempted to find out for getting a direction for future research in this topic. The implementation perspectives like the pattern recognition methods used in the existing literature have been explained in detail in the third section and the application areas along with the prospects of them will be marked out in the fourth section. Different sensors used to acquire taste-related data before analysis will be outlined in the fifth section. Finally, the concluding statements and future scope in this area have been presented in the final section.

12.4 RELATED WORKS

Accompanying the swift advancement in the application domain of Big Data and Artificial Intelligence within healthcare (Alotaibi et al., 2020), the research community has initiated different application-based studies on taste perception analysis. Tuwani et al. used an open source and proprietary software for generating molecular descriptors to classify sweet–bitter taste with real-life relevant applications (Tuwani et al., 2019). They collated all existing prediction models with their own model of bitter–sweet taste to eliminate redundancy by applying advanced pre-processing techniques. Wright et al. (1991) discussed a methodology to produce a confusion matrix of odour to evaluate different dimensions and standard connections of our olfactory system. Hettinger et al. (1999) discussed that confusion matrix is an outline of prediction outcomes on a classification problem based on taste training dataset. It is an application for animals to examine their taste perception differently from others. They have mentioned construction of other taste confusion matrices to address specific perceptual issues as the future scope. A combination of sweeteners or salts, for example, may be used to measure the perceived effects of taste-altering medications (Frank et al., 1992).

Yamamoto et al. have photographed faces of various participants by a cellphone camera after a while of consuming ten several solutions having five staple smacks and diverse voluptuary tones (Yamamoto et al., 2021). Each photograph was next copied to an artificially intelligent application, which generated scores ranging from 0 to 100 for eight emotions (surprise, happiness, fear, neutral, disgust, sadness, rage and shame). Each participant assessed the hedonics of each solution ranging between −10 (very uncomfortable) and +10 (very comfortable) for perceived evaluations (extremely pleasant). Multiple linear regression was executed then to forecast perceived hedonic ratings. The predicted result was promising as it had a strong correlation with perceived ratings. Dutta et al. have used a metal oxide sensor built electronic nose (Dutta et al., 2003) to determine taste and smell as for five types of tea samples by analysing their volatile and non-volatile organic compounds. The traditional methods are expensive in terms of manpower and time and sometimes they become inexact due to the lack of proper quantitative detail and sensitivity. Principal component analysis (PCA) and Fuzzy C means method was executed on the data and then the clustering was analysed using self-organizing plan, following classification by four artificial neural networks, namely multi-layered perceptron, learning vector quantization, radial basis function (RBF) and probabilistic neural network (PNN), which achieved a promising result. However, the electronic noise reliability in real-time scenario can be assessed in future and applying a commercial headspace auto sampler maybe beneficial to the sampling process by increasing the performance, leading to a more suitable system to profitable applications for nonstop monitoring.

Min et al. (2019) showed that on social network sites people have shared large amounts of food data, which has become the source of food computation and prediction. By this pattern of data, issues of national obesity and diabetes have been used to monitor human health. Beforehand work was done on food consumption, food taste, food choice and food safety by using small-scale data like question answer sessions and cookbook recipes. With rapid growth in different technologies, wide range of food data, food images and recordings are getting easily shared using Internet of Things (IoT), smartphones and social networks. This food data increases the breadth of knowledge and spans the theories of food perception, analysing cuisine art of food preparation in different cultures and to keep track of diet chart (Marks et al., 2007).

Differentiation between sweet and sour taste was the primary goal of Abidi et al. (2015) by extracting features from eight channel EEG signals. From every channel, two features, namely wavelet entropy and energy, achieved 98 percent accuracy in classifying the taste whereas kurtosis and skewness achieved only 60 percent of the same. The training dataset was too small as it only consisted of ten subjects and the feature set could be reduced to improve the performance. Time iron out multivariate pattern analysis using 128-channel electroencephalogram (EEG) was applied (Anderse et al., 2019) in order to distinguish the taste retaliation obtained from perceptually similar stimuli, which may cause a distinct cortical brain activation. A lot of research works are going on to find alternative sweeteners and salt having a good impact on health (Briand and Salles, 2016). The introduction of bitter blockers for reducing the bitterness of food items is another area of research, which helps to identify new food ingredients and reduce sodium and sugar content while preserving the other properties and food value as intact. Park et al. (2011) have compared the reaction to a taste against the response to an anamnesis to the same taste by monitoring the sensation elicited by taste using an electroencephalogram. The emotion characteristics were measured by hedonics score in response to each taste using common spatial pattern-based features.

Patients having altered taste perception following stroke suffer from malnutrition and often have related complications, which adversely affect the quality of life due to depressive thoughts slowing down the process of recovery (Dutta et al., 2013). Delta frequency range of EEG signal has been identified as responsible for taste perception in Wallroth et al. (2018). Spectral power and phase angle have been considered as the features to check the dependency of the timing of the fastest taste response on behavioural demands, and the decoding performance of two classes have been compared. The first one is fast-paced taste identification and the second one is a slow-paced taste categorization task and the experimental findings suggested that the taste stimuli's speed and degree are adjustable with the situation (Chandran and Perumalsamy, 2019).

The role of machine learning in the research on human body has been reviewed in Lotsch et al. (2019) where odour detection, pattern recognition in odoriferant phenotypes, implementation of intricate disorder biomarkers involving features of redolent and odour prediction from different features of volatile molecules were the main objectives. Deep learning has been applied in Teo et al. (2017) to classify the

TABLE 12.1
Comparative Study among Related Works

Ref.	Objective	Method used	Experimental Outcome	Shortcoming Found (if any)
(Li et al., 2020)	To gather information on the unions of persistent disease and tastes	Geo detector method	16 out of 71 chronic diseases have a connection with taste. Mixed tastes also give a sign of bad health	Quantitative approach is used here. Need more exploration of the health peril's part
(Guido et al., 2016)	To look into connection of savour, scent and meal choice	Single-centre sampling method and multivariate regression analysis	Food choices scaled in terms of gender, genetics, BMI, drugs consumption	Age shown a negative result on food perception
(El and Ziade, 2016)	To determine the need of proper physical activity with a person's diet	Survey-based statistical analysis	Association between obesity/overweight with dairy products intake noted	Create wrong value while experimenting with real weight
(Alvarez et al., 2020)	To regulate end user acceptance of new food products in the market	Random forest and convolution neural network (CNN)	Elucidated the panic, enjoyment, distaste, cortical, throb, emotions and facial expressions for human acceptance	Addition of other biometric signals like electroencephalogram (EEG) in the model and other flavours to bring out facial expressions
(Jilani et al., 2019)	To study sensory perception, delineation of taste phenotypes on food and beverages	Survey-based statistical study	Utilization of sweet and fatty foods increases as taste preference score also gets increased	The preference taste category is small for the children as it is slower in food propension scores
(Isezuo et al., 2007)	To determine the correlation between salt taste and blood pressure in type 2 diabetics	Univariate and multivariate regression model	Salt taste perception reduces the type 2 diabetics and relation with blood pressure	Survey on small population sample and without determining 24 h urinary sodium excretion
(Study Review)	To report and compare temporal resolution patterns for taste, smell with and without SARS-CoV-2 antibodies	Survey-based statistical study	To identify the existence of parosmia and female gender are at peril	Survey on female participants only and absence of general population without loss of smell and taste

Note: This table compares different existing literatures in terms of their objective, methodology, experimental result and drawbacks (if found any).

subject's preference based on real-time emotion monitoring by EEG electrode. The response patterns of two tasting persons are identical if their neural patterns are near in neural arena. Similarly, two tastes that are identical in perceptual field should have similar flavours while much work has gone into understanding how the chemical space is encoded within the neural space. The mapping of neural space into perceptual space – the function of neural response patterns in personalized gustatory perception and gustatory-related behaviour – has acquired little attention, despite identifying receptors for different tasting persons on the tongue and also characterizing response patterns of peripheral or cortical gustatory neurons (Crouzet et al., 2015).

Taste perception can be influenced by a number of external factors, such as emotional foozlition or the application of awful stress. Noel and Dando have obtained taste strength ratings and hedonic assessments from approximately 550 spectators (Noel and Dando, 2015) at men's hockey games during the 2013–2014 period to assess the impact of more commonplace day-to-day emotional variance on taste function, a period surrounding one tie, three losses, and four wins. Since various competitive sporting event results have been shown to elicit different affective responses, this field survey provided a unique setting for evaluating the impact of real-life emotional manipulations on our perception of taste, while antecedent studies had concentrated on irrelevant manipulation in a laboratory setting. Positive emotions were linked to increased sweetness and decreased sourness, while negative emotions were linked to increased sourness and decreased sweetness. Increasing sweet as well as decreasing sour taste strength will theoretically increase acceptance of a wide range of foods (see Table 12.1).

12.5 PATTERN RECOGNITION MODELS USED IN TASTE PERCEPTION ANALYSIS

In this section, different pattern recognition algorithms applied to analyse taste perception in the existing literature will be discussed and working mechanisms will be described.

12.5.1 SUPERVISED ALGORITHM

It is the machine learning process where machines are trained using well "labelled" training data and based on that training, the output prediction takes place (Singh et al., 2016). Based on the deviation from the actual and correct output, the performance of the prediction process gets evaluated by metrics like accuracy, precision, recall, error rate, F1 score, and area under the curve (AUC). Different classification and regression techniques used in taste perception are described below.

12.5.1.1 Logistic Regression

It is a binary classifier (Hosmer et al., 2013) as it deals at a time with sweet–bitter or sweet–sour like that. It pulls out the trait of taste buds and puts on the data accordingly. Logistic regression becomes wobbly when two taste categories sort out. Algorithm 1 and Figure 12.2 have shown the working mechanism of logistic regression where P and Q are two types of taste data taken. Using logistic regression, a, we initialize the whole taste data set. Now in data processing part, normalization of a data set is done using mean, mode and median value to fill the missing values, to maintain the range of taste intensity. The loop will repeat until we get our expected result, which is to segregate between mentioned datasets. Otherwise, it returns a.

12.5.1.1.1 Logistic Regression Hypothesis

The logistic regression classifier can be acquired by semblance to the linear regression hypothesis that is

$$h_\theta(x) = \theta^T x$$

Next, the logistic regression hypothesis theories originating from the linear regression hypothesis give rise to the logistic function

$$h_\theta(x) = g(\theta^T x)$$

$$g(z) = \frac{1}{1+e^{-z}}$$

where $g(z)$ is the logistic function, which is termed as sigmoid function. The obtained outcome of the logistic regression hypothesis is:

$$h_x(x) = \frac{1}{1+e^{-\theta Tx}}$$

12.5.1.1.2 Logistic Regression Decision Boundary

Our taste data has five features like sweet, sour, salt, bitter and umami, but we are using two features of taste perception. that is, sweet and sour then the logistic regression hypothesis is

$$h_\theta(x) = g(\theta_0 + \theta_1 x_1 + \theta_2 x_2)$$

The logistic regression classifier will predict sour if

$$\theta_0 + \theta_1 x_1 + \theta_2 x_2 \geq 0$$

This is caused by the logistic regression setting the threshold value, $g(z)$ at 0.5. To get access to the theta value, which is computed by scikit learn library. Firstly, manual prediction is done without using the predicting function (clf.predict) to calculate the taste feature product and to check if the obtained value is equal to zero (to predict sour) or otherwise (to predict sweet). A visualization of the decision boundary is given in which yellow points correlate to sour and the red one correlates sweet. Nonlinear functions as well as soaring stage polynomials are also logistic decision boundaries.

Application of Pattern Recognition in Taste Perception for Healthcare

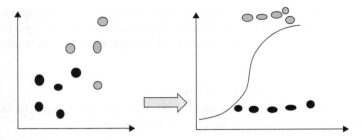

FIGURE 12.2 Working mechanism of logistic regression.

12.5.1.1.3 Computing the Logistic Regression Parameter

The Scikit–Learn library does a considerable project of detaching the numeration of the logistic regression parameter θ, and the way it is performed is by answering an enhancing problem. First, determine the logistic regression cost function for the two points of interest: y=1, and y=0, that is, when the theorem function predicts sweet or sour.

$$\cos t(h\theta(x), y) = \{-\log\log(h\theta(x))\}, \quad \text{if } y = 1$$

$$\cos t(h\theta(x), y) = \{-\log\log(1 - h\theta(x))\}, \quad \text{if } y = 0$$

Then, we lay hold of a convex amalgamation in y of these two taste perceptions to attain with the logistic regression cost function:

$$J(\theta) = -[y\log\log(h\theta(x)) + (1-y)\log(1-h\theta(x))]$$

Algorithm 1: Logistic Regression

Input: a, P, Q
Output: a
Initialize a= $\{1,\ldots,1\}^T$
Normalize P
Repeat until convergence
A = a + a/Mp^T(q-g(Pa))
Return a

12.5.1.2 Linear Discriminant Analysis

It is a dimensionality reduction methodology. This approach pares the number of taste features in a dataset while confining as much information is feasible to represent the whole dataset. It is a multiclass classifier (Izenman et al., 2013). Our taste buds not only cap to two types of tastes; when it undergoes multiple tastes, then it is extended to linear discriminant analysis (LDA). LDA is preferred more as the desired outcome is categorical (sweet, sour, salty, bitter, umami) (Marks et al., 2007). Algorithm 2 and Figure 12.3 have described the working mechanism of LDA where LDA approximates the likelihood of new taste data input that belongs to each class. LDA applies Bayes' theorem to estimate the probabilities. If the output class is (b) and the input is (a), here Bayes' theorem operates to evaluate the probability that the taste data belongs to that predicted class.

12.5.1.2.1 Data Preparation for LDA
Pattern recognition model performance works well after processing the taste data. Each taste feature is silhouetted like a 'bell-shaped' curve.

12.5.1.2.2 Outlier Treatment
Outlier updates us that the advertence of the taste dataset. There is no compulsory requirement about how much a data item requires to deviate for the existing dataset. For example, taking a dataset of food, which are all in salty range, and a chocolate, which belongs to sweet category. Outliers initiate unbalanced as a consequence of the numeration of mean and variance will be allured.

12.5.1.2.3 Equal Variance
We need to standardize the input taste data set where mean is zero and standard deviation is one. For example, taking a dataset having a chocolate (sweet intensity 80), an apple (sweet intensity 50) and a tea prepared with sugilite (sweet intensity 20) and to work on, we need to work with mean and standard deviation value of these parameters.

12.5.1.2.4 Impetus to Reduce Dimension in a Dataset
Searching any features in one-dimensional work is effortless because it is known the start state and the goal state. When we append another dimension (feature), then it enhances to a two-dimensional one; searching in it is more composite rather than in one-dimensional feature. With increase in dimensionality, the amount of data also enlarges and in two-dimensional features, there is also a tendency of some row or column having null value in it.

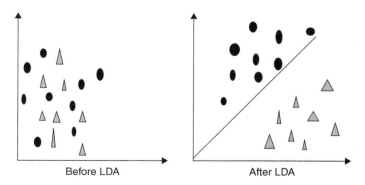

FIGURE 12.3 Working mechanism of linear discriminant analysis.

Algorithm 2: Linear discriminant analysis
Input: a
Output: b
Initialize: lda_pseudo(a, …)
lda_pseudo(a, b, first = NULL, sum = 1e-08, …)
Logic: lda_pseudo(equation, data, first = NULL, sum = 1e-08, …)
Prediction: predict(obj, newdata, …)
Return b

12.5.1.3 Soft Independent Modelling of Class Analogy

The principle of linear discriminant analysis (LDA) was developed by statisticians, but soft independent modelling of class analogy (SIMCA) was proposed by chemists. SIMCA gives priority to similar taste variables (Racz et al., 2018). There are a number of steps before modelling: mount up the variables (sweet, sour, salty, bitter, umami), way of identifying the number of principal components, enlarged range, weights of the classes and after auto-scaling regenerated weights (Forina et al., 2008). When there is an out-of-class variable or a new class found, SIMCA characterizes it. Algorithm 3 and Figure 12.4 have described the working mechanism of SIMCA where it needs a training taste dataset consisting of objects with a group of attributes and their category integration. SIMCA is actually a framework used to make a decision. If the taste data is graded as systematic, it will put on the mentioned class, otherwise make a new class. That's how fitness of a model is calculated.

Algorithm 3: Soft Independent Modelling of Class Analogy
Initialize_tastedata
Preprocessing_tastedata
//sublayer loops
for h=1 to 20
for i=1 to h
Create_neural_network(net)
//training loop
For k=1 to 100
Train(net)
Run_Regression-test(net,training_set)
Simulate(net)
Run_Regression-test(net,validation_set)
Avoid_overfitting
end for//k
end for//i
end for//h
Get_best_net
Display_results

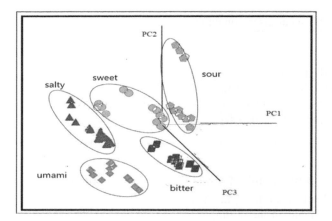

FIGURE 12.4 Working mechanism of soft independent modelling of class analogy.

12.5.2 Unsupervised Algorithm

Unsupervised algorithms of machine learning are basically concerned with modelling itself to find some hidden patterns and insights from the given data instead of supervision using labelled data (Kassambara et al., 2017). If an input dataset having several categories of cats and dogs' images is given to the unsupervised learning technique, it never gets trained upon the given dataset, having no idea about the dataset features. The unsupervised learning algorithm's goal is to recognize image features on its own. Unsupervised learning algorithms can complete this task by grouping the image dataset into groups based on image similarities.

12.5.2.1 Hierarchical Clustering

It is used in the classification of unauthenticated datum into a clutch, having two types of outlooks (Johnson et al., 1967). This is also known as distance-based algorithm as we calculate the distances between the clusters. Two types of approaches are commonly applied while we perform hierarchical clustering.

a. *Agglomerative*: It is a bottom-up approach. Designate each taste point to a separate cluster in the agglomerative hierarchical clustering technique. Considering there are four taste data spikes. We would allocate each of this point to different clusters and will discover four clusters in the starting. Then at each monotony, we amalgamate the closest pair of clusters and redo this step until there is only a solitary cluster. It is also termed as additive hierarchical clustering.
b. *Divisive*: It is a top-down approach. Divisive hierarchical clustering projects work in another way. In spite of beginning with in clusters (in case n taste observations), we establish with an exclusive cluster and allocate all the taste datapoints to that cluster. Now at each rotation, we divide the most distant

Application of Pattern Recognition in Taste Perception for Healthcare

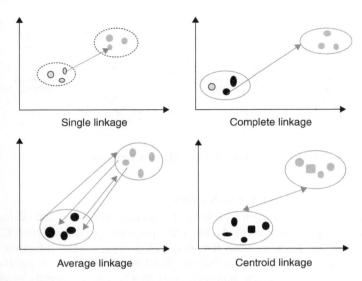

FIGURE 12.5 Working mechanism of hierarchical clustering.

point in the cluster and redo these steps until each cluster has only one solo point. This is called divisive hierarchical clustering because we are dividing the clusters at each step.

With the implementation of this model, the taste pattern of a person is recognized and preferred food is offered. Algorithm 4 and Figure 12.5 have described the working mechanism of hierarchical clustering where b1 to bN are taste datasets. By traversing two for loops, we calculate distance between taste datasets. Hierarchical clustering separates data into groups based on some measure of similarity. After that, datasets are reduced and split into clusters. Then allocate all the positions to the nearest clutch of midpoint. Calculate the midpoint of the recently formed bundle. Redo steps three and four until cluster form.

Algorithm 4: Hierarchical Clustering
Initialize a dataset (b1, b2, b3, ..., bN) of size N
for a=1 to N:
for b=1 to a:
dis_mat[a][b] = distance[ba, bb]
each data point is a singleton cluster
repeat
merge the two clusters having minimum distance
update the distance matrix
until only a single cluster remains

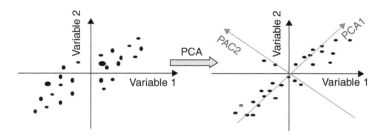

FIGURE 12.6 Working mechanism of principal component analysis.

12.5.2.2 Principal Component Analysis

It is a dimensionality reduction tool, not a classifier. All classifiers and estimators are responsible for predicting methods, but principal component analysis (PCA) does not. We need an appropriate classifier on the PCA-module data. It is used to bring down the variables present in the dataset (Wold et al., 1987). PCA signifies association between two variables, scales from −1 to +1, denotes directly proportional and −1 indicates inversely proportional to each other. Algorithm 5 and Figure 12.6 have described the working mechanism of PCA where $x1$ to xN are taste datasets. Evaluate the covariance matrix for feature extraction. By standardizing the dataset, we form the initial matrix where the diagonal element is 1. Reckon the eigenvalue and eigenvector for the resultant matrix. Calibrate the eigenvalues and their consonant eigenvectors. Eigen value and eigen vector are calculated. After sorting these values, remodel the original matrix.

Algorithm 5: Principal Component Analysis
Input: Data matrix $X = (x1, \ldots, xN) \in R^{M \times N}$ Output: Eigenvectors W   Parameters: Step size a, Number of epoch P, Number of Dimension Q   Initialize Matrix $W0 = (w1, \ldots, wQ)$ (diagonal element is 1, otherwise 0) $\in R^{M \times Q}$ Compute the diagonal matrix $D = \text{Diag}(Q, \ldots, 1) / 105 \in R^{Q \times Q}$ Calculate the output eigenvectors $W \in R^{Q \times M}$

12.6 APPLICATION AREAS

On a comparison of olfactory systems, taste perception allows us not only quality but also different impetus names and keeps away from the limitation to basic four attributes of taste interpretation (Tuwani et al., 2019). It will be useful to study the gustatory system by selecting a preferred stimulus. The experiment constitutes an identification procedure to create a taste confusion matrix which determines the pattern of confusion produced by a group of taste stimuli and also collates the show gauges. The experiment examined the pattern by using hierarchical analysis, a cluster analysis that describes relative similarities and to determine whether they are in distinct groups.

12.6.1 APPLICATION IN HEALTHCARE

Almost 10 million people from 1.7 million people in Instagram shared their daily food posts by using #foodporn hashtags. By mining that data using pattern recognition algorithms, foodborne illness has been prevented. By using pattern recognition methods on Twitter data set, public health issues means what people face in the nation can be detected. Rehabilitation is the technique of restoring someone's original physical/mental ability (Saha and Ghosh, 2019; Ghosh and Saha, 2020). Some of the existing literature has also worked on recovering taste and smell functions of persons suffering from diseases like laryngectomy (Caldas et al., 2012), stroke (Dutta et al., 2013), and COVID-19 (Niklassen et al., 2021). The goal of health labels like decreased iodine or the healthy preferences badge is to make it easier for consumers to make good food choices. However, they may operate as a caution flag for those end users who are more bothered with the flavour of the products, rather than healthy lifestyle. Liem et al. (2012) explored how consumers' expectations and actual perceived taste quality of a chicken soup were affected by front-of-pack health labelling.

12.6.2 APPLICATION IN CULTURE

Food practice tells a lot about culture. By exploring food cultures, personalized food recommendations application has been made in different urban areas. This culture-based study helped to a greater extent in tourism too and promoted their regional food courts too. People who check-in into food courts find strong temporal and spatial correlation between their cultural preferences and eating drinking habits (Trachootham et al., 2018). Trachootham et al. have compared different taste thresholds in Thai and Japanese people by recognizing the threshold of five different tastes by filter paper disc. The preference of spicy was measured by calibrated questionnaires. For the Japanese and Thai, the average thresholds for bitter, sour, salty and sweet were two and four, respectively, where the mean umami threshold were found to be three and five, respectively. Furthermore, the Thai people exhibited a higher predilection for spicy cuisine with 70 percent mild- to moderate-spicy eaters and 10 percent strong spicy eaters, compared to over 90 percent non- or mild-spicy eaters in Japan. To determine the impact of food culture on flavour perception, more large-scale international surveys are required.

12.6.3 APPLICATION IN AGRICULTURE

Food processing is used in cultivation systems. The implementation of automated agriculture and nutrition security are examined using food image analysis (Senthilnath et al., 2016). Pattern recognition systems predict the shape and size of fruits and vegetables, which helps to identify the food-based health. By applying the same, more new recognizing techniques are established. By integrating taste perception with pattern recognition, ingredients of an unknown dish get distinguished (Min et al., 2019).

12.6.4 APPLICATION IN FOOD SCIENCE

Food science can be termed as basic science and it is a study of physical, chemical and biochemical aspects of food. The study to appraise food by human sense styled as sensory analysis, such as magnetic resonance imaging (MRI) (Killgore and Yurgelun, 2005). Food sense perception is multi-modal as it includes senses that are visual, auditory, taste, smell and palpable (Min et al., 2019). Food recommendation task also has been performed based on consumer's taste preference based on their culture. Naresh et al. (2020) have represented a menu of food items each having six taste values (salty, sweet, bitter, umami, astringency, sour) in Manhattan cut-off based six-dimensional graphical space having each food item as its one node and each taste as that particular node's dimension. Different factors can be taken into account in the prediction task including the food item colour, chemical compounds associated with it, relative amount of each compound and type of the food. Cardiovascular disease, diabetes and obesity are all nutriment-related chronic diseases that have become an acute public health concern. Dietary control is crucial in the prevention and remedy of many disorders. Conventional food diary systems need manual recording of the types and portions of food consumed, making accuracy difficult to ensure. Novel food detection systems for dietary assessment have been enabled by the increasing usage of cell phones and developments in computer vision. We can then undertake numerous health-related analyses, such as calorie intake estimation, nutrition analysis, and eating habits analysis, once we've identified the meal's category or ingredients (Min et al., 2019).

12.7 SENSORS USED TO ACQUIRE TASTE PERCEPTION

A sensor is a tool, which dredges and transfers taste data that is revealed by the five human senses. Of the five senses, a photometer replicates vision, microphone instead of hearing and touch is stimulated by thermometer. There is no machine-made sensor which can acquire taste perception. In this section, different sensors used to acquire taste perception data (Wang et al., 2015) will be described along with their working mechanisms.

12.7.1 INTRODUCTION TO ELECTROENCEPHALOGRAM

Since the human brain cells communicate with each other by electrical impulses, the electroencephalogram (EEG) is an electrical activity recording and monitoring method for the brain using small metallic discs called electrodes attached to the scalp. Among different regions of our brain shown in Figure 12.7, namely occipital, parietal, frontal, cerebellum and temporal, the parietal lobe is responsible for the generation of taste-based stimulation and the three channels recording the corresponding activities are C3, C4 and Cz (Sanei and Chambers, 2013). Figure 12.7b has shown the International 10-20 system of EEG electrode placement where the 10 and 20 refer to the distance between adjacent nodes that are 10 or 20 percent of the entire scalp. Every electrode is represented by a letter followed by a digit where the letter F, P, O, C, T stand for the frontal, posterior, occipital, central and temporal region, respectively,

Application of Pattern Recognition in Taste Perception for Healthcare 223

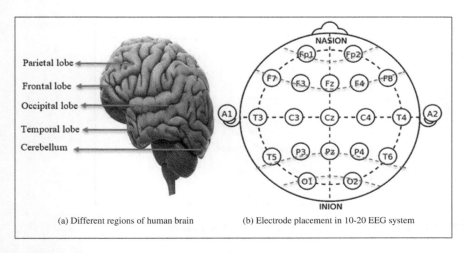

(a) Different regions of human brain   (b) Electrode placement in 10-20 EEG system

FIGURE 12.7   Pictorial overview of electroencephalogram electrode detail.

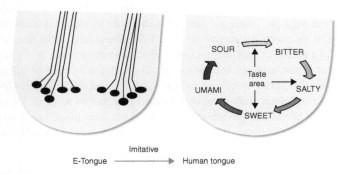

FIGURE 12.8   Pictorial overview of electronic tongue.

and the digit is odd representing the left hemisphere and even representing the right one (Wasim et al., 2018; Hussin and Sudirman, 2013).

12.7.2   Introduction to Electronic Tongue

A multisensor system incorporates an amount of low and selective sensor and cross sensitivity to incompatible species and it is a pertinent method of pattern recognition for data pre-processing is known as electronic tongue (Vlasov et al., 2005). E-Tongue fetches the signals that are pre-processed in consecutive steps comprising drift compensation, scaling and lineage of indicative variables (Prieto et al., 2012). In spite of cognitive conditions or personal choices, e-tongue comes up with positive responses to taste perception (Rodriguez et al., 2016). Artificial electronic tongue with natural human features is a sought-after pioneering tool for the biological, medical and industrial researchers (Ishihara et al., 2005) (Figure 12.8).

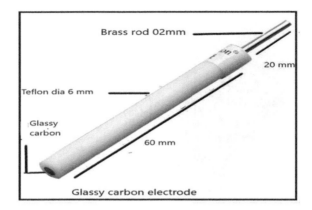

FIGURE 12.9 Pictorial overview of glassy carbon electrode.

12.7.3 Introduction to Glassy Carbon Electrode (GCE)

The basic proposition of electroanalytical chemistry is straight conversion of chemical to electrical signals, where an analyte is diffused in electrolyte emulsion in an electrochemical solution (Braungardt et al., 2015). The signal of the interaction helps in detection of tastes. As they are focussed on electrochemicals, a slight change in food molecules is also noticeable (Radh et al., 2013). Combination of glassy and ceramic properties of graphite is termed as glassy carbon electrode (GCE) (Jakubowska et al., 2016). To make accurate measurements of free amino acids (FAA), contained in different types of tea such as green tea, oolong tea, and black tea (Ouyang et al., 2020), GCE also gets applied. It really helps a lot to detect bad smells of food, but using GCE is expensive (March et al., 2015) (Figure 12.9).

12.7.4 Multichannel Sensor

A multichannel taste sensor converter is unruffled of respective types of lipoid elastomer membranes with dissimilar features, which can detect human taste buds. When we receive some taste, it comes through membranes as an electrical signal and formulated patterns. A multichannel sensor distinguishes repression of bitterness of quinine and a nostrum chattel by sucrose. This method customarily can evaluate the level of bitterness (Takagi et al., 1998). Figure 12.10 has shown different components of the sensor and the working mechanism in taste perception.

12.7.5 Smart Chemical Sensor

An innovative and acute, chemical taste sensor that pragmatically imitates conduct of human exteroception system is termed as smart chemical sensor. Smart chemical sensors take input as an array of electrochemical sensors that illustrates the taste receptors on the human tongue and a two-phase optimized radial basis function

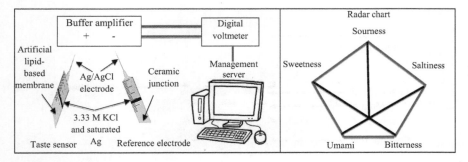

FIGURE 12.10 Pictorial overview of multichannel sensor.

network (RBFN) to express the human brain, which describes the sense trigger and explains the entire tastes (Ishihara et al., 2005).

12.7.6 INTRODUCTION TO ELECTROGUSTOMETRY

Electrogustometry is the quantification of taste skill by advancing controlled anodal current along the articulation (Tomita and Ikeda, 2002). It is useful for masking the malformations of taste. It is an assessment of facial nerve function, that is, chorda tympani nerve test. A solution made of salt, sugar, citrate, quinine or electrical simulation are used to perform the test. It balances the amount of current vital for a retaliation on each side of the tongue. If the resultant value is less than 20uAmp, then the facial nerves are normal and when it is more than 25 percent, it is considered as aberrant testers.

12.7.7 INTRODUCTION TO FUNCTIONAL MAGNETIC RESONANCE IMAGING (fMRI)

The subtle changes in blood flow that occur with brain activity are measured using functional magnetic resonance imaging (fMRI). It may be applied to look at the functional structure of the brain (to see which parts of the brain are in charge of which functions), to assess the effects of a stroke or other illness, or to help with brain care. fMRI can detect anomalies in the brain that other imaging methods cannot. To assess the taste perception ability also, some recent literature has used fMRI (Barros et al., 2012; Kuhn and Gallinat, 2013) where the brain activation regions after hearing taste-related words have been identified as thalamus, lateral orbitofrontal gyrus, frontal operculum, anterior insula along with posterior middle, superior temporal gyri and left inferior frontal. Figure 12.11 shows different components of the fMRI-based experimental setup.

12.8 CONCLUSION AND FUTURE SCOPE

This chapter gives a demonstration of the power of technological innovations to identify taste perception ability. Thus, in a clinical setting, a relatively small battery

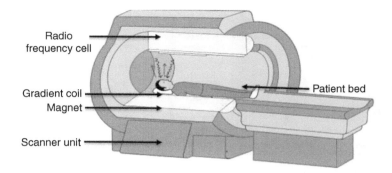

FIGURE 12.11 Pictorial overview of functional magnetic resonance imaging (fMRI) device.

of tasting persons may suffice to provide useful diagnostic information on the functioning of the gustatory system. Exposure to such environmental toxins as lead, mercury, insecticides and solvents (e.g., paint thinner) also can damage taste buds and sensory cells in the nose or brain. Many people face taste disorders due to accidents, chronic disease and in the near future we can collaborate with medical science to cure them. Future research to identify taste-enhancing therapies will help improve health and quality of life in patients affected by various types of taste disorders.

REFERENCES

Abidi, I., Farooq, O., & Beg, M. M. S. (2015, December). Sweet and sour taste classification using EEG based brain computer interface. In *2015 Annual IEEE India Conference (INDICON)* (pp. 1–5). IEEE.

Alotaibi, S. R. (2020). Applications of artificial intelligence and big data analytics in m-health: a healthcare system perspective. *Journal of Healthcare Engineering*, https://doi.org/10.1155/2020/8894694.

Álvarez-Pato, V. M., Sánchez, C. N., Domínguez-Soberanes, J., Méndoza-Pérez, D. E., & Velázquez, R. (2020). A multisensor data fusion approach for predicting consumer acceptance of food products. *Foods, 9*(6), 774.

Andersen, C. A., Kring, M. L., Andersen, R. H., Larsen, O. N., Kjær, T. W., Kidmose, U., ... & Kidmose, P. (2019). EEG discrimination of perceptually similar tastes. *Journal of Neuroscience Research, 97*(3), 241–252.

Barros-Loscertales, A., González, J., Pulvermüller, F., Ventura-Campos, N., Bustamante, J. C., Costumero, V., ... & Ávila, C. (2012). Reading salt activates gustatory brain regions: fMRI evidence for semantic grounding in a novel sensory modality. *Cerebral Cortex, 22*(11), 2554–2563.

Berrueta, L. A., Alonso-Salces, R. M., & Héberger, K. (2007). Supervised pattern recognition in food analysis. *Journal of Chromatography A, 1158*(1–2), 196–214.

Braungardt, C. B. (2015). Evaluation of analytical instrumentation. Part XXVI: instrumentation for voltammetry. *Analytical Methods, 7*(4), 1249–1260.

Briand, L., & Salles, C. (2016). Taste perception and integration. In *Flavor* (pp. 101–119). Woodhead Publishing.

Caldas, A. S. C., Facundes, V. L. D., & Silva, H. J. D. (2012). Rehabilitation of smell and taste functions in total laryngectomy: systematic review. *Revista CEFAC, 14*, 343–349.

Callejon-Leblic, M. A., Moreno-Luna, R., Del Cuvillo, A., Reyes-Tejero, I. M., Garcia-Villaran, M. A., Santos-Peña, M., ... & Sanchez-Gomez, S. (2021). Loss of smell and taste can accurately predict COVID-19 infection: a machine-learning approach. *Journal of Clinical Medicine*, *10*(4), 570.

Chandran, K. S., & Perumalsamy, M. (2019). EEG–Taste classification through sensitivity analysis. *The International Journal of Electrical Engineering & Education*, 0020720919833036.

Crouzet, S. M., Busch, N. A., & Ohla, K. (2015). Taste quality decoding parallels taste sensations. *Current Biology*, *25*(7), 890–896.

Dutta, R., Hines, E. L., Gardner, J. W., Kashwan, K. R., & Bhuyan, M. (2003). Tea quality prediction using a tin oxide-based electronic nose: an artificial intelligence approach. *Sensors and Actuators B: Chemical*, *94*(2), 228–237.

Dutta, T. M., Josiah, A. F., Cronin, C. A., Wittenberg, G. F., & Cole, J. W. (2013). Altered taste and stroke: a case report and literature review. *Topics in Stroke Rehabilitation*, *20*(1), 78–86.

El-Kassas, G., & Ziade, F. (2016). Exploration of the dietary and lifestyle behaviors and weight status and their self-perceptions among health sciences university students in North Lebanon. *BioMed Research International*, https://doi.org/10.1155/2016/9762396

Forina, M., Oliveri, P., Lanteri, S., & Casale, M. (2008). Class-modeling techniques, classic and new, for old and new problems. *Chemometrics and Intelligent Laboratory Systems*, *93*(2), 132–148.

Frank, R. A., Mize, S. J., Kennedy, L. M., de los Santos, H. C., & Green, S. J. (1992). The effect of Gymnemasylvestre extracts on the sweetness of eight sweeteners. *Chemical Senses*, *17*(5), 461–479.

Ghosh, A., & Saha, S. (2020). Interactive game-based motor rehabilitation using hybrid sensor architecture. In *Handbook of Research on Emerging Trends and Applications of Machine Learning* (pp. 312–337). IGI Global.

Guido, D., Perna, S., Carrai, M., Barale, R., Grassi, M., & Rondanelli, M. (2016). Multidimensional evaluation of endogenous and health factors affecting food preferences, taste and smell perception. *The Journal of Nutrition, Health & Aging*, *20*(10), 971–981.

Hettinger, T. P., Gent, J. F., Marks, L. E., & Frank, M. E. (1999). Study of taste perception. *Perception & Psychophysics*, *61*(8), 1510–1521.

Hosmer Jr, D. W., Lemeshow, S., & Sturdivant, R. X. (2013). *Applied Logistic Regression* (Vol. 398). John Wiley & Sons.

Hussin, S. S., & Sudirman, R. (2013). Sensory response through EEG interpretation on alpha wave and power spectrum. *Procedia Engineering*, *53*, 288–293.

Isezuo, S. A., Saidu, Y., Anas, S., Tambuwal, B. U., & Bilbis, L. S. (2008). Salt taste perception and relationship with blood pressure in type 2 diabetics. *Journal of Human Hypertension*, *22*(6), 432–434.

Ishihara, S., Ikeda, A., Citterio, D., Maruyama, K., Hagiwara, M., & Suzuki, K. (2005). Smart chemical taste sensor for determination and prediction of taste qualities based on a two-phase optimized radial basis function network. *Analytical Chemistry*, *77*(24), 7908–7915.

Izenman, A. J. (2013). Linear discriminant analysis. In *Modern Multivariate Statistical Techniques* (pp. 237–280). Springer.

Jakubowska, M., Sordoń, W., & Ciepiela, F. (2016). Unsupervised pattern recognition methods in ciders profiling based on GCE voltammetric signals. *Food Chemistry*, *203*, 476–482.

Jilani, H., Pohlabeln, H., De Henauw, S., Eiben, G., Hunsberger, M., Molnar, D., ... & Hebestreit, A. (2019). Relative validity of a food and beverage preference questionnaire to characterize taste phenotypes in children adolescents and adults. *Nutrients*, *11*(7), 1453.

Johnson, S. C. (1967). Hierarchical clustering schemes. *Psychometrika*, *32*(3), 241–254.
Kassambara, A. (2017). *Practical Guide to Cluster Analysis in R: Unsupervised Machine Learning* (Vol. 1). Sthda.
Killgore, W. D., & Yurgelun-Todd, D. A. (2005). Body mass predicts orbitofrontal activity during visual presentations of high-calorie foods. *Neuroreport*, *16*(8), 859–863.
Kühn, S., & Gallinat, J. (2013). Does taste matter? How anticipation of cola brands influences gustatory processing in the brain. *PLoS One*, *8*(4), e61569.
Li, H., Jia, P., & Fei, T. (2021). Associations between taste preferences and chronic diseases: a population-based exploratory study in China. *Public Health Nutrition*, *24*(8), 2021–2032.
Liem, D. G., Aydin, N. T., & Zandstra, E. H. (2012). Effects of health labels on expected and actual taste perception of soup. *Food Quality and Preference*, *25*(2), 192–197.
Lötsch, J., Kringel, D., & Hummel, T. (2019). Machine learning in human olfactory research. *Chemical Senses*, *44*(1), 11–22.
Makaronidis, J., Firman, C., Magee, C. G., Mok, J., Balogun, N., Lechner, M., ... & Batterham, R. L. (2021). Distorted chemosensory perception and female sex associate with persistent smell and/or taste loss in people with SARS-CoV-2 antibodies: a community based cohort study investigating clinical course and resolution of acute smell and/or taste loss in people with and without SARS-CoV-2 antibodies in London, UK. *BMC Infectious Diseases*, *21*(1), 1–11.
March, G., Nguyen, T. D., & Piro, B. (2015). Modified electrodes used for electrochemical detection of metal ions in environmental analysis. *Biosensors*, *5*(2), 241–275.
Marks, L. E., Elgart, B. Z., Burger, K., & Chakwin, E. M. (2007). Human flavor perception: application of information integration theory. *Teorie&modelli*, *1*(2), 121.
Min, W., Jiang, S., Liu, L., Rui, Y., & Jain, R. (2019). A survey on food computing. *ACM Computing Surveys (CSUR)*, *52*(5), 1–36.
Naresh, A., Shaastry, M. S. S., Yadav, B. P., & Bhaskar, K. (2020). Understanding user taste preferences for food recommendation. *International Journal of Engineering Research & Technology, 9*(6), 6–12.
Niklassen, A. S., Draf, J., Huart, C., Hintschich, C., Bocksberger, S., Trecca, E. M. C., ... & Hummel, T. (2021). COVID-19: Recovery from chemosensory dysfunction. A multicentre study on smell and taste. *The Laryngoscope*, *131*(5), 1095–1100.
Noel, C., & Dando, R. (2015). The effect of emotional state on taste perception. *Appetite*, *95*, 89–95.
Ouyang, Q., Yang, Y., Wu, J., Chen, Q., Guo, Z., & Li, H. (2020). Measurement of total free amino acids content in black tea using electronic tongue technology coupled with chemometrics. *LWT*, *118*, 108768.
Park, C., Looney, D., & Mandic, D. P. (2011, January). Estimating human response to taste using EEG. In *2011 Annual International Conference of the IEEE Engineering in Medicine and Biology Society* (pp. 6331–6334). IEEE.
Prieto, N., Rodriguez-Méndez, M. L., Leardi, R., Oliveri, P., Hernando-Esquisabel, D., Iñiguez-Crespo, M., & De Saja, J. A. (2012). Application of multi-way analysis to UV–visible spectroscopy, gas chromatography and electronic nose data for wine ageing evaluation. *Analytica Chimica Acta*, *719*, 43–51.
Rácz, A., Gere, A., Bajusz, D., & Héberger, K. (2018). Is soft independent modeling of class analogies a reasonable choice for supervised pattern recognition? *RSC Advances*, *8*(1), 10–21.
Radhi, M. M., Al-Damlooji, N. K., Jobayr, M. R., & Dawood, D. S. (2013). Electrochemical sensors of cyclic voltammetry to detect Cd (II) in blood medium. *Sensors & Transducers*, *155*(8), 150.

Rodríguez-Méndez, M. L., De Saja, J. A., González-Antón, R., García-Hernández, C., Medina-Plaza, C., García-Cabezón, C., & Martín-Pedrosa, F. (2016). Electronic noses and tongues in wine industry. *Frontiers in Bioengineering and Biotechnology*, *4*, 81.

Saha, S., & Ghosh, A. (2019, December). Rehabilitation using neighbor-cluster based matching inducing artificial bee colony optimization. In *2019 IEEE 16th India Council International Conference (INDICON)* (pp. 1–4). IEEE.

Sanei, S., & Chambers, J. A. (2013). *EEG Signal Processing*. John Wiley & Sons.

Senthilnath, J., Dokania, A., Kandukuri, M., Ramesh, K. N., Anand, G., & Omkar, S. N. (2016). Detection of tomatoes using spectral-spatial methods in remotely sensed RGB images captured by UAV. *Biosystems Engineering*, *146*, 16–32.

Singh, A., Thakur, N., & Sharma, A. (2016, March). A review of supervised machine learning algorithms. In *2016 3rd International Conference on Computing for Sustainable Global Development (INDIACom)* (pp. 1310–1315). IEEE.

Takagi, S., Toko, K., Wada, K., Yamada, H., & Toyoshima, K. (1998). Detection of suppression of bitterness by sweet substance using a multichannel taste sensor. *Journal of Pharmaceutical Sciences*, *87*(5), 552–555.

Teo, J., Hou, C. L., & Mountstephens, J. (2017, October). Deep learning for EEG-based preference classification. In *AIP Conference Proceedings* (Vol. 1891, No. 1, p. 020141). AIP Publishing LLC.

Tomita, H., & Ikeda, M. (2002). Clinical use of electrogustometry: strengths and limitations. *Acta Oto-Laryngologica*, *122*(4), 27–38.

Trachootham, D., Satoh-Kuriwada, S., Lam-Ubol, A., Promkam, C., Chotechuang, N., Sasano, T., & Shoji, N. (2018). Differences in taste perception and spicy preference: a Thai–Japanese cross-cultural study. *Chemical Senses*, *43*(1), 65–74.

Tuwani, R., Wadhwa, S., & Bagler, G. (2019). BitterSweet: building machine learning models for predicting the bitter and sweet taste of small molecules. *Scientific Reports*, *9*(1), 1–13.

Vlasov, Y., Legin, A., Rudnitskaya, A., Di Natale, C., & D'amico, A. (2005). Nonspecific sensor arrays ("electronic tongue") for chemical analysis of liquids (IUPAC Technical Report). *Pure and Applied Chemistry*, *77*(11), 1965–1983.

Wallroth, R., Höchenberger, R., & Ohla, K. (2018). Delta activity encodes taste information in the human brain. *Neuroimage*, *181*, 471–479.

Wang, P., Liu, Q., Wu, C., & Hsia, K. J. (Eds.). (2015). *Bioinspired Smell and Taste Sensors*. Springer.

Wasim, M., Sajjad, M., Ramzan, F., Khan, U. G., & Mahmood, W. (2018). A review and classification of widely used offline brain datasets. *International Journal of Advanced Computer Science and Applications*, *9*(2), 399–408.

Wold, S., Esbensen, K., & Geladi, P. (1987). Principal component analysis. *Chemometrics and Intelligent Laboratory Systems*, *2*(1–3), 37–52.

Wright, H. N., Sheehe, P. R., & Leopold, D. A. (1991). The odorant confusion matrix as an aid to diagnosis. *Chemical Senses*, *16*, 601.

Yamamoto, T., Mizuta, H., & Ueji, K. (2021). Analysis of facial expressions in response to basic taste stimuli using artificial intelligence to predict perceived hedonic ratings. *Plos One*, *16*(5), e0250928.

13 New Perspectives for Knowledge Management in Telemedicine

Preliminary Findings from a Case Study in the COVID-19 Era

Francesca Dal Mas, Helena Biancuzzi,
Maurizio Massaro, Lorenzo Cobianchi,
Rym Bednarova and Luca Miceli

CONTENTS

13.1 Introduction 231
13.2 Literature Review 232
13.3 Methodology 235
13.4 Findings 235
13.5 Discussions and Conclusions 237
References 237

13.1 INTRODUCTION

Telemedicine is usually referred to as the application of electronic information and communication technologies to provide and support healthcare at a distance (Nikolian et al., 2018). The recent COVID-19 pandemic (WHO, 2020) has catalysed an unprecedented need to provide care remotely, following the social distancing criteria and the disruption in hospitals and clinics worldwide (Cobianchi et al., 2020). This compelling factor has aided the rapid expansion of telemedicine devices and technical solutions in a variety of medical specialties, including radiology, neuroscience, dermatology, cardiology, oncology, and internal medicine, to meet patients' ambulatory care needs (Sorensen et al., 2020). Many people assume that virtual visits will continue to occur in the post-COVID "new normal" (Cobianchi et al., 2020), since telemedicine has been shown to enhance treatment access, improve resource efficiency, and reduce costs as compared to traditional in-person hospital or ambulatory visits (Nikolian et al., 2018). Furthermore, studies show that many patients are afraid of being injured or infected while visiting a hospital (Acar and Kaya, 2019).

Patients may be provided with virtual appointments without needing to access the hospital facilities thanks to telemedicine equipment.

The recent literature has examined the public's attitudes, interactions, and satisfaction with telemedicine (Grenda et al., 2020; Reed et al., 2019), with many reflecting on the technological solutions used in response to the need for user-friendly yet secure platforms and resources (Jain and Kaur, 2018). For sure, telemedicine brings new ways of interaction and knowledge sharing between the healthcare professionals and the patients, requiring new knowledge management and knowledge-sharing mechanisms. Starting from the preliminary results of a case study (Yin, 2014), this chapter deepens some of the new perspectives on knowledge management in telemedicine.

13.2 LITERATURE REVIEW

The implementation of emerging technologies in healthcare practices is radically changing the whole scenario, both in terms of clinical practice and the educational needs of clinicians (Au-Yong-Oliveira et al., 2021; Sousa et al., 2021). Modern operating robots influence not only the surgical practice but also the surgeons' and medical doctors' ideal skillset and how such technical and non-technical skills should be learned. Big data can be used to train the physicians-to-be and improve current healthcare techniques (Sousa et al., 2019). Internal processes, such as team management and stakeholder partnerships, are impacted by emerging technologies and data availability. Scientific clinical research is advancing in a variety of fields, such as transplantation and organ regeneration (Cobianchi et al., 2009; Croce et al., 2019; Hogan et al., 2012; Marzorati et al., 2009), calling for a fast translation to clinical practice to improve the patients' outcomes (McAneney et al., 2010). The recent COVID-19 pandemic has exacerbated these challenges. In this evolving scenario, managing knowledge and the relationships among the various actors, especially patients and clinical professionals (Dal Mas, Biancuzzi, Massaro, Barcellini, et al., 2020; Dal Mas, Garcia-Perez, et al., 2020; Ferreira Polonia and Coutinho Gradim, 2020; Martins et al., 2020), looks essential (Pateiro Marcão et al., 2020; Therkildsen Sudmann et al., 2020; Vold and Haave, 2020).

The need for social distancing generated by the global pandemic COVID-19 has pushed e-health to become an increasingly concrete and widespread reality (Gadeikienė et al., 2021; Grenda et al., 2020; Ritchey et al., 2020; Sorensen et al., 2020).

Even today, the terms telemedicine and e-health are often considered synonyms. These phenomena are in continuous advancement in parallel with the constant evolution of information technologies. Still, e-health turns out to be a development and expansion of telemedicine. In any case, these concepts are based on the idea that it is no longer the patients who move but the information concerning them.

Telemedicine refers to the medical staff's activity that allows them to communicate with and monitor the patients without being physically in the same place through the use of technology. It is, therefore, not a new medical specialty but the provision of a health service that exploits the tools made available by information and communication technology (ICT). The aim is to improve the service and reduce costs, always guaranteeing the patient's centrality and the assistance they need. The terms e-health or telehealth have a much broader scope, going beyond the mere concept of geographical distance between the patient and medical staff, conceptually involving the entire healthcare system in a reorganization and rationalization of the whole sector (Della Mea, 2001).

The technologies described range from the exchange of information via telephone to computer and network applications that store/forward messages or ensure connections in real time, up to the most advanced virtual reality technologies.

The increasing use of ICT and the internet in healthcare have made it extremely difficult to define the boundaries of telemedicine interventions, which, as anticipated, refers to a limited number of technologies and processes.

According to Eysenbach (2001), e-health can be defined as:

> an emerging field in the intersection of medical informatics, public health and business, referring to health services and information delivered or enhanced through the Internet and related technologies. In a broader sense, the term characterizes not only a technical development, but also a state-of-mind, a way of thinking, an attitude, and a commitment for networked, global thinking, to improve health care locally, regionally, and worldwide by using information and communication technology.

There are different ways of categorizing the various areas and interventions of e-health. One of the first studies on the topic, that of Taylor (1998), suggested that telemedicine services could be divided into three distinct categories:

1. treatment services, such as telesurgery;
2. diagnostic services and their related management;
3. information and educational services, such as distance learning (divided into four sub-categories: teleconsultation, teleconference, telereporting, telemonitoring).

The guidelines of the Italian Ministry of Health as of October 27, 2020 (Ministero della Sanità, 2020) defined the indications for the provision of the services in telemedicine. These are divided as follows:

1. Tele-visit, namely the medical act in which the professional interacts remotely and in real time with the patient, possibly with a caregiver's support. This cannot be the only relationship between doctor and patient, nor a substitute for the first medical examination in the presence. It is up to the clinician to decide which measures and remote method to use in favour of the patient. The possibility of exchanging clinical data, medical reports, images, audio-video relating to the patient in real time must always be guaranteed.
2. Medical teleconsultation, namely the medical act in which professionals interact remotely to discuss the clinical situation of a patient, based primarily on the sharing of clinical data, reports, images, and audio-video concerning the specific case. This method can also take place asynchronously. If the patient is present at the teleconsultation, this must be carried out in real time. The

goal is to share the medical choices concerning a patient by the professionals involved.
3. Clinical teleconsultation, namely the specific activity of the healthcare professionals, not necessarily medical doctors, performed at a distance by two or more people with different responsibilities with respect to the specific case treated. It consists of a request for support during healthcare activities, followed by a video call in which the consulted professional provides the other with indications for the correct execution of assistance actions aimed at the patient. The teleconsultation can be carried out in a deferred manner or, if in the presence of the patient necessarily in real time.
4. Remote assistance by health professionals, happening when a professional provides remote support to the patient/caregiver through a video call. Such activity can require the sharing of data, reports, or images if necessary. The professional can also propose the use of suitable apps for administering questionnaires or sharing images and videos. The aim is to facilitate the correct performance of assistance activities, which can be carried out mainly at home. Teleservice is mainly programmed and repeatable on the basis of the accompanying programmes provided for the patient.
5. Telerefertation is a report issued by the professional who offered the patient a clinical or instrumental examination. This documentation is produced and transmitted using digital systems. The remote report can be issued within a remote management process of the clinical or instrumental examination (remote management), in which the medical doctor who performs the remote report, who is not in the same place where the examination is performed, can make use of the collaboration of a healthcare professional located near the patient, can communicate with them in real time via computer or by telephone.
6. Virtual triage, namely a virtual consultation carried out by health personnel to patients to indicate the most appropriate diagnostic or therapeutic path.

According to the literature, telemedicine can contribute to generating value for the various stakeholders involved (Gadeikienė et al., 2021; Miceli et al., 2021). Still, new managerial practices are needed to reach the potential of such a value fully. Indeed, while ICT has allowed virtual collaborative practices, increasing the number of the available information tools, virtual environments pose new challenges in how such collaborations and knowledge-sharing activities can take place, including the way knowledge is translated, transferred, shared, and created (Dal Mas, Garcia-Perez, et al., 2020). The design of ICTs is typically focused on implicit assumptions that are consistent with models of rational decision making and explicit knowledge transfer. This makes it much more difficult to express tacit knowledge through narration and demonstration (Paul, 2006). Since knowledge translation, transfer, and sharing look fundamental when it comes to the outcome, namely, the care of the patient (Brunoro-Kadash and Kadash, 2013; Dal Mas, Biancuzzi, Massaro, Barcellini, et al., 2020), new ways of managing such practices are needed. Therefore, following the expansion due to the COVID-19 pandemic, telemedicine opens today several challenges in defining the best ways knowledge should be managed in the relationship between the clinical staff involved and the patient.

13.3 METHODOLOGY

The methodology used is based on a case study (Massaro et al., 2019; Yin, 2014). The case study was used to analyse a new initiative by the National Cancer Institute – IRCCS CRO of Aviano (Italy), one of Europe's most renowned institutes and research centres in the area of oncological surgery and cancer therapies (Dal Mas, Biancuzzi, Massaro, Barcellini, et al., 2020). The institute recently unveiled a new telemedicine program named "Doctor@Home (D@H)" (Miceli et al., 2021).

The D@H project, which began in the pain therapy department (Miceli et al., 2017), is now used for all of the institute's follow-up visits. Although an ethical committee request for further investigation and patient satisfaction is still pending as of March 2021, the project has defined its steps and milestones.

Data collection was made by interviewing the scientific chief and principal investigator of the initiative (LM) and two executives belonging to the Regional Government healthcare management team.

13.4 FINDINGS

The program is developed through specific steps. At the hospital, the clinician and the patient meet for the first time in person. The patient can choose to continue with the brand-new telemedicine program or schedule in-person follow-up appointments. The online meeting takes place on an online platform. To allow the medical staff to submit the e-invitation for the tele-visit, the patient must own a valid email address. The medical staff schedules the appointment on the "LifeSize" approved portal and gives the patient the access code on the scheduled date and time. The patient may access the link sent via email or use the software on his or her mobile phone. The device recognizes the patient's identity through the unique social security number badge. At the end of the appointment, the medical staff creates an encrypted digital report sent to the patient via email, together with an unlock code. If a medical prescription is requested, the patients are emailed an identification code that allows them to pick up the paperwork and prescribed medication at any pharmacy.

Despite the fact that the D@H project is one of the several ones in the telemedicine scenario, the National Cancer Institute of Aviano has chosen to concentrate the protocol and ongoing experience on two fundamental and possibly understudied aspects of telehealth knowledge management: service co-production and continuous learning.

In healthcare, co-production refers to the patient's active role in achieving the medical outcome by behaving in specific ways or carrying out particular tasks in cooperation with the clinical staff (Batalden et al., 2016; Dal Mas, Biancuzzi, Massaro and Miceli, 2020; Elwyn et al., 2020). Since physical contact is not permitted in the D@H program, doctors and patients must devise new ways of communicating symptoms and concerns. Clinical professionals play an important role in guiding patients in this process, assisting them in becoming more familiar with the potential of technology. Specific knowledge translation techniques, including the use of non-technical skills (Dal Mas, Bagarotto, et al., 2021; Yule and Smink, 2020), may be used to provide certain patients with guidance about the prescribed behaviours or necessary actions, as the presence of a caregiver supporting them.

While the ethical committee is being adopted at a regional level, most of these recommendations are still missing; thus, learning is crucial in raising consciousness about knowledge management best practices.

According to the National Cancer Institute of Aviano, the learning component of the program is the second pillar of the D@H initiatives. The clinical team is using a learning-by-doing approach to help handle the patient during the e-visit. In this type of "learning-on-the-job," which is conducted in an interprofessional manner involving medical practitioners and patients, co-learning is used to facilitate the co-production of care and knowledge flows. To better understand how to handle the remote visit and the patient's relationship, medical doctors involved in the program may seek help from colleagues from different specialties, nurses, and psychologists (Dal Mas, Biancuzzi, Massaro, Barcellini, et al., 2020). With the support, help, and advice of the nursing team, patients must learn how to take advantage of telemedicine technology, enhancing the knowledge flows to and from the clinical staff.

To merge co-production and co-learning, the D@H project uses four distinct phases (Elwyn et al., 2020; Miceli et al., 2021):

1. *Co-assess*: Clinical staff at D@H will assist patients in correctly self-assessing and describing their symptoms. Self-evaluation should be documented in a sort of "diary" even between follow-ups, to be shared with the medical doctor during the e-meeting. In order to document the health status and issues in a consistent manner, adequate shared methodologies and instructions should be provided, like how to measure pain. Tacit knowledge of the patient must then be translated into explicit knowledge, to be shared with the clinical team taking care of the situation.
2. *Collaborate*: D@H clinical staff should use communication and translation tools to pass medical information to patients and learn about their wishes, goals, and priorities (Angelos, 2020). Understanding the patient's needs will make collaborative decision-making (Woltz et al., 2018) about the next steps and rehabilitation much easier.
3. *Co-design*: Once D@H clinical staff and the patient have a greater understanding of one another, there is room for one step ahead in the knowledge process: co-design the care plan. The co-design action (Osei-Frimpong *et al.*, 2018) includes the telehealth component as well as the efforts and behaviours needed to enhance the medical outcomes.
4. *Co-deliver*: Although the patient must follow the three previous steps, telemedicine will aid D@H clinical staff in tracking the patient's progress and assisting or directing the patient as needed, even if (s)he is far away. Clinicians will be able to learn from the program's aggregate data in order to strengthen their medical research and knowledge about specific diseases and treatment strategies.

13.5 DISCUSSIONS AND CONCLUSIONS

Telemedicine is meant to soon become part of physicians' and patients' everyday practices in today's healthcare environment.

Telehealth tools can provide a range of benefits to the patients, including the ability to avoid going back and forth from and to the hospitals. Such an advantage looks particularly useful for specialized hubs like the National Cancer Institute of Aviano, where patients come from all over Italy as well as from abroad. Furthermore, during difficult times such as the COVID-19 pandemic, social distancing measures are required, especially for vulnerable patients such as oncological ones. At the same time, telemedicine proved to offer a significant response to patients' ambulatory needs in a variety of specialties (Grenda et al., 2020; Ritchey et al., 2020; Sorensen et al., 2020).

However, the increasing use of telemedicine and e-health tools opens up new issues regarding how knowledge should be translated, transferred, and shared in the virtual relations between clinicians and patients, to maximize the medical outcomes. Moreover, it is crucial to understand how telemedicine and the related aggregate data collection can facilitate the creation of new knowledge for a single clinical case and also to study a disease or the results of different treatment strategies. While explicit knowledge can be more easily shared, taking advantage of protocols and procedures, open questions still exist when it comes to tacit knowledge (Gadeikienė et al., 2021; Paul, 2006).

Although telemedicine appears to have distinct advantages, guidance and understanding of how to use such tools are required. The National Cancer Institute of Aviano's D@H initiative serves as a pilot study to explore how knowledge can be translated, transferred, shared, and created in the interaction between clinicians and patients. We argue that the initial study and design of the D@H initiative identified the co-production and co-learning pillars as the means to maximize the benefits of modern technologies. In our preliminary findings, co-production and co-learning will enable maximum participation, knowledge translation, and interaction between clinical professionals and patients.

The four phases of co-assess, collaborate, co-design, and co-deliver constitute different levels of knowledge flows between the clinical staff, which can be multidisciplinary, and the patients and their caregivers. Adequate tools and methods must be identified to understand and facilitate such knowledge dynamics. Non-technical skills like communication and empathy have already proved to play a central role in such a process (Dal Mas, Bagarotto, et al., 2021; Dal Mas, Biancuzzi, et al., 2021).

Although this study is at an early stage, its preliminary results look promising in providing new perspectives into knowledge management in telemedicine.

Further investigation within the D@H initiative and its development will allow the dissemination of knowledge and best practices to other healthcare organizations.

REFERENCES

Acar, M. and Kaya, O. (2019), "A healthcare network design model with mobile hospitals for disaster preparedness: A case study for Istanbul earthquake", *Transportation Research Part E: Logistics and Transportation Review*, Vol. 130, pp. 273–292.

Angelos, P. (2020), "Interventions to improve informed consent perhaps surgeons should speak less and listen more", *JAMA Surgery*, Vol. 155, No. 1, pp. 13–14.

Au-Yong-Oliveira, M., Pesqueira, A., Sousa, M.J., Dal Mas, F. and Soliman, M. (2021), "The Potential of Big Data Research in HealthCare for Medical Doctors' Learning", *Journal of Medical Systems*, Vol. 45, p. 13.

Batalden, M., Batalden, P., Margolis, P., Seid, M., Armstrong, G., Opipari-arrigan, L. and Hartung, H. (2016), "Coproduction of healthcare service", *BMJ Quality & Safety*, Vol. 25 No. 7, pp. 509–517.

Brunoro-Kadash, C. and Kadash, N. (2013), "Time to care: A patient-centered quality improvement strategy", *Leadership in Health Services*, Vol. 26, No. 3, pp. 220–231.

Cobianchi, L., Pugliese, L., Peloso, A., Dal Mas, F. and Angelos, P. (2020), "To a new normal: Surgery and COVID-19 during the transition phase", *Annals of Surgery*, Vol. 272, pp. e49–e51.

Cobianchi, L., Zonta, S., Vigano, J., Dominioni, T., Ciccocioppo, R., Morbini, P., Bottazzi, A., et al. (2009), "Experimental small bowel transplantation from non-heart-beating donors: a large-animal study", *Transplantation Proceedings*, Vol. 41 No. 1, pp. 55–56.

Croce, S., Peloso, A., Zoro, T., Avanzini, M.A. and Cobianchi, L. (2019), "A hepatic scaffold from decellularized liver tissue: Food for thought", *Biomolecules*, Vol. 9, No. 12, doi:10.3390/biom9120813.

Dal Mas, F., Bagarotto, E.M. and Cobianchi, L. (2021), "Soft skills effects on knowledge translation in healthcare. Evidence from the field", in Lepeley, M.T., Beutell, N., Abarca, N. and Majluf, N. (Eds.), *Soft Skills for Human Centered Management and Global Sustainability*, Routledge, New York, doi:10.4324/9781003094463-7-11.

Dal Mas, F., Biancuzzi, H., Massaro, M., Barcellini, A., Cobianchi, L. and Miceli, L. (2020), "Knowledge translation in oncology: A case study", *Electronic Journal Of Knowledge Management*, Vol. 18, No. 3, pp. 212–223.

Dal Mas, F., Biancuzzi, H., Massaro, M. and Miceli, L. (2020), "Adopting a knowledge translation approach in healthcare co-production. A case study", *Management Decision*, Vol. 58, No. 9, pp. 1841–1862.

Dal Mas, F., Biancuzzi, H. and Miceli, L. (2021), "The importance of soft skills in the co-production of healthcare services in the public sector: The oncology in motion experience", in Lepeley, M.T., Beutell, N., Abarca, N. and Majluf, N. (Eds.), *Soft Skills for Human Centered Management and Global Sustainability*, Routledge, New York, doi:10.4324/9781003094463-8-12.

Dal Mas, F., Garcia-Perez, A., Sousa, M.J., Lopes da Costa, R. and Cobianchi, L. (2020), "Knowledge Translation in the Healthcare Sector: A Structured Literature Review", *Electronic Journal of Knowledge Management*, Vol. 18, No. 3, pp. 198–211.

Della Mea, V. (2001), "What is e-health (2): The death of telemedicine?", *Journal of Medical Internet Research*, Vol. 3 No. 2, p. e22. doi: 10.2196/jmir.3.2.e22

Elwyn, G., Nelson, E., Hager, A. and Price, A. (2020), "Coproduction: When users define quality", *BMJ Quality and Safety*, Vol. 29 No. 9, pp. 711–716.

Eysenbach, G. (2001), "What is e-health?", *Journal of Medical Internet Research*, Vol. 3 No. 2, p. e20. doi: 10.2196/jmir.3.2.e20; www.jmir.org/2001/2/e20

Ferreira Polonia, D. and Coutinho Gradim, A. (2020), "Innovation and knowledge flows in healthcare ecosystems: The Portuguese case", *Electronic Journal of Knowledge Management*, Vol. 18, No. 3, pp. 374–391.

Gadeikienė, A., Pundzienė, A. and Dovalienė, A. (2021), "How does telehealth shape new ways of co-creating value?", *International Journal of Organizational Analysis*, Vol. 29 No. 6, pp. 1423–1442. doi:10.1108/IJOA-07-2020-2355.

Grenda, T.R., Whang, S. and Evans, N.R. (2020), "Transitioning a surgery practice to telehealth during COVID-19", *Annals of Surgery*, Vol. 272, No. 2, pp. e168–e169.

Hogan, A.R., Doni, M., Damaris Molano, R., Ribeiro, M.M., Szeto, A., Cobianchi, L., Zahr-Akrawi, E., et al. (2012), "Beneficial effects of ischemic preconditioning on pancreas cold preservation", *Cell Transplantation*, Vol. 21, No. 7, pp. 1349–1360.

Jain, P. and Kaur, A. (2018), "Big data analysis for prediction of coronary artery disease", *2018 4th International Conference on Computing Sciences (ICCS)*, Jalandhar, pp. 188–193.

Martins, M., Costa, M., Gonçalves, M., Duarte, S. and Au-Yong-Oliveira, M. (2020), "Knowledge creation on edible vaccines", *Electronic Journal of Knowledge Management*, Vol. 18, No. 3, pp. 285–301.

Marzorati, S., Bocca, N., Molano, R.D., Hogan, A.R., Doni, M., Cobianchi, L., Inverardi, L., et al. (2009), "Effects of systemic immunosuppression on islet engraftment and function into a subcutaneous biocompatible device", *Transplantation Proceedings*, Vol. 41 No. 1, pp. 352–353.

Massaro, M., Dumay, J. and Bagnoli, C. (2019), "Transparency and the rhetorical use of citations to Robert Yin in case study research", *Meditari Accountancy Research*, Vol. 27, No. 1, pp. 44–71.

McAneney, H., McCann, J.F., Prior, L., Wilde, J. and Kee, F. (2010), "Translating evidence into practice: A shared priority in public health?", *Social Science and Medicine*, Vol. 70, No. 10, pp. 1492–1500.

Miceli, L., Bednarova, R., Di Cesare, M., Santori, E., Spizzichino, M., Di Minco, L., Botti, R., et al. (2017), "Outpatient therapeutic chronic opioid consumption in Italy: A one-year survey", *Minerva Anestesiologica*, Vol. 83, No. 1, pp. 33–40.

Miceli, L., Dal Mas, F., Biancuzzi, H., Bednarova, R., Rizzardo, A., Cobianchi, L. and Holmboe, E.S. (2021), "Doctor@Home: Through a telemedicine co-production and co-learning journey", *Journal of Cancer Education*, Vol. 3 No. 2, doi:10.1007/s13187-020-01945-5.

Ministero della Sanità. (2020), *Indicazioni nazionali per l'erogazione dei servizi di telemedicina*, Italy, pp. 1–19.

Nikolian, V.C., Williams, A.M., Jacobs, B.N., Kemp, M.T., Wilson, J.K., Mulholland, M.W. and Alam, H.B. (2018), "Pilot study to evaluate the safety, feasibility, and financial implications of a postoperative telemedicine program", *Annals of Surgery*, Vol. 268, No. 4, pp. 700–707.

Osei-Frimpong, K., Wilson, A. and Lemke, F. (2018), "Patient co-creation activities in healthcare service delivery at the micro level: The influence of online access to healthcare information", *Technological Forecasting and Social Change*, Vol. 126, pp. 14–27.

Pateiro Marcão, R., Pestana, G. and Sousa, M.J. (2020), "Knowledge management and gamification in pharma: An approach in pandemic times to develop product quality reviews", *Electronic Journal of Knowledge Management*, Vol. 18, No. 3, pp. 255–268.

Paul, D.L. (2006), "Collaborative activities in virtual settings: A knowledge management perspective of telemedicine", *Journal of Management Information Systems*, Vol. 22, No. 4, pp. 143–176.

Reed, M.E., Huang, J., Parikh, R., Millman, A., Ballard, D.W., Barr, I. and Wargon, C. (2019), "Patient–provider video telemedicine integrated with clinical care: Patient experiences", *Annals of Internal Medicine*, Vol. 171, No. 3, pp. 222–224.

Ritchey, K.C., Foy, A., McArdel, E. and Gruenewald, D.A. (2020), "Reinventing palliative care delivery in the era of COVID-19: How telemedicine can support end of life care", *American Journal of Hospice and Palliative Medicine*, Vol. 37, No. 11, pp. 992–997.

Sorensen, M.J., Bessen, S., Danford, J., Fleischer, C. and Wong, S.L. (2020), "Telemedicine for Surgical consultations – Pandemic response or here to stay?", *Annals of Surgery*, Vol. 272, No. 3, pp. e174–e180.

Sousa, M.J., Dal Mas, F., Pesqueira, A., Lemos, C., Verde, J.M. and Cobianchi, L. (2021), "The potential of AI in health higher education to increase the students' learning outcomes", *TEM Journal*, Vol. 10 No. 2, 488–497. https://doi.org/10.18421/TEM102-02.

Sousa, M.J., Pesqueira, A., Lemos, C., Sousa, M. and Rocha, A. (2019), "Decision-making based on big data analytics for people management in healthcare organizations", *Journal of Medical Systems*, Vol. 43, No. 9, p. 290.

Taylor, P. (1998), "A survey of research in telemedicine. 2: Telemedicine services", *Journal of Telemedicine and Telecare*, Vol. 4, No. 2, pp. 63–71.

Therkildsen Sudmann, T., Haukeland Fredriksen, E., Træland Børsheim, I. and Heldal, I. (2020), "Knowledge management from senior users of online health information point of view", *Electronic Journal of Knowledge Management*, Vol. 18, No. 3, pp. 325–337.

Vold, T. and Haave, H.M. (2020), "Relevance of adult higher education on knowledge management in the healthcare sector", *Electronic Journal of Knowledge Management*, Vol. 18, No. 3, pp. 236–254.

WHO. (2020), "Coronavirus disease (COVID-19) pandemic", *Health Topics*, available at: www.who.int/emergencies/diseases/novel-coronavirus-2019 (accessed 8 April 2020).

Woltz, S., Krijnen, P., Pieterse, A.H. and Schipper, I.B. (2018), "Surgeons' perspective on shared decision making in trauma surgery: A national survey", *Patient Education and Counseling*, Vol. 101, No. 10, pp. 1748–1752.

Yin, R.K. (2014), *Case Study Research: Design and Methods*, Sage, Thousand Oaks, CA.

Yule, S. and Smink, D.S. (2020), "Non-technical skill countermeasures for pandemic response", *Annals of Surgery*, Vol. 272, No. 3, pp. e213–e215.

14 Prioritization of Quality of Public Health Services in the Sector of Graphic Methods
University Hospital

Roger da Silva Wegner, Leoni Pentiado Godoy, Taís Pentiado Godoy, Maria José Sousa and Luciana Aparecida Barbieri da Rosa

CONTENTS

14.1 Introduction 241
14.2 Theoretical Framework 243
    14.2.1 Hospital Management with an Emphasis on the Quality of Public Health Services 243
14.3 Methodological Procedures 245
    14.3.1 Data Collection Procedures 246
    14.3.2 Data Analysis Procedures 247
14.4 Discussion of Results 248
    14.4.1 Profile of Professionals 248
    14.4.2 AHP Method Prioritizing the Actions of the Investigated Sector 249
14.5 Final Considerations 252
References 253

14.1 INTRODUCTION

Companies increasingly need to engage in people management to be able to implement strategies in order to develop employees' knowledge in order to deliver quality goods and services. The way employees are managed reflects on the company's success and on the development of professionals working in the organization. Working conditions influence both the professional's satisfaction and the activities performed (Gomes et al., 2011). The high performance of employees is a determining factor for the production of quality goods and services (Calixto et al., 2011).

In this way, organizations, especially those in the public sector, need to develop goods that present a differential in terms of service and quality. Because it is through user satisfaction that hospitals guarantee their reference aiming at quality. Health organizations are increasingly seeking the continuous development of their employees, because it is through them that services are developed.

Hospital management seeks excellence in the services provided and, consequently, the satisfaction of its users. The employees' experience regarding the activities developed results in their good performance and, as a result, the services tend to improve (Fukuda et al., 2014; Guerreiro; Barroso; Rodrigues, 2016). According to the authors (Silva, Matsuda, and Waidman, 2012; Rodrigues Barrichello and Morin, 2016), the human resources area employs all the organization's efforts regarding its employees.

However, evaluating the quality of services, whether in public or private, for-profit or non-profit organizations, is a key factor in the sense of accurately analyzing the real conditions of service provision, in addition to enhancing means for excellence in carrying out these activities (Morais, 2016). The interaction and relationship between them ensure the efficiency of services with motivation, leadership, teamwork, internal communication and remuneration, which are some of the factors that contribute to the employees' performance.

The quality developed by health organizations originated with the purpose of developing goods and services with excellence (Papadomichelaki et al., 2013) meeting the needs and desires of users. High-quality performance in activities not only leads to patient satisfaction but also ends up making life easier for employees.

The tools used in the healthcare area guarantee the excellence of the services delivered. SERVQUAL has the function of measuring the quality of services (Parasuraman et al., 1985; Deb; Lomo-David, 2014), exposes the users' perception regarding the expectation, and experience of the services received. And through this, it is possible to identify the problems that affect the quality of services (Peixoto et al., 2020). The analytic hierarchy process (AHP) method meets the search for improvements (Saaty, 2008; Min, 2010). This methodology allows the solution of complex problems, indicating the hierarchy of possible alternatives.

In this context, this research sought to evaluate the relationship between people management and the quality of public services in the Graphic Methods sector of a university hospital located in southern Brazil through the integration of the following methodologies and tools: the 4Ps of services (profile, processes, procedures, and people) related to the SERVQUAL tool and the AHP method to analyze and assess the perception of respondents. The importance of using the methods mentioned to achieve results relevant to the public health sector, referring to a teaching hospital, is justified, because through this, it is possible to understand the position of priorities and the alternatives raised in order for them to come to contribute to the performance of employees.

Multi-criteria methods add significant value in decision-making, as they not only allow for the approach of complex problems, but also provide clarity and, consequently, transparency to the decision-making process. This study presents, in addition to the introduction, the theoretical framework, addressing hospital management with an emphasis on service quality and the AHP method contributing to the health area. Next, it presents the methodological procedures on reporting the tools and methods used, discusses the results found and presents the final considerations.

14.2 THEORETICAL FRAMEWORK

Literature reviews are presented as an important activity to identify, know, and monitor the development of research; in addition, it contributes to the identification of perspectives for future research.

14.2.1 Hospital Management with an Emphasis on the Quality of Public Health Services

The actions of hospital services (Souza, Scatena, 2010; Regis, 2011; Chakravarty, 2011) and the search for the efficiency of these organizations is a much-debated subject. Hospital management has some objectives, which are equality, efficiency, effectiveness, quality and satisfaction of consumers/patients. The dimensions of quality, present in hospital management, are associated with the balance between the needs of patients and the way professionals are delivering services (Reis, et al., 2020). The service performed efficiently ensures progress and loyalty of customers. According to the authors (Saaty et al., 2012; Reis, et al., 2020). Health care professionals need to be prepared through training to face the working conditions that hospitals offer. Great interactions with the patient result in the quality of services provided (MENDES et al., 2013; Handayani et al., 2014) because quality care has become a fundamental requirement in the health area (Shieh; Wu; Huang, 2010; Chen, H. R.; Cheng, 2012; Saloner et al., 2018).

It is important to emphasize that all the dedication coming from the professionals in the area contributes to the progress of patients. Management plays a participative role with regard to the administration of human resources, catalyzing them to achieve good results. The leadership that acts on human resources management must influence the team to perform the activities in the best possible way; thus, everyone involved tends to grow professionally and deliver services with excellence to society (Vendemiatti et al., 2010).

University hospitals are health centers that serve the general public and are linked to universities, where employees and students participating in the institution perform activities in favor of knowledge (Lim et al., 2018). Some are highly regarded for offering the community treatment of high quality without reimbursement as is the case of the hospital under study.

In the public hospital institution, the demand for quality services is unquestionable since its primary mission is focused on the human being. Thus, the movement for quality in healthcare services is currently necessary and incorporated into healthcare management, in order to ensure risk-free care for the user. This implies the awareness of everyone involved in the system because of its importance and the value of their actions (Rodrigues, 2015; Santos, 2020).

Thus, there must be the commitment of people in addition to continuous improvement so that the results desired by the patient are achieved. In health quality management, the critical point is the culture of organizations, the set of beliefs and values that define how members of an organization comment on the facts and the need to change this reality (Donabedian, 2003; Donabedian, 2010). The theme of the quality of public services offered to society represents a crucial aspect for managers and administrators of public health services. In this context, public hospitals have been the object of many studies, and the perception of clients/patients and employees/public servants are very important elements to determine the quality of delivered goods.

TABLE 14.1
Saaty's Scale

Scale	Evaluation	Reciprocal
Extremely preferred	9	1/9
Very strong to extreme	8	1/8
Very strongly preferred	7	1/7
Strong to very strong	6	1/6
Strongly preferred	5	1/5
Moderate to strong	4	1/4
Moderately preferred	3	1/3
Equal to moderate	2	1/2
Equally preferred	1	1

Source: Saaty (2008).

Initially, a hierarchy is created that aims to describe the problem. The global objective is located at the top, followed by the criteria and alternatives (Zatta et al., 2019). Peer comparisons express linguistic-verbal terms and are converted into numerical values using the Saaty scale for comparative judgments, see Table 14.1. The value of this scale is given in 9 points (Table 14.1). The scale aims to measure the degree of importance of the elements surveyed.

After establishing the weights of the variables, the consistency index is analyzed and the consistency ratio of the data judgments determined. Consistency is given through an amount of raw data from a given base; all other data can be logically deduced from it. In some cases, the values may be inconsistent.

Saaty developed some procedures to assess the consistency of judgments: the first is based on the consistency index (CI), which assesses the degree of inconsistency of the matrix carried out pairwise, using Equation (14.1):

$$IC = (\lambda m\acute{a}x - n)/(n - 1)$$

And the second is the consistency ratio (RC): it is a calculation that allows evaluating the inconsistency in terms of the judgments made, through Equation (14.2).

$$RC = (IC/IR)$$

where IR is the random index.

The random index is perceived as the consistency index obtained for a reciprocal random matrix, presenting non-negative elements, for different sizes of matrix N. Table 14.2 shows the order of the matrices and their respective IR values.

For the realization of an acceptable matrix, the researcher needs to have a consistency ratio less than or equal to 10 percent.

Among the studies in the health area that address the AHP method, we can briefly mention the "Evaluation of Health Care Providers by Users Using the Analytic

TABLE 14.2
Random Consistency Index

MATRIX DIMENSION	1	2	3	4	5	6	7	8	9	10
Random inconsistency	0	0	0,58	0,9	1,12	1,24	1,32	1,14	1,45	1,49

Source: Adapted from Saaty e Shih (2009).

Hierarchy Process Method", developed by Wollmann, Steiner, Vieira and Steiner in the Revista Saúde Pública in 2012. This study sought to understand the quality of services offered by health insurance companies, considering the perception of customers. In the methodology, a cross-sectional study was used with 360 users, with seven health plan operators analyzed. A questionnaire containing questions related to the quality of services was also applied. To analyze the results, the AHP method was used, in order to understand the decision and planning of multiple criteria (Zatta et al., 2019).

The results revealed that users prioritize the "price" criterion. The organizations investigated were grouped into two groups, based on the relationship of their attributes. It was noted that two had less preference and the others indicated greater preference. Based on the results reported, the research provided strategic information, so that processes, pricing structures, and credential networks could be reformulated with the intention of meeting the needs of users and improving market positioning (Wollmann et al., 2011; Zatta et al., 2019).

Another study to be briefly mentioned is "Information technology as a tool for optimizing the quality of healthcare services in Manaus-AM", developed by Barreiros, Neto, Kuwahara and Gonçalves, published in 2011 in the *Iberoamerican Journal of Industrial Engineering*. The aim of this study was to inform that IT is important in decision making, considering the quality of health services in the nursing sectors of UBS as per Dr. Geraldo Magala. The study population consisted of 24 service users, and the AHP method was used to better analyze the problem and generate information for decision-making. The criteria used were tangibility, reliability, responsiveness, courtesy, communication, and credibility. The alternatives analyzed were preparation, nursing office, dressing, inhalation, and vaccination. Data were analyzed using Experthoice 11 software. The results obtained highlight that the best alternative for the user was the nursing office and the prioritized subcriteria were courtesy and reliability (Barreiros et al., 2011). Through the AHP method, health professionals identified the priorities highlighted by users, in order to improve the quality of services provided.

14.3 METHODOLOGICAL PROCEDURES

The present study was carried out in the Graphic Methods sector of a university hospital located in southern Brazil. This hospital, founded in 1970, is a health reference for the entire region. The services to the community are provided in the 291 beds of the Inpatient Unit and 37 beds in the Intensive Care Unit, in addition to the 53 outpatient rooms, 11 emergency care rooms, the six rooms of the Surgical Center and the two rooms of the Obstetrics Center. Various forms of health procedures are diagnosed and treated, always seeking to combine technology with a team of trained and updated professionals and students.

The limitation of the research is the interview, as all professionals in the sector were questioned, that is, 27 employees, performing the activities of nurses, doctors, and receptionists. This limitation is due to the fact that only employees in the sector were interviewed, with the aim of prioritizing improvement actions and determining the strengths and weaknesses of this area in the short term. It can be said that the study has a scientific character considering the experience of professionals in the sector under study. To delineate the scope of the methodology, the phases of the methodology adapted for this study will be presented.

As for the nature of the study, it presents concrete questions and shows a measurable part of what was proposed. This search for knowledge for practical application and solution of specific problems in the sector was under study, resulting only in the solution of specific problems, at the moment and within a given context (Gil, 2010).

As for the approach to the problem, the research is mixed, as it uses both qualitative and quantitative analyses (Marconi; Lakatos, 2010; Lakatos, 2021). In order to know information that allows highlighting the problem of the research sector, and subsequently provided through the application of the proposed model, it is a system to aid in decision-making by managers.

Regarding the objectives, the research is considered exploratory and descriptive, based on the need to get to know in greater depth the concepts of the subject and the characteristics of the Graphic Methods laboratory of a public hospital. Research 34 is exploratory because of its exploratory nature and because it does not contain hypotheses, that is, its function is to discover without the intention of testing specific research hypotheses. It is descriptive because it exposes characteristics of the population and is not committed to explaining the phenomena it describes (Gil, 2010; Marconi; Lakatos, 2010).

The method is characterized as inductive, which considers that knowledge is based on experience, not taking into account the pre-established principles, that is, the generalization derives from observations of cases of concrete reality (Gil, 2010).

14.3.1 Data Collection Procedures

As a data collection technique, SERVQUAL was used, addressing 22 questions related to the methodology of the 4Ps of services. SERVQUAL is a research method, which intends to assess the quality of service under the perception and experience of consumers. It is based on the implementation of the concept of quality failure that may be present in the delivery of the service. The 4Ps of services need to be known by the employees in order to improve the service provided. The "P" for Profile refers to visual communication, such as cleanliness, furniture arrangement, physical layout, employee appearance, and uniform. The "P" for Process refers to the processes that are performed during the provision of the service. The "P" for Procedure refers to the procedures that are performed during the services and the "P" for People refers to the employees who are performing the services (LAS CASAS, 2007). Thus, having knowledge about what the 4Ps represent, it is possible to develop quality services and, as a consequence, meet customer expectations.

14.3.2 Data Analysis Procedures

For the analysis of this research, the analytic hierarchy process (AHP) method was used. The method takes a systematic approach aimed at solving complex problems, helping the manager in decision making. The steps used in the application of the AHP method will be described to make its application easier for the reader.

The first step sought to model the problem, structuring it through a hierarchical model. Thus, it begins with the formulation of the general problem, which was based on the search to improve the quality of services by employees, seeking to identify the best alternative through the perception of employees together with the human resources manager of the Graphic Methods sector of a university hospital.

The second stage was execution, in which the criteria, subcriteria, and alternatives comparison matrices were developed, with the intention of constructing the local and global priority vectors. The criteria were composed of a 4×4 matrix covering the dimensions of the 4Ps of services (profile, process, procedure, and people). The subcriteria were composed of two 4×4 matrices and two 7×7 matrices, with questions referring to the dimensions mentioned. The alternatives were composed of four 6×6 matrices, reporting possible solutions for the company's reality. Table 14.3 shows the six alternatives used in the study and their contributions.

Thus, the variables mentioned were evaluated according to the binary combination, using the Saaty scale of importance, which includes nine points and their reciprocals, with the weights 3, 5, 7, and 9 representing the order of magnitude, the weights 2, 4, 6, and 8 representing intermediate values and weight 1 represents equal importance. Therefore, comparison matrices were developed through the judgment of the participants.

The third step was the analysis, which was carried out through the judgments of employees, seeking to identify the priorities highlighted. The priority vector points to the hierarchy of variables and their consistency. The random inconsistency index of the matrices were 0.90, 1.24 and 1.32; this is justified by the composition of the matrices. The method extension adopts an error up to 10 percent consistency.

The fourth step was the classification of criteria, subcriteria, and alternatives through local and global analysis. Figure 14.1 shows the hierarchical structure of the method correlated to the investigated topic.

The hierarchical structure of the method presented the overall objective, criteria, subcriteria, and alternatives. The mentioned subcriteria are presented as follows: Scr1 – updated equipment; Scr2 – physical facilities; Scr3 – employees' clothing; Scr4 – appearances of the premises; Scr5 – trusted sector; Scr6 – carrying out activities; Scr7 – updated records; Scr8 – punctuality in processes; Scr9 – reliable services; Scr10 – security in agreements; Scr11 – true collaborators; Scr12 – punctuality in procedures; Scr13 – friendly employees; Scr14 – exact information; Scr15 – individual attention; Scr16 – willing employees; Scr17 – emergency requests;

TABLE 14.3
Alternatives and Their Contributions

Alternatives	Contributions
Physical Layout	The physical organization is considered an extremely important factor, a well-presented layout contributes to the performance of employees, when the environment is organized, unnecessary activities are eliminated.
Training	Investments in employee training and development contribute to the organization's performance. When professionals are trained, the goals that management proposes are easier to be achieved. The use of knowledge results in benefits for all involved.
Internal Communication	This factor is linked to the efficiency of passing information to everyone involved. The encouragement of internal communication takes place through conversations, meetings and organizational instruments, provides interaction between professionals in the face of the transfer of information.
Remuneration	The compensation system that offers added value to the actions carried out contributes to the performance of all the professionals involved. There are several ways of remuneration such as: company shares, retirement plans and creative alternatives.
Contracting/ Outsourcing	The hiring of new professionals in some cases contributes to the human resources area, adopting outsourcing in the organization implies reducing costs and reducing the efforts made by employees
Technology	Technology is an important investment in the business environment, as it directs efforts at both the strategic and operational levels. The use of technology offers opportunities for the organization to develop its activities more efficiently, taking advantage of all possible resources.

Source: Survey data (2016).

Scr18 – educated employees; Scr19 – support to employees; Scr20 – special attention; Scr21 – knowledge of customer needs and Scr22 – convenient times. Through this method, it was possible to understand the judgment of the participants and identify the priorities of the sector under study. Data were analyzed using the software Microsoft Excel.

14.4 DISCUSSION OF RESULTS

14.4.1 Profile of Professionals

The hospital is currently a reference in health for the southern region located in the state of Rio Grande do Sul, Central Region. Its structure is integrated into a federal university that is perceived as a teaching hospital, with the objective of developing teaching, research, and health care.

Briefly, it was found that the majority of employees are female (61.5 percent). Thus, it can be noted that more and more women are engaged in the labor market.

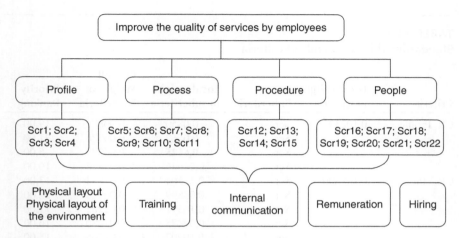

FIGURE 14.1 Hierarchical structure of the AHP adapted to the study.
Source: Survey data (2016).

The predominant age is between 31 and 40 years (42.3 percent). Regarding education, higher education (80.8 percent) stood out, as their high level of education is perceived as a positive point. To carry out the activities, it is necessary for employees to have advanced knowledge, which contributes to the quality of services. Regarding marital status, it was found 61.5 percent were single and the dominant remuneration above four salaries (69.3 percent). Most employees carry out their activities within a period of five to ten years. After knowing the profile of the respondents, the AHP method was used through their judgments to identify the priorities of the sector under study.

14.4.2 AHP METHOD PRIORITIZING THE ACTIONS OF THE INVESTIGATED SECTOR

The analysis of the AHP method occurred through the judgment of the manager of the people management area and some employees, seeking to find the alternative that will contribute to the excellence of the services offered. Based on this, through Table 14.4, it is possible to view the criteria and their related subcriteria.

When performing the pairwise comparison, it was found that Cr3 obtained 0.5483, that is, the highest priority of local average; this criterion represents the "Procedure" since the sector always seeks to pass the necessary information to the patient before and after the exam, in addition to seeking to be friendly and receptive individually.

From the local priorities, it is possible to see that in Cr1, Scr3 (0.6125) stood out; it refers to the clothing of employees in relation to the position held. Employees' clothing is a very important factor in the health area; white is a characteristic of the sector, representing cleanliness and hygiene, which are fundamental elements for these professionals, in addition to easily identifying any type of dirt, as it stands out in white color.

TABLE 14.4
Standardized criteria and subcriteria

Criteria	Local Weight of Criteria	Subcriteria	Local weight of subcriteria	Overall Weight of Subcriteria	Priority Ranking
CR$_1$ Profile	0,102530	Scr$_1$	0,0850	0,00872	19,00
		Scr$_2$	0,0562	0,00576	21,00
		Scr$_3$	**0,6125**	0,06279	5,00
		Scr$_4$	0,24636	0,02526	10,00
CR$_2$_ Process	0,138000	Scr$_5$	0,13504	0,01864	13,00
		Scr$_6$	0,03899	0,00538	22,00
		Scr$_7$	0,16391	0,02262	11,00
		Scr$_8$	0,09275	0,01280	16,00
		Scr$_9$	0,10907	0,01505	15,00
		Scr$_{10}$	0,05624	0,00776	20,00
		Scr$_{11}$	**0,40399**	0,05575	6,00
CR$_3$_ Procedure	**0,548360**	Scr$_{12}$	**0,42545**	0,25000	**1,00**
		Scr$_{13}$	0,38595	0,21164	2,00
		Scr$_{14}$	0,11598	0,06360	4,00
		Scr$_{15}$	0,07262	0,03982	7,00
CR$_4$_ People	0,211110	Scr$_{16}$	0,15266	0,03223	9,00
		Scr$_{17}$	0,16264	0,03433	8,00
		Scr$_{18}$	0,09722	0,02052	12,00
		Scr$_{19}$	0,04241	0,00895	18,00
		Scr$_{20}$	0,07658	0,01617	14,00
		Scr$_{21}$	**0,41025**	0,08661	3,00
		Scr$_{22}$	0,05823	0,01229	17,00

Source: Survey data (2016).

In Cr2, Scr11 (0.40399) is evidenced, the same concerns the employees being true to the customers; this means that the customer feels trust in the professional, developing a relationship that will contribute to their satisfaction.

In Cr3, the Scr12 (0.42545) was over, both in local and global weight; the question seeks to know the punctuality of procedures. In this regard, it is clear that professionals are punctual with the delivery of exams on time, which was determined when the patient used the service in addition to prioritizing care for patients who are hospitalized.

In Cr4 it was Scr21 (0.41025), which refers to the issue of the sector's employees knowing what the patients' needs are. Employees know the needs due to the constant training carried out by the hospital, as the profile of this organization is to develop education.

Table 14.5 shows the consistency index and consistency ratio of the matrices pointed out to prove the evaluations carried out.

Through the analysis carried out from the local priorities, it is possible to see that in Cr1 (Profile), the alternative that stood out was "A" with (0.33325). This

TABLE 14.5
Consistency index and consistency ratio of the criteria and subcriteria.

Criteria	Índice de Consistência e Razão de Consistências dos Critérios	Subcriteria	Consistency Index and Subcriteria Consistency Ratio
CR_1 _ Profile	λ MAX = 4,0977 CI =0,032566 **CR =0,03619**	Scr_1	λ MAX = 4,24773 CI =0,082576 **CR =0,09175**
		Scr_2 Scr_3 Scr_4	
CR_2 _ Process		Scr_5	λ MAX = 7,49566 CI =0,08261 **CR =0,06258**
		Scr_6 Scr_7 Scr_8 Scr_9 Scr_{10} Scr_{11}	
CR_3 _ Procedure		Scr_{12}	λ MAX = 4,23546 CI =0,78486 **CR =0,08721**
		Scr_{13} Scr_{14} Scr_{15}	
CR_4 _ People		Scr_{16}	λ MAX = 7,64835 CI =0,10805 **CR =0,08186**
		Scr_{17} Scr_{18} Scr_{19} Scr_{20} Scr_{21} Scr_{22}	

Source: Survey data (2016).

alternative concerns investments in the physical layout of the sector. This is justified by the fact that the sector needs up-to-date equipment so that services are performed efficiently.

In the Process (0.28365) and also in the Procedure (0.21422), alternative "B" was the one that obtained the highest degree of importance; it refers to the training of employees. These professionals prioritize this alternative fundamental for the execution of its activities, mainly in the care of patients who seek services in the Graphic Methods sector. In the People dimension, there was a tie between alternatives "B" and

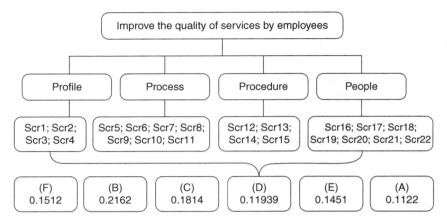

FIGURE 14.2 Hierarchical structure of the AHP showing final result.
Source: Survey data (2016).

"D" with (0.23598); the alternative "D" refers to the remuneration of employees. After analyzing the consistency ratio indices of the alternatives, it is possible to see that they all had a value of less than 0.10. Thus, it is clear that the judgments are consistent.

In order to find the best alternative for the above criteria, it was necessary to perform the calculation proposed by Martins, Souza and Barros (2009), which is as follows: PG (a1) = PML (Cr1)*PML(a1)Cr1 + PML (Cr2)*PML(a1)Cr2 + ... + PML (Cr4)*PML(a1)Cr4. In Figure 14.2, it is possible to understand which alternative is prioritized according to the perception of these professionals.

According to the reported data, the alternative that stood out was "B", as it obtained 0.2162 as a global priority, the highest among the others, referring to training. Thus, it is clear that investment in training results in development, contributing to the improvement of skills and knowledge of these professionals; the result of this action implies in quality services. When it comes to health, this issue must be prioritized, as poorly performed services tend to compromise the lives of patients.

The second alternative was "D" with (0.1939), which refers to remuneration, which means that, in this case, this issue represents a motivational factor, also contributing to the performance of employees in this sector. A subject that is also related to aspects of an economic nature, with professionals seeking to invest in knowledge with the intention of obtaining personal and financial fulfillment. And the third alternative highlighted was internal communication (0.1814); employees perceive that through this, information channels are developed, providing transparency in actions, interaction between the team, and exchange of knowledge.

14.5 FINAL CONSIDERATIONS

This research aimed to evaluate the relationship between people management and the quality of public services in the Graphic Methods sector of a university hospital located in the central region of Rio Grande do Sul. The 4Ps of services were used (profile, process, procedure, people) related to the SERVQUAL instrument (experience) and the AHP decision-making method, in order to find the best alternative that will contribute in the area of human resources and consequently in public health services.

The method helped in prioritizing criteria, subcriteria and alternatives; the structure proved to be flexible, being recommended for the health area, seeking to improve the quality of services provided by employees. More than determining the correct decision, the AHP model brings subsidies to justify the choices. In this case, the importance scale developed by Saaty was used, which allows a rational and consistent comparison of the studied elements. Through this consistency and rationality of judgments, the AHP method differs from other decision-making methods.

The AHP extension adopts an error of up to 10 percent consistency, as the objective of this method was to prioritize the alternative that most contributes to the reality of the sector. The criterion prioritized by the collaborators was "Procedure" and the sub-criterion that got the most relevance was "Puctuality in the procedures"; this service offers the performance and delivery of exams in the allotted time, prioritizing the attention of patients. These factors contribute to the quality of services provided.

The alternative highlighted through the professionals' judgments was "Training", since it is very important to identify this perception, as it is perceived that the professionals are aware that the effect of training is an action that contributes to improving the performance of the sector. Training is relevant to the health area, and a systematic approach must be used; its contributions come from the training of technical skills, pointing to motivation, and developing learning. The result of these actions supports the professional's behavior through the activities performed.

Learning this process allows the sector to develop its strengths and work on its weaknesses, with the intention of meeting the needs of patients, thus becoming a reference hospital for the entire community. The results presented were adequate to meet what was proposed. The main limitation of this study is centered on the results obtained, indicating only the reality of the investigated sector, which makes extrapolation of information impossible.

But in the application one can observe the relevance of this method for complex and subjective decisions. This assessment is not restricted to the health sector, as all sectors can acquire a management system. Thus, this method is recommended for other organizations that face the challenge of making complex, subjective decisions which have a great impact on the future of business.

In the short term, as future research, it is intended to understand with this model the perception of patients using this service. Given the above, studies of applications of fuzzy/AHP methods in the human resources sector are suggested, aiming to improve the quality of services in public health. Consequently, in order to generalize the conclusions, it is recommended to study the other sectors of the university hospital.

REFERENCES

Barreiros, N. R. et al. (2011). A tecnologia de informação como ferramenta para otimização da qualidade nos serviços de saúde em Manaus-AM. *Iberoamerican Journal of Industrial Engineering*, 3: 34-48.

Büyüközkan, G.; Çifçi, G.; Güleryüz S. (2011). Strategic analysis of healthcare service quality using fuzzy AHP methodology. *Expert Systems with Applications*, 38: 9407–9424.

Chakravarty A. C. (2011). Evaluation of service quality of hospital outpatient department services. *Medical Journal*, Armed Forces India, 67: 221–224.

Chen, H. R.; Cheng, B. W. (2012). Applying the ISO 9001 process approach and service blueprint to hospital management systems. *The TQM Journal*, 24: 418–432.

Deb, M.; Lomo-David, E. (2014). Evaluation of retail service quality using analytic hierarchy process. *International Journal of Retail & Distribution Management*, 42: 521–541.

Donabedian, A. (1990). The seven pillars of quality. *Archives of Pathology and Laboratory Medicine*, Northfield, 114: 1115–1118.

Donabedian, A. (2003). *An Introduction to Quality Assurance in Health Care*. New York: Oxford University Press.

Fukuda, H; Okuma, K; Imanaka, Y. (2014). Can experience improve hospital management? Peer Reviewed, open Access Journal, 9: 1–7. e106884. doi:10.1371/journal.pone.0106884

Gil. A. C. (2010). *Como elaborar projetos de pesquisa*. 5. ed. São Paulo: Atlas.

Guerreiro, M. D.; Barroso, A. M. M.; Rodrigues, E. A. A. (2016). Organizações saudáveis e qualidade do trabalho na Europa. Desafios para organizações e profissões no setor público de saúde. *Organizações & Sociedade*, 23: 421–437.

Handayani, P. W. et al. (2014). Strategic hospital services quality analysis in Indonesia. *Expert Systems with Applications*, 42: 3067–3078.

Ho, W.; Dey, P. K.; Lockström, M. (2011). Strategic sourcing: A combined QFD and AHP approach in manufacturing. *Supply Chain Management: An International Journal*, 16: 446–461.

Lakatos, E. M. (2021). *Fundamentos de Metodologia Científica*. Atlas. Grupo GEN. 1ª ED. 2021. ISBN: 9788597026566

Las Casas, A. L. L. (2007). *Marketing de Serviços*. 5a Ed. Atlas.

Lee, S.-H. (2010). Using fuzzy AHP to develop intellectual capital evaluation model for assessing their performance contribution in a university. *Expert Systems with Applications*, 37: 4941–4947.

Marconi, M. A.; Lakatos, E. M. (2010). *Fundamentos de Metodologia Científica*. 5 Ed. Atlas: 2010. ISBN: 9788522473922

Mendes, A. C. G, et al. (2013). Condições e motivações para o trabalho de enfermeiros e médicos em serviços de emergência de Alta complexidade. *Rev. bras. Enferm*, 66: 161–166.

Miguel, P. A. C. et al. (2012). *Metodologia de Pesquisa em Engenharia da Produção e Gestão de Operações*. 2nd ed. Elsevier.

Min, H. (2010). Evaluating the comparative service quality of supermarkets using the analytic hierarchy process. *Journal of Services Marketing*, 24: 283–293.

Pan, F. F. C. (2014). Using analytic hierarchy process to identify the nurses with high stress-coping capability: model and application. *Iranian Journal of Public Health*, 43: 273–281.

Papadomichelaki, X. et al. (2013). An analytic hierarchy process for the evaluation of E-government service quality. *International Journal of Electronic Government Research*. 9: 19–44.

Parasuraman, A. et al. (1985). Conceptual model of service quality and its implications for future research. *Journal of Marketing*. 49 : 441–450.

Peixoto, M. R,; Rocha, A. M. C.; Oliveira, L. R.; Costa, C. T. Q. (2020). Avaliação da Utilização das Escalas SERVQUAL e SERVPERF no Brasil: uma análise da adequação do uso de escalas existentes em pesquisas empíricas. XXIII SEMEAD Seminários em Administração. Novembro. ISSN 2177-3866. www.researchgate.net/publication/349700499

Regis, M. F. De A. (2011). O Serviço Social e a área de gestão de pessoas: mediações sintonizadas com a Política Nacional de Humanização no Hospital Giselda Trigueiro. *Serv. Soc. Soc*, 107: 482–496.

Reis, B. C. M.; Bortoluzzi, M. B. O.; Boas, T. S. V.; Nascimento, E. S. (2020). Aplicação da ferramenta SERVQUAL para a análise da qualidade em serviços: o caso de uma franquia odontológica. *GEPROS. Gestão da Produção, Operações e Sistemas*, 15(3): 230–254.

Rodrigues A. L.; Barrichello, A.; Morin, E. M. (2016). Os Sentidos Do Trabalho Para Profissionais De Enfermagem: Um Estudo Multimétodos. *Revista de Administração de empresas*, 56: 192–208.

Rodrigues, M. K. (2015). *Análise da organização formal dos processos de trabalho dos Hospitais Universitários Federais do RS: segurança dos pacientes e dos profissionais. Dissertação de Mestrado em Engenharia de Produção*. Universidade Federal de Santa Maria.

Roesch, S.; Maria, A. (2013). Projetos de estágio e de pesquisa em Administração: guia para estágios, trabalhos de conclusão, dissertações e estudos de caso. 3 ed. Atlas.

Saaty Karassavidou, E.; Glaveli, N.; Zafiropoulos, K. (2012). Assessing hospitals' readiness for clinical governance quality initiatives through organisational climate. *Journal of Health Organizationand Management*, 25: 214–240.

Saaty, T. L. (2008). Decision making with the analytic hierarchy process. *International Journal of Services Sciences*, 1: 83–98.

Saaty, T. L.; Shih, H.-S. (2009). Structures in decision making: On the subjective geometry of hierarchies and networks. *European Journal of Operational Research*, 199: 867–872.

Saloner, B.; Akosa Antwi, Y.; Maclean, J. C.; Cook, B. (2018). Access to health insurance and utilization of substance use disorder treatment: Evidence from the Affordable Care Act dependent coverage provision. *Health Economics*, 27(1): 50–75.

Santos, H. L. P. C.; Maciel, F. B. M.; Junior, G. M. S.; Martins, P. C.; Prado, N. M. L. B. (2020). Gastos públicos com internações hospitalares para tratamento da covid-19 no Brasil em Rev Saude Publica. 2021; 55: 52. https://doi.org/10.11606/s1518-8787.2021055003666

Shieh, J.-I.; Wu, H. H.; Huang, K. K. (2010). A DEMATEL method in identifying key success factors of hospital service quality. *Knowledge-Based Systems*, 23: 277–282.

Silva, E. L. Da; Menezes, E. M. (2005). *Metodologia da pesquisa e elaboração de dissertação, UFSC*, 4. ed. Florianópolis.

Silva, L. G.; Matsuda, L. M.; Waidman, M. A. P. (2012). A estrutura de um serviço de urgência público, na ótica dos trabalhadores: perspectivas da qualidade. *Texto contexto – enferm*, 21(2): 320–328.

Souza, P. C.; Scatena, J. H. G. (2010). É economicamente viável regionalizar a atuação de um hospital público de médio porte? *Physis*, 20(2): 571–588.

Vendemiatti, M. et al. (2010). Conflito na gestão hospitalar: o papel da liderança. *Ciênc. Saúde coletiva*, 15: 1301–1314.

Wollmann, D. et al. (2012). Avaliação de operadoras de saúde por usuários pelo método *Analytic Hierarchy Process*. Rev Saúde Pública, 46: 777–783.

15 Telework and Conflict (Work–Family and Work–Family)
What Is the Effect on Occupational Stress?

Ana Moreira, Mónica Salvador, Alexandra de Jesus, Catarina Furtado and Madalena Lopez-Caño

CONTENTS

15.1 Introduction 257
15.2 Literature Review 258
    15.2.1 Work–Family Conflict, Family–Work Conflict and Performance 258
    15.2.2 Work–Family Conflict, Family–Work Conflict and Occupational Stress 259
    15.2.3 Occupational Stress and Performance 259
    15.2.4 Mediating Effect of Occupational Stress 259
15.3 Method 260
    15.3.1 Data Collection Procedure 260
    15.3.2 Participants 260
    15.3.3 Data Analysis Procedure 261
    15.3.4 Measures 261
15.4 Results 262
15.5 Discussion and Conclusions 265
15.6 Final Considerations 266
References 267

15.1 INTRODUCTION

According to the World Health Organization (2002), stress is a worldwide epidemic with high individual, family, organisational, community and socio-political costs. According to Stanton et al. (2001), occupational stress is perceived as something threatening, giving employees a negative and uncomfortable feeling. According to Murta and Tróccoli (2004), work may become a source of illness, since there are risk factors and the employee does not have strategies to adaptively cope with them. In addition to the fact that, in current times, the labour market is characterised by feelings of uncertainty, high control and precarious working conditions (Linhares & Siqueira, 2014), in times of the SARS-CoV-2 pandemic, stress symptoms have worsened.

The work–family conflict is a concept that refers to a conflict between the work and family dimensions (Bruck et al., 2002). This type of conflict results from pressures in the domains of work and family, becoming irreconcilable at a given moment (Greenhaus & Beutell, 1985). The work–family conflict is associated with depression, dissatisfaction at work and/or with life and/or marital (Netemeyer et al. 1996), while family experiences (i.e. responsibility for children, family conflicts) interfere with the work aspect.

For Abbad (1999), work performance reflects the total production achieved during the workday. This author also mentions that performance may be conceptualised by behaviours, activities, roles, standards and/or goals existing at work.

Work–family and family–work conflict are one of the causes of higher levels of work stress (Netemeyer et al., 2005) and lower levels of performance (Allen, et al., 2000). In turn, decreased levels of performance are also a consequence of work stress (Nouri, 2017).

This study aims to investigate the effect of work–family and family–work conflict on perceived performance and whether this relationship is moderated by occupational stress. It also aims to study the effect of work–family and family–work conflict on occupational.

15.2 LITERATURE REVIEW

15.2.1 Work–Family Conflict, Family–Work Conflict and Performance

The relationship between work and family has been much studied in recent years. Studies have focused on the mechanisms that link these two factors.

The negative paradigm of work–family relations is a conflict of roles, in which the tension and pressure required by performance at both work and family levels are incompatible (Greenhaus & Beutell, 1985).

Greenhaus and Beutell (1985) identify three types of conflict, which are determined by the origin of the conflict: the temporal conflict, which arises when the time required for the performance of one role interferes with the performance of another role; the pressure conflict that arises when the pressure caused by one role influences the other role; the behavioural conflict arises due to the incompatibility of behaviours desired in the performance of each of the roles.

According to Sonnentag and Frese (2002), the concept of performance can be studied from various perspectives, since it has a multidimensional nature. Performance can also be defined as "a set of actions that involve a complex interaction with reality, and the integration of different psychological, interpersonal and material facets" (Bendassolli, 2012, p. 173).

The work–family conflict has a significant effect on the work environment, causing absences, leaves or delays (Oliveira et al., 2013), which leads to a decrease in individual performance (Silva, 2017). With regard to family–work conflict, according to Allen et al. (2000), this also has a negative effect on performance. Thus, the first hypothesis of this study is put forward.

Hypothesis 1: Work–family conflict and work–family conflict have a significant and negative effect on perceived performance.

15.2.2 WORK–FAMILY CONFLICT, FAMILY–WORK CONFLICT AND OCCUPATIONAL STRESS

According to the Portuguese Psychologists' Association (2020), occupational stress occurs when the individual perceives that the demands of their job are greater than their abilities and may be perceived as something threatening, giving the employee a negative and uncomfortable feeling (Stanton et al., 2001). Scientific studies are increasingly addressing occupational stress due to the high losses caused by it, both at a personal and organisational level (Lyonette et al., 2007).

According to several authors, including Netemeyer et al. (2004), there is a positive relationship between work–family conflict and occupational stress, that is, high levels of work–family conflict are associated with high levels of work stress. In turn, Russo (2013) argues that occupational stress is a consequence of the family–work conflict. Therefore, the second hypothesis was formulated.

Hypothesis 2: Work–family conflict and work–family conflict have a significant and positive effect on occupational stress.

15.2.3 OCCUPATIONAL STRESS AND PERFORMANCE

Occupational stress has serious financial implications for organisations, which are related to productivity losses, high levels of absenteeism, presenteeism and intentions to leave the organisation. According to Leite and Uva (2010) absence or leaving work due to stress-related illness or injury has a negative and significant effect on employee performance. Based on the assumption that occupational stress has a negative and direct effect on employees' perceived performance, the third hypothesis was formulated.

Hypothesis 3: Occupational stress has a significant and negative effect on perceived performance.

15.2.4 MEDIATING EFFECT OF OCCUPATIONAL STRESS

From Oliveira et al.'s (2013) perspective, work–family conflict has a significant impact on job satisfaction, organisational commitment, absenteeism, employee leaves or delays, which according to Silva (2017) has a significant and negative effect on individual performance. Allen et al. (2000), also refer that low performance can be related to a high family–work conflict. Russo (2013) also mentions that there are behavioural consequences resulting from both work–family conflict and family–work conflict, and that among these consequences is occupational stress. In turn, occupational stress has a negative effect on performance (Nouri, 2017). This is the reasoning that leads us to formulate the following hypotheses:

FIGURE 15.1 Research model.

Hypothesis 5: Occupational stress is the mechanism that explains the relationship between work–family conflict and performance.

Hypothesis 6: Occupational stress is the mechanism that explains the relationship between family–work conflict and performance.

In order to integrate the various hypotheses formulated, a theoretical model was developed (Figure 15.1), which aims to synthesise the expected relations and associations between the constructs.

15.3 METHOD

15.3.1 DATA COLLECTION PROCEDURE

A total of 504 participants collaborated in this study on a voluntary basis. However, only 495 participants were considered in the subsequent statistical analyses, since nine participants did not meet the conditions for participation in this study (having worked for at least six months in organisations based in Portugal). The sampling procedure was non-probability, convenience and intentional snowball sampling (Trochim, 2000).

The questionnaire, which was placed online on the Google Docs platform, contained information about the purpose of the study. It was also expressed that the confidentiality of the answers would be ensured. The questionnaire was composed of six questions for sample characterisation (age, gender, academic qualifications, seniority in the organisation, seniority in the position and type of employment contract) and three scales (work–family and family–work conflict, perceived performance and occupational stress). Data collection occurred during the month of April 2021.

15.3.2 PARTICIPANTS

The sample of this study comprises 495 participants, 350 (70.7 percent) being female and 145 (29.3 percent) are male, aged between 18 and 66 years, with an average of 39.26 years (SD = 10.96). With regard to academic qualifications, 240 (48.5 percent) participants have completed at least the 12th grade, 179 (36.2 percent) have a bachelor's degree and the remaining 79 (15.4 percent) a master's degree or higher. With regard to seniority in the organisation and in the job, the values vary between 50 and 46 years, with a mean seniority in the organisation of 10.17 (SD = 9.65) and a mean seniority in the job of 10.63 (SD = 9.49). Regarding the work sector, 365

(73.7%) of the participants work in the private sector and 130 (26.3 percent) in the public sector.

15.3.3 Data Analysis Procedure

The first analysis was to test the metric qualities of the instruments used in this study. To test the validity of the instruments used in this study, confirmatory factor analyses were performed for each instrument using the AMOS 24 for Windows software. The procedure was according to a "model generation" logic (Jöreskog & Sörbom, 1993), considering in the analysis of their adjustment, interactively the results obtained: for the chi-square (χ^2) ≤ 5; for the Tucker Lewis index (TLI) > .90; for the goodness-of-fit index (GFI) > .90; for the comparative fit index (CFI) > . 90; for the root mean square error of approximation (RMSEA) ≤ .08. The internal consistency of each scale was then analysed by calculating Cronbach's alpha, whose value should vary between "0" and "1", not assuming negative values (Hill & Hill, 2002) and being higher than .70, the minimum acceptable in organisational studies (Bryman & Cramer, 2003). As for the sensitivity study, the different measures of central tendency, dispersion and distribution were calculated for the different items of the scales used, thus carrying out the normality study for all items and the various scales. The hypotheses formulated in this study were tested through multiple linear regressions.

15.3.4 Measures

The questionnaire used in this study consisted of six sociodemographic questions and the work–family and family–work conflict, perceived performance and occupational stress scales.

In order to measure the work–family and family–work conflict, we used the instrument developed by Netemeyer et al. (1996), translated and adapted to the Portuguese population by Santos and Gonçalves (2014), consisting of ten items classified in a 7-point Likert-type rating scale (from 1 "strongly disagree" to 7 "strongly agree"), which assess the two-way component of the work–family (items 1, 2, 3, 4 and 5) and family–work (items 6, 7, 8, 9 and 10) conflict. Their validity was tested through a two-factor confirmatory factor analysis (CFA). The adjustment indices obtained are adequate (χ^2/gl = 2.57; GFI =.99; CFI =.98; TLI =.98; RMSEA =.056), which indicates that this instrument is composed of two factors. With regard to internal consistency, a Cronbach's alpha of .88 was obtained for both work–family conflict and family–work conflict.

Perceived performance was measured through the short version of the instrument developed by Williams and Anderson (1991). This instrument is composed of four items, rated using a 5-point Likert-type rating scale (from 1 "strongly disagree" to 5 "strongly agree"). The validity of this instrument was tested through a one-dimensional CFA. The adjustment indices obtained are adequate (χ^2/gl = .47; GFI =.99; CFI =.98; TLI =.99; RMSEA =.010). With regard to internal consistency, a Cronbach's alpha of .86 was obtained.

To measure occupational stress, we used Gomes's Occupational Stress Questionnaire – General Version (QSO-VG) (2010) developed by Gomes (2010) based on studies in different professional areas, which gave it credibility in the study of occupational stress. This instrument identifies and assesses the potential sources of stress (i.e. stressors) in the exercise of professional activity. The questionnaire is composed of 24 items, classified in a 5-point Likert-type rating scale (from 0 "no stress" to 4 "very stress"). The theoretical model consists of seven factors corresponding to the following dimensions: relationship with patients (items 2, 8, 13 and 21); relationship with superiors (items 12, 20 and 24); relationship with colleagues (items 4, 17 and 22); overwork (items 5, 10, 11 and 16); career and remuneration (items 1, 6, 15 and 19); family problems (items 3, 14 and 23); working conditions (items 7, 9 and 18). The validity of this instrument was tested through a seven-factor AFC. The adjustment indices obtained are adequate (χ^2/gl = 2.76; GFI =.91; CFI =.95; TLI =.94; RMSEA =.060), which indicates that this instrument is composed of seven factors. As for the internal consistency, the Cronbach's alpha values of the dimensions vary between .80 and .92. The scale presents a Cronbach's alpha of .95.

Regarding the sensitivity of the items that compose the instruments, it was found that none of the items has the median close to one of the extremes, all items have responses in all points, the absolute values of asymmetry and kurtosis are below 3 and 7, respectively, which indicates that no item grossly violates normality (Kline, 1998).

15.4 RESULTS

Initially, we tested the intensity and direction of the association between the variables under study using Pearson's correlations. The mean and standard deviation were also calculated for each of the variables (Table 15.1). Performance is negatively and significantly correlated with work–family conflict ($r = -.11$; $p = .014$), with family–work conflict ($r = -.366$; $p < .001$) and with occupational stress ($r = -.14$; $p = .003$). Occupational stress is positively and significantly associated with work–family conflict ($r = .58$; $p < .001$) and family–work conflict ($r = .38$; $p < .001$).

The participants in this study perceived their performance as high (t (494) = 50.26; $p < .001$; $M = 4.33$; SD = .59), as the mean is significantly above the central point. As for occupational stress (t (494) = 26.43; $p < .001$; $M = 3.06$; SD = .89), it is also significantly above the centre point (t (494) = 26.43; $p < .001$; $M = 3.06$; SD = .89). With regard to both work–family conflict (t (494) = –3.32; $p = .001$; $M = 3.77$; SD = 1.52) and family–work conflict (t (494) = –26.28; $p < .001$; $M = 2.47$; SD = 1.30).

The results indicate that only the family–work conflict has a negative and significant effect on the perceived performance ($\beta = -.37$; $p < .001$), the higher the family–work conflict, the lower the perceived performance (Table 15.2). A coefficient of determination of .13 was obtained, which indicates that the model accounts for 13 percent of the variability in perceived performance (Table 15.2). The model is statistically significant (F (2, 492) = 36.17; $p < .001$).

Hypothesis 1 was partially confirmed.

The results indicate to us that both work–family conflict ($\beta = .51$; $p < .001$) and family–work conflict ($\beta = .17$; $p < .001$) have a positive and significant effect on occupational stress, that is, the higher the work–family and family–work conflict the higher the occupational stress (Table 15.3). A coefficient of determination of .36 was obtained, which indicates that the model accounts for 36 percent of the

TABLE 15.1
Association Between Variables, Means and Standard Deviation Next, the Hypotheses Formulated in This Study Were Tested

	Mean	SD	1	2	3.1	3.2
1. Perceived performance	4,33	,59	--			
2. Occupational stress	3,06	,89	-,14**	--		
3.1. Work–family conflict	3,77	1,52	-,11*	,58***	--	
3.2. Family–work Conflict	2,47	1,30	-,36***	,38***	,41***	--

Note: * $p \leq .05$; ** $p < .01$; *** $p < .001$.

TABLE 15.2
Multiple Linear Regression Results with Perceived Performance as Dependent Variable

	F	p	R2a	β	t	p
Work–family conflict	36.17***	< .001	.13	.04	.94	.346
Family–work Conflict				-.37***	-8.09***	< .001

Note: *** $p < .001$.

TABLE 15.3
Results of the Multiple Linear Regression with Occupational Stress as the Dependent Variable

	F	p	R2a	B	t	p
Work–family conflict	140.34***	< .001	.36	.51***	12.98***	< .001
Family–work conflict				.17***	4.31***	< .001

Note: *** $p < .001$.

variability in occupational stress (Table 15.3). The model is statistically significant ($F (2, 492) = 140.34; p < .001$).

Hypothesis 2 was confirmed.

Occupational stress has a negative and significant effect on perceived performance ($F (1, 493) = 9.12; R^2 = .02; β = -.13; p = .003$), the higher the levels of occupational stress the lower the perceived performance (Table 15.4). The model accounts for 2 percent of the variability in perceived performance (Table 15.4).

Hypothesis 3 was confirmed.

In order to test hypotheses 4 and 5, the procedures of Baron and Kenny (1986) were followed as it concerns a mediating effect and were confirmed using Pearson

TABLE 15.4
Simple Linear Regression Results with Perceived Performance as Dependent Variable

	F	R2	β	t	p
Occupational stress	9.12**	.02	−.13**	−3.02	.003

Note: ** $p < .01$.

TABLE 15.5
Results of the Mediating Effect of Occupational Stress on the Relationship Between Work–Family Conflict and Perceived Performance

Independent Variables	Step1	Step2
Work–family conflict	−.11*	−.05
Occupational stress		−.11*
F	6.07*	4.94**
R^2_a	.01	.02
R^2 Change		.01*

Note: * $p \leq .05$; ** $p < .01$.

TABLE 15.6
Results of the Mediating Effect of Occupational Stress on the Relationship Between Family–Work Conflict and Perceived Performance

Independent Variables	Step1	Step2
Work–family conflict	−.36***	−.36***
Occupational stress		−.01
F	71.46***	35.66***
R^2_a	.13	.13
R^2 Change		.00

Note: *** $p < .001$

correlations (Table 15.1). Next, multiple linear regressions were carried out in two steps, with the predictor variables being introduced in the first step as independent variables and the mediator variable in the second step, the results of which are presented in Tables 15.5 and 15.6.

There is a full mediation effect of occupational stress on the relationship between work–family conflict and perceived performance, because when the mediating variable was introduced in the linear regression model, the effect of work–family conflict on perceived performance ($\beta 1 = -.11$; $\beta 2 = -.05$) was no longer significant. There is also a significant increase of 1 percent ($p = .05$) in the value of the coefficient of determination. Sobel's test was then performed, using the interactive instrument of Preacher and Leonardelli (2001), in which we obtained a $Z=-3.02$ with a $p=.001$, which proved both the total mediation effect.

Hypothesis 4 was proved.

The mediating effect of occupational stress on the relationship between work–family conflict and perceived performance was not verified as occupational stress has no significant effect on perceived performance ($\beta = -.01$; $p > .05$).

Hypothesis 5 was not proved.

15.5 DISCUSSION AND CONCLUSIONS

The main objective of this study was to investigate the effect of work–family and work–family conflict on perceived performance and whether these relationships are mediated by occupational stress.

In terms of H1, the negative and significant effect of the family–work conflict on the perceived performance was confirmed. These results are in line with the literature, since, according to Allen et al. (2000), the family–work conflict has a negative effect on performance. The effect of work–family conflict on perceived performance was found to be non-significant. These results go against what the literature suggests. For Oliveira et al. (2013), work–family conflict has a significant effect on performance by causing employees to miss work, leave early or be late, which has a negative effect on performance (Silva, 2017). These results may be a reflection of the data being collected during the month of April 2021, when Portugal was in lockdown due to the COVID-19 (the disease caused by novel corona virus, SARS-CoV-2). Due to this fact many employees were teleworking, so they did not miss work, did not arrive late or leave early. However, many of these employees have small children, many of them of school age, who were also at home at the time, which could often interfere with their work. Perhaps this is why only the family–work conflict was found to have a significant negative effect on perceived performance.

The positive and significant effect of work–family and family–work conflict on occupational stress was confirmed (hypothesis 2). These results are in line with what was reported by Netemeyer et al. (2004) and Russo (2013), since, according to these authors, high levels of work–family and family–work conflict are associated with high levels of work stress. Other studies show that the relationship between professional and personal life can lead to conflicts that, consequently, negatively affect individuals, which may lead to depression, extreme tiredness, anxiety, lower performance, as well as lower job satisfaction and greater stress (Priyadharshini & Wesley, 2014). In this study, it was found that the effect of work–family conflict on occupational stress is stronger than family–work conflict. Once again, the fact that we were in this pandemic context may have enhanced this effect, since, for certain employees, it became very difficult to work from home and this situation often led to working more hours than those established in the work schedule. In fact, all these abuses led to the emergence of legislation on teleworking and the respective schedules.

A negative and significant effect was also found in relation to the effect of occupational stress on the perceived performance (hypothesis 3), which is in line with the literature. Occupational stress leads to absenteeism, presenteeism and increases the intentions to leave the organisation, which leads to major financial losses related to the loss of productivity (Leite and Uva, 2020).

At last, occupational stress was found to be the mechanism that explains the relationship between work–family conflict and perceived performance (hypothesis 4). According to Oliveira et al. (2013), work–family conflict has a significant effect on employee absenteeism, leaves or delays, which has a significant and negative effect on individual performance. Russo (2013) also mentions that work–family conflict has, among others, as a behavioural consequence, occupational stress and this, in turn, has a significant negative effect on perceived performance (Nouri, 2017). Indeed, it was found that work–family conflict has a significant effect on occupational stress and this has a significant effect on perceived performance, while the relationship between work–family conflict and perceived performance became non-significant. This situation may have worsened with confinement and telecommuting.

The mediating effect of occupational stress on the relationship between family–work conflict and perceived performance was not confirmed. These results are possibly due to the fact that the relationship between family–work conflict and performance is stronger than the relationship between family–work conflict and occupational stress.

It should be noted that when calculating the mean and standard deviation of the variables under study, it was found that the work–family and family–work conflicts were significantly below the midpoint of the scale. However, the type of conflict that was perceived as higher by the participants in this study was the work–family conflict. As for the perceived performance, participants also perceived their performance as significantly above the central point of the scale, which is not surprising since it is a self-perception. With regard to occupational stress, it was also significantly above the central point, which indicates that participants showed high levels of occupational stress. The fact that participants revealed high levels of occupational stress may be related to the context in which they were at the time of data collection (April 2021), since we had been in confinement (the second) since the 15th of January.

The limitations include the data collection process, as well as the fact that the questionnaires were self-report instruments, with closed and mandatory questions, which may condition the answers given by the participants, and the fact that this was a cross-sectional study, which did not allow establishing causal relationships between the variables.

15.6 FINAL CONSIDERATIONS

This study and those conducted by others recently remind us to pay close attention to the mental health of employees. In the current context, the number of individuals who have sought psychological help has increased exponentially, especially among health professionals.

As for the organisations, they should be concerned with the mental health of their employees, assuming human resources management practices that allow the prevention of their occupational stress. In what concerns teleworking, the employees' schedules should be respected, not requesting work or information at any time, which may cause work–family conflict.

REFERENCES

Abbad, G. (1999). *Um modelo integrado de avaliação do impacto do treinamento no trabalho – IMPACT*. Tese de Doutoramento. Instituto de Psicologia, Universidade de Brasília, Brasília.

Allen, T. D., Herst, D. E., Bruck, C. S., et al. (2000). Consequences associated with work-to-family conflict: A review and agenda for future research. *Journal of Occupational Health Psychology*, 5(2), 278–308.

Baron, R.M. and Kenny, D.A. (1986). The moderator-mediator variable distinction in social psychological research: Conceptual, strategic, and statistical considerations. *Journal of Personality and Social Psychology*, 51, 1173–1182.

Bendassolli, P. F. (2012). Desempenho no trabalho: Revisão da literatura. *Psicologia Argumento*, 30(68), 171–186. https://doi.org/10.7213/psicol.argum.5895

Bryman, A., & Cramer, D. (2003). *Análise de dados em ciências sociais. Introdução às técnicas utilizando o SPSS para windows* (3ª ed.). Oeiras: Celta.

Bruck, C. S., Allen, T. D., Spector, P. E. (2002). The relation between work–family conflict and job satisfaction: A finer-grainer analysis. *Journal of Vocational Behavior*, 60, 336–353.

Greenhaus, J. H., & Beutell, N. J. (1985). Sources of conflict between work and family roles. *The Academy of Management Review*, 10(1), 76. doi:10.2307/258214

Gomes, A. R. (2010). *Questionário de stress ocupacional – versão geral (QSO-VG)*. Braga: Universidade do Minho.

Hill, M., & Hill, A. (2002). *Investigação por Questionário*. Lisboa: Edições Sílabo.

Jöreskog, K. G., & Sörbom, D. (1993). *LISREL8: Structural equation modelling with the SIMPLIS command language*. Chicago, IL: Scientific Software International.

Kline, R. B. (1998). *Principles and practice of structural equation modeling*. New York: The Guilford Press.

Leite, E. S., & Uva, A. D. S. (2010). *Stress (relacionado com o trabalho) e imunidade*. Sociedade Portuguesa de Medicina do Trabalho.

Linhares, A. R. P., & Siqueira, M. V. S. (2014). Vivências depressivas e relações de trabalho: Uma análise sob a ótica da psicodinâmica do trabalho e da sociologia clínica. *Cadernos Ebape. BR*, 12(3), 719–740.

Lyonette, C., Crompton, R., & Wall, K. (2007). Gender, occupational class and work–life conflict. *Community, Work & Family*, 10(3), 283–308. doi:10.1080/13668800701456245

Murta, S. G., & Tróccoli, B. T. (2004). Avaliação de Intervenção em Estresse Ocupacional [Evaluation of Occupational Stress Intervention]. *Psicologia: Teoria e Pesquisa*, 20(1), 039–047. https://doi.org/10.1590/S0102-37722004000100006

Netemeyer, R. G., Brashear-Alejandro, T. & Boles, J. S. (2004). A cross-national model of job-related outcomes of work role and family role variables: A retail sales context. *Journal of the Academy of Marketing Science*, 32, 49–60.

Netemeyer, R., McMurrian, R. & Boles, J. (1996). Development and validation of work–family conflict and family–work conflict scales. *Journal of Applied Psychology*, 4, 400–410.

Netemeyer, R.G., Maxham, J.G. & Pullig, C. (2005). Conflicts in the work–family interface: links to job stress, customer service employee performance, and customer purchase intent. *Journal of Marketing*, 69(2), 130–143. https://doi.org/10.1509/jmkg.69.2.130.60758

Nouri, B. A. (2017). Effective factors on job stress and its relationship with organizational commitment of nurses in hospitals of Nicosia. *International Journal of Management, Accounting and Economies*, 4(2), 100–118.

Oliveira, L. B. de, Cavazotte, F. de S. C. N., & Paciello, R. R. (2013). Antecedentes e consequências dos conflitos entre trabalho e família. *Revista de Administração Contemporânea*, 17(4), 418–437. https://doi.org/10.1590/S0102-37722005000200007

Ordem dos Psicólogos Portugueses (2020). Prosperidade e Sustentabilidade das Organizações. Relatório do Custo do Stresse e dos Problemas de Saúde Psicológica no Trabalho, em Portugal. Lisboa.

Priyadharshini, R. & Wesley, R. (2014). Personality as a determinant of work family conflict. *Journal of Industrial Engineering and Management*, 7(5), 1037–1060.

Relatório Mundial Da Saúde (2002). Saúde mental: nova concepção, nova esperança. [consultado em 2 de dezembro de 2021]. www.who.int/whr/2001/en/whr01_po.pdf

Russo, M. (2013). Reducing the effects of work–family conflict on job satisfaction: The kind of commitment matters. Department of Management and Strategy, Rouen Business School. *Human Resource Management Journal*, 91–108 https://doi.org/10.1111/j.1748-8583.2011.00187.x

Santos, J. & Gonçalves, G. (2014). Contribuição para a Adaptação Portuguesa das Escalas de Conflito Trabalho-Família e Conflito Família-Trabalho.

Silva, T. (2017). O conflito trabalho-família e família-trabalho e a sua relação com o sentimento de culpa, satisfação com a vida e a satisfação e paixão pelo trabalho – Estudo empírico. Dissertação de Mestrado em Gestão de Recursos Humanos. Universidade do Algarve, Portugal.

Sonnentag, S., & Frese, M. (2002). Performance concepts and performance heory. In S. Sonnentag (Ed.), *Psychological management of individual performance* (pp. 3–26). Baffins Lane, Chichester: John Wiley & Sons, Ltd.

Stanton, J. M., Balzer, W. K., Smith, P. C., Parra, L. F., & Ironson, G. (2001). A general measure of work stress: The stress in general scale. *Educational and Psychological Measurement*, 61(5), 866–888. doi:10.1177/00131640121971455.

Trochim, W. (2000). *The research methods knowledge base* (2nd edition). Cincinnati, OH: Atomic Dog Publishing.

Williams, L., & Anderson, S. (1991). Job satisfaction and organizational commitment as predictors of organizational citizenship and in-role behaviors. *Journal of Management*, 17, 601–617.

Index

A

absenteeism, presenteeism, intentions to leave 259, 266
accountability 34
actor network 171–4, 187, 190
acute care 42
acute patients 53
adapt 87–9, 91–3
adaptability 123
adaptation to requirements 123
advanced technology 170–87
age of disruption 22, 29
agriculture 207, 221
AIDS 123
AI-powered robots 159–60
algorithm 174, 180, 185, 188–9
applications 134
artificial intelligence (AI) 21–5, 27–8, 153, 157–9, 166, 170, 173, 179–81, 186–7, 192
attractor 56–7
automation 22, 25, 29
autonomy 172, 176–7, 181, 187, 189, 191
availability 42

B

behavior 86–8, 89–96
beneficence 176–8, 180
benefits 86–7, 90, 92–3
big data 153, 155, 157, 162–6, 170, 174, 177, 179, 187, 189, 190, 192
bitter 209–15, 217, 221–4
blockchain 153, 157, 164–6
body 88–9, 90, 92
bottleneck, 40, 45
business value 62, 65, 82

C

capabilities 22
capacity 42–3, 50–3, 58
care 47, 54, 56, 58–9, 61
case reporting 202
case study 235
challenges 85–7, 90, 92
change 85–7, 89, 90, 94, 95, 96
channels 62, 64–5, 67, 69, 74–5, 78, 81
China 118–21
chronic diseases 3
classifier 213–15, 220

clinic 49, 52, 60
clinical and health psychology 86, 92–3
clinical managers 51
clinical solution 1
clinician 170–86
cloud services 153, 164–6
clustering 207, 210, 218–19
co-assess 236
co-deliver 236
co-design 236
co-learning 236
collaborate 236
collaboration with others 123
commercial excellence 62–4, 67, 74
commercial strategies 62
common care wards 42
communication 21–2, 24–5, 29–34, 44, 88–90, 93
community 46–7, 171–7, 182–3
competencies 29, 33–4
complex 38, 45–6, 60; systems theory 46
compliance 62, 66, 71, 75
confidentiality 260
confirmatory factor analysis, measures 261
continuous improvement 38–9, 42, 48–9, 51
control 47, 54, 56–7, 60–1; of impacts 63, 65
coordination 44, 48, 51–2, 56
COPI regulations 177
co-production 236
coronavirus restriction 202–3
cost 42, 47
COVID-19 pandemic 16, 64–79, 114, 182, 188, 190–2, 195–6, 231
crisis 85, 91
cross-clinical flows 52
culture 221–2
cure 46, 47, 58
customer 39, 41, 42–3, 58; engagement 62, 64, 66, 74; journeys 62, 74, 81, 83
cyber resilience 7
cyber security 7

D

data analytics 169, 170, 187, 190, 192
data-driven (advanced) technology 171, 175–86, 187, 190, 192
data mining 50
data privacy 62, 68, 75
data science 174–5, 177–8, 186
data scientist 170, 174–5, 177–8, 181, 184–6

269

dataset 174, 178
decision-making 159, 161, 165–6
decision process 42, 47
decision-support system 173–5, 177–8, 182, 184
depression 258, 265
developer 173–5, 177, 181, 183
developing 25, 32–3
diagnosis, 200, 202
digital 129–31, 133–7, 140–2; communications 62; disruption 21, 22, 26–7; HRM in healthcare 21; 25–7; leaders 7; skills 7; technologies 155, 157, 165; transformation 6
digitalisation 5, 129, 131–2, 134, 141
digital-savvy professionals 62
disadvantages 86, 89, 93
disease 194, 197–8
dissatisfaction, behaviours, workday 258
distance 90–2, 94
Doctor@Home (D@H) 235
drones 153, 160–1

E

eHealth 87, 94, 232, 237
electroencephalogram 207, 211–12, 222–3
electrogastrometry 225, 229
electronic tongue 207, 223, 228
emergency ward 42, 49, 52
engagement 32–4, 130
engaging 32–3
ENISA 5
ethics 91–4, 170–1, 176–9, 181, 184–6, 187–91
european countries 3
evaluation 91–4
explainability 180, 182, 186
explainable AI 179–81, 186–7, 189, 191–2

F

face-to-face interactions 62, 65, 81
family–work conflict 258, 266
fee-for-service 121
flow dynamics 53
flow efficiency 38, 40–2, 44–6, 50–3, 57
flow management 38, 40–4, 47, 49, 51–4
flow units 39–40
focus group method 63
focus group review 63
follow-up 235
food science 207, 222
forecasts 42–3
four principles of medical ethics 176–8, 185, 188–90
four Vs 177, 179
fourth industrial revolution 21
frontline health workers 194, 203

functional magnetic resonance imaging 225, 226
future 92, 93, 130, 142–3; scenarios 12

G

glassy carbon electrode 207, 224
government 114–26; agencies 16
guidelines 47, 90–1

H

HCP engagement 74
health 85, 86–9, 91–3, 130, 257, 266–7; organizations 22–4, 32, 34; professionals 23, 27; sector 203; tech 89
healthcare 4.0 157, 197, 207, 209–10, 221, 226; logistics 153, 155, 159; management systems 62; organization 22, 27, 31, 34
healthcare professionals (HCPs): key opinion leaders (KOLs) 61, 82
help 86, 92
hierarchical classification 130
HOBE PLUS 124–5
hospital 38, 40, 42–5, 49–57, 59–61; beds 5; clinical risk 3; facilities 3; group 49; planning 12; process 49, 53
human and social capital 21
human resource systems 129–30, 133, 135–8, 140–2
human resources management 21, 24–5, 31

I

individual care plan 43
industry 4.0 9, 153, 157, 160
inflow variation 43
innovation 2, 9, 22, 28, 30, 31, 34, 129–32, 134–41, 143
instrument 261–2, 265
intangible: flows 150, 152, 160
integration 44, 47–8, 51–2, 56–8; care approaches 3; people management model 21–2, 31–2
intensive care units 43, 49
interdisciplinary approaches 6, 86, 93
internal variation 43
internet 89–94, 201–2
Internet of Things (IoT) 157, 161–6
intersubjectivity 87–8
intervention 87–9, 90, 94
investments 5

J

Job Characteristics Model 181, 189, 191
Job Demand-Control Model 181, 189
job satisfaction 259, 265
journey 156–7, 159–60, 162–3, 165
justice 176–9

Index

K

KIBS 129, 133, 140–1
knowledge management 231
knowledge sharing 237
knowledge transfer 130, 234
knowledge translation 237

L

laptop 195, 200–1
leadership 12, 32–3
Lean 39, 43, 58–9
learning 30, 33–4, 236; management 123
legal 62, 66, 71, 75
linear discriminant analysis 207, 215–17, 227
Little's law 40
lockdown 98, 201, 202
logistics 153–4, 156, 165–6; in healthcare 153–4; regression 207, 213–15, 227

M

machine learning (ML) 153, 158, 160–1, 170, 173–4, 187
management 129, 134; of healthcare 24, 27; models 14; principles 46
managing people 33
maturity 130, 133, 139–41
mechanism 258, 260, 266
mediated 265
medical education 202, 206
medical ethics 176, 185, 187–91
medical outcome 235, 237
medical practice 13, 172–6, 181, 189
medical research 172, 174–5
Medline 120
meta-analysis 6, 130–1, 141, 144
mHealth 197
microsystems 40–1, 47, 56–7, 60
MIND 123
mobile application 193, 202–3
mobile device 194
mobile phone 200–1
mobile technology 195–6
model 130, 133, 142, 170, 174–5, 178–9, 181–4
monitoring app 169, 172–5, 177–8, 182–3
multi-professional teams 44
multichannel sensor 207, 224–5, 229

N

National Cancer Institute (CRO) 237
Nigeria Centre for Disease Control (NCDC) 193–4
non-maleficence 176–8
non-technical skills 236
normalisation process theory 185

O

occupancy 42, 45, 49, 51, 54, 59
occupational stress 257–67
online 85, 86, 90–7
online psychology 90–3
operation theatres 43
opportunities 129, 133–4, 136
organizational capabilities 6
organizational culture 25, 29, 30–3
Osakidetza 124–5

P

PAEHRs 124
pandemic 85, 86, 88, 92, 93, 94, 175, 177, 182–3, 190
participants 260–2, 266
patient engagement: behavioural changes 62
patients 22, 24, 28, 30, 169–81, 183–8, 191, 196, 200–1; experience 1; flows 38, 43–5, 49–52, 54, 150, 152, 155, 166
pattern recognition 207–11, 213, 216, 221, 223, 226–7
PCP 42–3, 53
people 129, 131, 135
people management in healthcare 21, 25–30, 32–4
performance 258, 266; monitoring 123
person-centered 113
pharmaceutical industry: business models 62–3, 67, 79, 82
physician 196, 198–9
planning horizon 42–3
planning levels 42, 43
policy 13, 16
practice 86, 90–3, 95
primacy 176
primary health care 113, 115–16, 118–19, 122–4
principal component analysis 207, 210, 217, 220, 229
prisma 130–1, 133–4
processes 22, 24, 29, 32–3, 41, 46, 61–82; development 41, 45; flow 44; management 39; managers 42, 51–2; maturity 42; measurement system 54; models 40
production and capacity planning 42, 43, 50–3
professional 86, 87, 91–3; bureaucracy 28
protocols 237
psychology 85–7, 89–94
psychotherapy 86–7, 90, 92–3

Q

quality 39, 42, 46, 49–50, 53–4, 57–9, 61
quasi-public 125
queues 41, 45, 53, 58

R

radiology departments 43
Rawls' theory of justice, utilitarian or virtue ethics 176
regulator 172–5, 177
regulatory 62, 66
relational 87, 89
relationship 85, 87–9, 90–3, 258–60, 262, 264–6
reliance 171, 174–5, 177, 180, 185
remote assistance 234
remote clinic 195–6, 201
resource efficiency 40–1, 53
resourcing 32–3
RFID: tags 157, 162, 166; technology 155
robotics 157–9, 161
robots 3, 6, 159–60, 166

S

salty 208, 215–17, 221–3
SARS-COV-2 177, 182–3
scenarios 93–4, 143
scientific research 130
scope 130, 133, 135, 136, 140–1, 143
security 91–3
self-guided vehicles 153, 159, 161
sensor 207, 210, 222–9
Six Sigma 39
skills 22, 24, 29, 31, 34
smart hospitals 1, 6–7
smartphone 194, 198
society 5.0 22–3, 34
socio-technical system 171, 174–6, 181, 182
soft independent modelling of class analogy 207, 217–18, 228
sour 208, 213–15, 217
spinner flow 129, 132–3, 141–2
spinner innovation 129–34, 138–9, 141–2
stakeholders 7, 13, 120, 124
strategic human resources management 25
strategy 25–7, 32–4
structure 23, 26–7, 32–4
super smart society 23
supervised algorithm 207, 213, 226, 228
sustainability 3, 22, 32–4, 42, 47, 59, 86, 93
sweet 208, 221–3, 226, 229

T

talent 22, 29–30, 32
tangible: flows 150–1, 154, 157, 160, 163, 165
technologies applied in healthcare 157, 165
transition 153, 166
taste perception 208–15, 220

tax-funded 125
technology 85, 87, 90–4, 196; development 23–4; 30; readiness levels 131, 145; solutions 232; transformation 2
telecommuting 266
teleconsultation 233–4
telemedicine 157, 163–4, 197, 231, 235, 237
teletherapy 93
teleworking 265
therapy 87–9, 90–2; alliance 87–8; relationship 87–8, 90, 93
3D printing 157, 159, 162, 165–6
throughput rate 43
throughput time 40–1, 57
total quality management 39, 59
track and trace 175, 178
traditional 86, 87, 90; commercial models 62
training 12, 170–2, 176–7, 180
transference 87–8
trust 171, 173–7, 181–3, 185, 189–90, 192
trustworthiness 170, 172, 173, 175, 177, 180, 188

U

ubiquitous computing 23, 27
umami 214–15, 217, 221–3, 225
unified health system 114
unsupervised algorithm 207, 218
unwanted variation 43
upstream 43, 54
use of technology 91, 92, 94
utilitarianism 176

V

validation 177, 185
value 176–9, 181, 186, 190; -based healthcare 1, 22, 31; co-creation 3
variation 41, 43, 45, 53–4, 61
variety 177–9, 181
VCoP 124–5
velocity 177–9
viability 32, 34
virtual/virtualization 22, 33, 88–4, 93–4
virtue ethics 176
volume 177, 179

W

well–being 86
work-family conflict 258–7
World Health Organization 257

Z

Zoom 202